Evidence-Based Inquiries in Ethno-STEM Research

Investigations in Knowledge Systems Across Disciplines and Transcultural Settings

FOREWORD

Studies on indigenous knowledge systems (IKSs) have been growing up in the last three decades at different parts of the world. Particularly studies on the relation between the nature of mathematics and the nature of science and the sociocultural heritages from various civilizations —a lot of these just recognized as such, after the colonial discriminating regimes fell. Whatever these studies and reflections about that relationship are named, there is a common understanding that is now conceptualized as ethnomathematics, ethnobotany, ethnoscience, ethnotechnology, and so forth. Critical debates are taking place all over the world as we can witness with the five International Congresses on Ethnomathematics already organized, each time on a different continent.

Like several study groups on ethno-STEM (science, technology, engineering and mathematics) do, we are now being offered by these anthologies of IKS-studies coming from Morocco, South Africa, India, Brazil, the United States of America, but also including reflections on mathematics and science from an old center of mathematical development—Ancient China. These authors bring stimulating results of their studies addressing today's hot issues in STEM education. It is important for every society to create a reasonable learning environment for their citizens from generation to generation so that they can feel themselves part of the global world, if we accept that both "Western" and "non-Western" systems of knowledge are (or should be) acting to ensure human dignity. The debates on "post-

Evidence-Based Inquiries in Ethno-STEM Research: Investigations in Knowledge Systems Across Disciplines and Transcultural Settings, pp. ix–x modern integrative paradigm shift" referred in this book are examples of evidence that approach the still conflicting "Western" and "non-Western" IKSs, contributing in that way for more investigations in knowledge systems across disciplines and transcultural settings.

The present book is indeed a great contribution for these debates going on among students, teachers, researchers, and policymakers in the field of ethno-STEM education. The various chapters provide insight into epistemological border-crossing between Western STEM and IKSs, and research in various fields. Please, take the opportunity to enjoy what the authors are stimulating us to reflect on. Enjoy!

Marcos Cherinda

PREFACE

Traditionally, the teaching of scientific disciplines particularly STEM (Science, Technology, Engineering, and Mathematics) has been characterized as being disembodied, decontextualized, and deeply concentrated in specific domains. Such instructional approaches are principled by the assumptions that the cognitive capabilities of the mind are trivially dependent upon the environment (Richardson et al., 2008).

The inspiration of this book is the mundane observation that each and every learner as cognitive agent has a body, inhabits an environment, and lives by the constraints of both (Warren, 2006). In this volume, we reflect an increasing tendency to regard this observation as the appropriate starting point for an understanding of cognition as neither disembodied nor decontextualized and far from being centralized. We have come to this realization after years of involvement in researching scientific and mathematical thought inherent in cultural practices and manifested in indigenous knowledge systems (IKS). We are especially dedicated to showing that IKS should draw its foundation from a system-based thinking that is rooted in embodied, situated, and distributed approaches to cognition (Chahine, 2013). As new research in the cognitive neuroscience domain is emerging providing new insights that focus on the role that immersion in inquiries within indigenous, authentic contexts play in enhancing scientific thinking, it becomes critically important to not only define IKS but also to provide examples of what it "looks like" in scientific practices and how it might

Evidence-Based Inquiries in Ethno-STEM Research: Investigations in Knowledge Systems Across Disciplines and Transcultural Settings, pp. xi–xiv impact policy decisions about teaching and learning STEM concepts. While the focus on STEM, nonetheless, in this volume several chapters also refer to tenets of indigenous knowledge such as its creative, oral and cooperative nature, and the epistemological border-crossing that is being advocated would move us closer towards STEAM education (Fulton & Simpson-Steele, 2016). The "A" represents "arts," and pedagogies that also engage students as *Homo ludens* (the playing human, also engaging with art) (Huizinga, 1955) might make STEM subjects more accessible to students.

We define IKS as comprising a constellation of indigenous practices and local empirical knowledges and rituals that are developed and appropriated by the people to sustain the wellbeing of their societies. Despite the multiplicity of definitions prevalent in the literature for IKS, nevertheless, there is an underlying shared understanding that characterizes the epistemology of such dynamic systems, namely that IK is embodied in local practices, situated in everyday experiences, and is orally transmitted and distributed within a semantic space. In our approach to understanding IKS, we invoke a principle of great generality, namely, nonlinear dynamic systems which concerns, in its most generic form, how complex structures emerge from the concurrence of several individual parts (Thelen & Smith, 2008). From a system's perspective, we argue then that an examination of IKS could help understand how people construct knowledge from daily practices and how this knowledge continuously changes and develops over time.

In the context of IKS, we advance a new panoramic view of STEM disciplines that encompasses localized knowledge that is practiced by several cultural groups across various contexts. We borrow the term "ethno" to signify an attachment to cultures and hence we propose the term ethno-STEM (ethno-Science, ethno-Technology, ethno-Mathematics, ethno-Engineering). As such, we seek to demonstrate that ethno-STEM are relativistic and culturally motivated fields of study that are negotiated through sociological and anthropological descriptions. Our aim is not to disrupt the normalized ideas related to STEM but rather to encourage a broader conceptualization of what counts as scientific knowledge that is informed by cultural considerations. We are well aware that controversies and skepticism will come along regarding the legitimacy of ethno-STEM and STEM, however, our intention is to help the reader realize the synergistic potential that these fields of study share.

This book volume will furnish the reader with investigations across the various ethno-STEM disciplines. In particular, our purpose is to provide an international lens to examine evidence-based investigations in ethno-STEM research. These themes grew out of multi-national, multi-institutional, and multi-disciplinary efforts to preserve as well as epitomize the role that IKS play in cognitive development and its vital contributions to successful and meaningful learning in conventional and nonconventional contexts. Principled by the embodied, situated, and distributed cognition (ESDC) (Chahine, 2013), our focus is to provide evidence supporting the embeddedness of a thinking-in-acting model as a fundamental framework that explains and supports students' acquisition of scientific knowledge.

Research such as the ROSE (Relevance of Science Education) study in the Scandinavian countries showed us that it is important that students experience the science they learn in school as relevant (De Beer & Whitlock, 2009). So often "Western" science curricula are experienced as irrelevant, since it does not take cognizance of the daily experiences and world in which the learner finds himself. This book takes a sociocultural look at indigenous knowledge systems and applies research in neuroscience to make a case for the incorporation of IKS in the STEM classroom.

Although the value of IKS is often acknowledged in curriculum policy documents (e.g., in the National Curriculum Statement [NCS] and Curriculum and Assessment Policy Statement [CAPS] in South Africa), teachers are most often not trained in incorporating IK in the classroom. Teachers' lack of the necessary pedagogical content knowledge (PCK) in effectively incorporating IK in their classrooms is a tremendous problem internationally. The lack of PCK is probably the most important reason for what Odara-Hoppers (2004) calls "knowledge apartheid"—the marginalization of learners' cultural context and IK that they might bring to the classroom. Another problem is that IK is often perceived as "pseudoscience," and scholars advocating for the incorporation of IK in the school curriculum often do not contextualize their arguments within a convincing theoretical and conceptual framework (De Beer, 2016).

We envision the audience for this volume to include policymakers, deans, researchers, school administrators, and teachers. This book could also be a useful reference for both undergraduate education (in science, technology, and engineering), as well as for postgraduate studies in IKS, in several countries including Brazil, India, South Africa, and the United States. More specifically, we see this volume as particularly valuable for preservice and in-service teachers working in STEM fields as it affords opportunities to explore necessary pedagogical considerations for including IK in their classrooms.

Broadly, the book introduces a pragmatic approach to prospective studies on ethno-STEM based on sociocultural constructivism and ESDC. To ensure multiplicity of perspectives across disciplines, we have collected a representative sample of studies conducted in several countries, including the United States, South Africa, Brazil, and India. The contributions by outstanding specialists in their respective fields illustrate the diversity of sources of evidence that would support the value of implementing ethno-STEM in the schools. We hope that this book will serve as a landmark in the establishment of new pedagogies and methodologies to teaching ethno-STEM subjects in a wide range of contexts.

The idea of this book started as part of our research project on IKS in transcultural settings and continued with a partial support from the National Research Foundation in South Africa. A key milestone involved two 3-year funded research projects IKS150729132391 "STEM teachers learning from Indigenous Knowledge Systems practitioners" and CPRR150712125224 "The affordances of Indigenous Knowledge for Self-Directed Learning." Several of the chapters, particularly work conducted in South Africa, have grown out of the research projects' outcomes. We also invited additional authors to contribute to this volume. We are grateful to all the contributors for their interest, commitment and terrific work- and for trusting us and agreeing to become part of this enterprise. We would also like to acknowledge the support of the Fuchs Foundation in South Africa.

Chahine & De Beer
November 2019

REFERENCES

Chahine, I. C. (2013). The impact of using multiple modalities on students' acquisition of fractional knowledge: An international study in embodied mathematics across semiotic cultures. *The Journal of Mathematical Behavior, 32*(3), 434–449.

De Beer, J. (2016). Re-imagining science education in South Africa: The affordances of indigenous knowledge for self-directed learning in the school curriculum. *Journal for New Generation Sciences, 14*(3), 34–53.

De Beer, J., & Whitlock, E. (2009). Indigenous knowledge in the life sciences classroom: Put on your de Bono hats. *The American Biology Teacher, 71*(4), 209–216.

Fulton, L., & Simpson-Steele, J. (2016). Reconciling the divide: Common processes in science and arts education. *The STEAM Journal, 2*(2/3). https://doi.org/10.5642/steam.20160202.03

Huizinga, J. (1955). *A study of the play element in culture.* Beacon Press.

Odara Hoppers, C. A. (2004, March). *Culture, indigenous knowledge and development: The role of the university*. Occasional Paper No. 5, Centre for Education Policy Development. Braamfontein, Johannesburg.

Richardson, M., Skockley, K., Fajen, B., Riley, M., & Turvey, M. (2008). Ecological psychology: Six principles for an embodied-embedded approach to behavior. In P. Calvo & T. Gomila (Eds.), *Handbook of cognitive science* (pp. 161–182). Elsevier.

Thelen, E., & Smith, L. (2008). *Dynamic systems theories.* www.iub.edu/~cogdev/labwork/handbook.pdf

Warren, W. H. (2006). The dynamics of perception and actions. *Psychological Reviews, 11*, 358–389.

CHAPTER 1

EXPLORING LOCAL MATHEMATICAL KNOWLEDGE SYSTEMS BY APPLYING ETHNOMODELING RESEARCH

Milton Rosa and Daniel Clark Orey

ABSTRACT

The application of modeling methods often examines how members of distinct cultural groups have come to develop local mathematical knowledge. What is less evident is how mathematical thinking can be part of the way in which researchers, investigators, and educators reason and attempt to make sense of the underlying cultural framework within which mathematical ideas, patterns, and procedures are embedded. In this chapter, we present an argument that the linking of mathematics and culture is appropriate and necessary for an effective understanding of the development of local mathematical knowledge aimed at providing a holistic understanding of human behavior. Culture is a lens, shaping reality, as well, it often forms a blueprint specifying a plan of action, which is unique to a specific group of people, who grow, learn, and act everyday in distinct economic, social, cultural, political,

Evidence-Based Inquiries in Ethno-STEM Research: Investigations in Knowledge Systems Across Disciplines and Transcultural Settings, pp. 1–34

and natural environments in which language is the ultimate cognitive artifact, which allows the members of sociocultural groups to extend cognitive processes beyond the brain, the body, and the immediate environment. This means that cognition is a process that is not only embodied and situated, but also distributed because the members of sociocultural groups create, process, accumulate, and diffuse mathematical information conjointly. The main purpose of this chapter is to discuss the role of ethnomathematical theorizing and formal modeling in the developing theory of ethnomodeling. In this context, this chapter presents current research related to the use of both emic and etic knowledge bases by applying the dialectical perspective found in ethnomodeling research. Emic knowledge is essential for an intuitive and empathic understanding of local mathematical ideas of a culture and it is essential for conducting effective ethnographic fieldwork. Furthermore, emic knowledge is a valuable source of inspiration for etic hypotheses. Etic knowledge, on the other hand, can be an essential tool for cross-cultural comparisons. Both kinds of knowledge provide essential components of ethnographic perspectives where such comparison necessarily demands standard units and categories. Finally, we define ethnomodeling as the study of mathematical phenomena within a culture because it is a social construction and culturally bound.

INTRODUCTION

As researchers come to investigate the local mathematical knowledge of distinct cultural group members, they may be able to find characteristics of mathematical ideas, procedures, and practices that we refer to as ethnomodeling (Rosa & Orey, 2010a). However, an outsider's understanding of objectified cultural traits is always an interpretation that may emphasize inessential features and is in danger of misinterpreting local mathematical ideas and procedures. The challenge that arises from this perspective is how to extract culturally-bound, often local mathematical ideas, procedures, and practices while reducing the possibility of contamination by of the researchers, investigators, and educators (*outsiders*) cultural background. This interference or bias of the researchers, investigators, and educators will with no doubt color the findings gleaned from the culture of the members of the cultural group (*insiders*) under study.

This happens when the members of cultural groups share their own interpretation of their own culture (*emic*) opposed to an outsider's interpretation (*etic*). In this approach, the members of cultural groups have their own interpretation of their culture (*emic*) which can often be different from an outsider's interpretation (*etic*). Emic constructs are descriptions and analyses acceptable by members of the group as real, meaningful, and appropriate while etic constructs are categories and concepts used by the observers who can generate scientific theories. In our point of view, research strategies prioritize the study of etic phenomena over that of the emic.

In any ongoing ethnomodeling research strategies there are two approaches to be considered in order to investigate and study the mathematical ideas, procedures, and practices developed by the members of distinct cultural groups:

1. *Etic*, which may be defined as an external or outsider's view on beliefs, customs, and scientific and mathematical knowledge of the members of cultural groups. These individuals are considered as *culturally universal* (Sue & Sue, 2003).
2. *Emic*, which may be defined as an insider's view or the view from the members of cultural groups about their own customs, beliefs, and scientific and mathematical knowledge. These individuals are considered as *culturally specific* (Sue & Sue, 2003).

However, the use of emic or etic approaches depends upon the nature of the study itself. It is important here to mention that there are numerous important differences between the notions of culture and cultural differences and the ways they are studied (Li & Karakowsky, 2002). For example, culture, being universal, can be studied in every defined sociocultural group, but cultural differences can be studied only if there are at least two of these groups. This perspective has implications for research since every sociocultural group should be identified by taking in account their own cultural understandings. In this regard, while the effects of culture are testable from one culture alone, the effects of cultural differences are measurable only when individuals from different sociocultural groups are compared (Li & Karakowsky, 2002).

The terms emic and etic are neologisms that were initially coined by anthropologist Pike (1954) from phon***etic***, which refers to the sounds used in a particular language and phon***emic***, which provide the general aspects of vocal sounds and sound production in languages. All the possible sounds individuals make constitute the phon*etics* of the language. However, when people actually speak a particular language, they do not hear all its possible sounds because not all sounds *make a difference*. The sounds that are locally significant, as modeled by a linguist, are the phon*emics* of that language (Agar, 2005).

In the study of a language's sound system, there are two approaches in the investigation of cultural group systems that could be applied in the study of these groups, which provide the perspective of either the insiders or the outsiders (Pike, 1967). The concepts of emic and etic began to be more widely used both inside and outside of anthropology. In the late 60s, Berry (1969) transferred Pike's (1967) cross-cultural psychology using the term *etic* to analyze human behavior of people who focus on universals. Thus, *etic* behaviors are those that can be compared across cultures using common definitions and metrics. An *emic* analysis of these behaviors would focus on behaviors unique to a subset of cultures or on the diverse ways in which etic activities are carried out in a specific cultural setting. In this context, cross-cultural researchers make a distinction between culture-specific (emic) and culture general (etic).

The emic approach focuses on intrinsic cultural distinctions meaningful to the members of the sociocultural group whether the natural world is distinguished from the supernatural realm in the worldview of that specific culture. It attempts to describe a particular sociocultural group by investigating specific aspects of concepts or behaviors. This means that the members of cultural groups should, in the end be the primary judges of the validity of emic descriptions of their cultural, social, environmental, political, economic, and scientific and mathematical ideas, concepts, procedures, and practices (Rosa & Orey, 2012a). To reaffirm, the emic approach focuses on studying a construct from within a specific culture, and tries to understand that construct, as the people from within that culture understand it. On the other hand, etic approaches rely upon extrinsic concepts and categories that have meaning for scientific and researcher observers who are often the sole judges of the validity of an etic account. It involves developing an understanding of a construct by comparing it across cultures using predetermined characteristics.

The etic approach is equated with the objective explanation of sociocultural phenomena from external points of view while emic perspectives are identified with the sympathetic comprehension of subjective experience from the internal participants of a sociocultural group (Harris, 1980). On the other hand, from the dialectal perspective, both approaches, emic and etic, can be considered as two sides of the same coin. In so doing, by utilizing the research provided by both approaches, we gain a more complete understanding of knowledge and interests of the members of distinct cultural groups (Rosa & Orey, 2012a).

The distinction between etic and emic approaches was introduced into the discipline of cultural anthropology, which assumed a methodological dichotomy that corresponds to the opposition between the behavior of the cultural group members and their internal experience (Pike, 1967). The etic perspective is equated with the objective explanation of sociocultural phenomena from the external point of view while the emic perspective is identified with the sympathetic comprehension of subjective experience from the internal participants of a culture (Harris, 1980).

Researchers, investigators, educators who take on an emic perspective believe that many factors such as cultural and linguistic backgrounds, social and moral values, and lifestyle come into play when mathematical ideas, procedures, and practices are developed by the members of distinct cultural groups so they are able to understand and comprehend their own cultural, social, political, economic, and natural environments (D'Ambrosio, 1990).

THE CULTURAL INFLUENCE ON MATHEMATICS

The cultural influences placed upon mathematics are the interactions of culture and mathematics (Rosa& Orey, 2010a). The members of the sociocultural groups' mathematical knowledge are a result of their cultural value system developed in a particular context that developed over time as people become socialized into a particular sociocultural group. Society, culture, community, and family values and beliefs influence the formation of the members of cultural system as well as the acquisition of their scientific and mathematical knowledge (D'Ambrosio, 1990). Thus, the cultural value system includes cultural elements that members of these groups have in common within the sociocultural group to which they belong, as well as idiosyncratic values unique to each member. This is how culture manifests its influence on the mathematical ideas, procedures, and practices produced as part of its value system (Rosa & Orey, 2010b). According to this point of view, a sociocultural groups' mathematical practice may be viewed, imitated, and rejected by others as well as it may become the scientific and mathematical norms of these members (D'Ambrosio, 1990).

The development of mathematical ideas and procedures serve as a vehicle to transfer meanings from culturally constituted environment to academic mathematical activities. This means that the communication of these ideas represents or moderates the influence of culture on the mathematical concepts developed by members of distinct cultural groups (Rosa & Orey, 2010a). It is important to highlight that from an emic approach, culture may not be seen as a constructed apart from and causing the development of mathematical practices because emic researchers and investigators view culture as inseparable from the members of distinct cultural group as an inherent quality (Geertz, 1973). Culture is depicted as causing the development of mathematical ideas, procedures, and practices in order to develop a framework that enables the comparison of mathematical practices developed by members of distinct sociocultural groups and isolates the cultural causes of these practices (Rosa & Orey, 2008).

Culture influences mathematics through its manifestations such as symbols (Hofstede, 1997), which are the forms in which culturally mathematical knowledge is stored and expressed. These manifestations may be considered as the ideas and procedures that are organized, evaluated, and constructed in order to help members of distinct sociocultural groups to assign meaning to (mathematical) phenomena that occur in the environment that surrounds them. Thus, any sociocultural group possesses different mathematical manifestations that encompass most elements of their culture (Rosa, 2010). In this regard, emic research focuses on every meaning of symbols and cultural artifacts in the lives of the members of sociocultural groups explains how members organize and applies information to solve problems faced daily.

In our opinion, scientific and mathematical knowledge embodies and expresses cultural principles. For example, symbols are a broad category of processes and cultural artifacts that carry meaning and are unique to a particular sociocultural group (Geertz, 1973). Cultural artifacts are objects created by members of cultural groups, which provide clues and information about its creators and users (D'Ambrosio, 1993). Mathematical ideas and procedures are cultural artifacts that are socioculturally situated as well as distributed among members across generations. Embodied cognition includes any cognitive activities that use symbols as external resources that assist members of distinct cultural groups in the "mental representation and manipulation of things that are not present, and this includes purely internal uses of sensorimotor representations in the form of mental simulations" (Wilson, 2002). However, mathematical symbols may not exist in all sociocultural groups and their meaning may be different from one specific culture to another (Rosa & Orey, 2008).

Since mathematical ideas and symbolism are generated at the sociocultural level, they are expressed through mathematical practices that become charged with cultural meaning. In this regard, members of distinct sociocultural groups use mathematical ideas that may be charged with symbolic meaning and values having a central role amongst the other manifestations of their culture (Hofstede, 1997). Hence, symbolic language as well as symbology generally expresses cultural values through which language, mathematical ideas and procedures become part of the development of transmitted mathematical practices across generations. According to this context, researchers, investigators, and educators have to acknowledge and recognize that the members of distinct cultural groups possess valid scientifically mathematical knowledge (D'Ambrosio, 1990).

ETHNOMODELING RESEARCH

Ethnomodeling research is related to mathematical practices developed by the members of distinct sociocultural groups, which tend to privilege its organization and presentation in order to facilitate its communication and transmission through generations (emic). The representational idea of this local mathematical knowledge through scientific methods help researchers, investigators and educators to construct and understand the world (etic) by using small units of information called ethnomodels that compose its entire representation. Ethnomodels help to link the development of mathematical practices to the cultural heritage of the members of distinct cultural groups, who detain necessary information to solve problems and situations described in systems taken from their own reality (Rosa & Orey, 2012a).

The emphasis of ethnomodeling research takes into consideration the processes that help the construction and development of local mathematical knowledge systems, which includes collectivity, creativity, and inventivity (Ascher & Ascher, 1981; Eglash, 1999). According to this approach, it is impossible to imprison mathematical ideas, procedures, and practices in registers of univocal designation of reality because there are distinct systems that provide an unambiguous representation of reality as well as universal explanations (Craig, 1998). This means that mathematics cannot necessarily be conceived as a universal language because its principles, concepts, and foundations are not always the same everywhere around the world (Rosa & Orey, 2010b). The production process of mathematical ideas, concepts, procedures, and practices operates within the register of interpretative singularities regarding possibilities for symbolic construction of local mathematical knowledge (Rosa & Orey, 2007).

Ethnomodeling studies local mathematical processes developed by the members of distinct sociocultural groups. Many interesting models are formulated by using data obtained from studies related to ethnomathematics, which proposes the rediscovery of knowledge systems adopted by the members of these groups (Babbitt, Liles, & Eglash, 2012; Bassanezi, 2002; Caldeira, 2007; Rosa & Orey, 2013a). When this knowledge uses mathematical ideas and procedures via ethnomodels, we are able to understand the origin of mathematical practices more efficiently.

From an ethnomodeling perspective, emic constructs are accounts, descriptions, and analyses of mathematical ideas, procedures, and practices expressed in terms of conceptual schemes and categories that are regarded as meaningful and appropriate by members of the cultural group under study. This means that emic constructs are in accordance with the perceptions and understandings deemed appropriate by the insider's culture. The validation of the emic knowledge comes with a matter of consensus of local people who must agree that emic constructs match shared perceptions that portray characteristics of their culture (Lett, 1996). It is paramount to note that the particular research technique used in acquiring mathematical knowledge has nothing to do with the nature of that knowledge. Emic mathematical knowledge can be obtained through elicitation and observation because observers infer local perceptions. In taking an emic approach, researchers, investigators, and educators try to put aside bias, prior theories and assumptions in order to let the members of distinct cultural groups and provide mathematical themes, patterns, and concepts to emerge (Lett, 1996). Some of its strength lies in its appreciation of the uniqueness of the context being studied in its respect for local viewpoints, and potential to uncover unexpected findings.

In an ethnomodeling perspective, "the etic constructs are accounts, descriptions, and analyses of mathematical ideas, procedures, and practices expressed in terms of conceptual schemes and categories that are regarded as meaningful and appropriate by the community of scientific observers" (Lett, 1990, p. 130). An etic approach uses as its starting point theories, hypothesis, perspectives, and concepts from outside of the setting being studied that were developed by researchers, investigators and educators, and must be precise, logical, comprehensive, replicable, and observer-researcher independent (Rosa & Orey, 2010b). The validation of etic knowledge becomes a matter of logical and empirical analysis, in particular, the logical analysis of whether the construct meets the standards of comprehensiveness and logical consistency. It is important to emphasize that the particular research technique that is used in the acquisition of scientific and mathematical knowledge have no bearing on the nature of that knowledge. Etic knowledge may be obtained at times through elicitation as well as observation. One of the strengths of the etic approach is that it allows for comparison across contexts and populations, and the development of more general cross-cultural concepts (Morris et al., 1999).

While some researchers, investigators, and educators heavily privileging one approach over the other, many still work within the tension between these two extremes. In this regard, it is necessary to point out that a completely etic approach risks blinding these professionals to potentially perceive new and innovative ideas, procedures, and concepts. At the same time, they come with previous ideas, perspectives, and commitments and so it may be impossible to be purely emic. For example, the study conducted by Rosa and Orey (2013b) showed that a quilt theme is an interesting way for teachers to develop their pedagogical work by integrating mathematics, art, history, and reading in an interdisciplinary approach by elaborating lesson plans that combine an ethnomathematical-historical perspective that proposes a history project related to the *Underground Railroad*.[1] This approach allows teachers to develop classroom activities and projects that help students to better understand history and geometry, especially concepts of symmetry and transformations through ethnomodeling by incorporating emic and etic ethnomodels in the curricular mathematics activities. One major implication of this project for the mathematics classroom is to stimulate student's creativity and interest, because quilts may be considered as cultural and mathematical expressions of student's daily life.

Finally, ethnomodeling emphasizes the organization and presentation of mathematical ideas and procedures developed by the members of distinct cultural groups in order to facilitate its communication and transmission across generations. In this regard, these members construct ethnomodels of mathematical practices found in sociocultural systems, which link cultural heritage with the development of mathematical practices (Rosa & Orey, 2010a). It is our understanding that this approach helps the organization of pedagogical action in classrooms by using emic and etic aspects of mathematical knowledge through the development and elaboration of ethnomodels.

ETHNOMODELS

Traditional mathematical models, more often than not, do not fully take into account the implications of the varied cultural aspects of human social systems. The cultural component in this process is critical because its accounts "emphasize the unity of culture, viewing culture as a coherent whole, a bundle of practices and values" (Pollak & Watkins, 1993, p. 490) that are incompatible with the rationality of the elaboration of traditional mathematical models. In the context of mathematical knowledge, what is meant by the cultural component varies widely and ranges from viewing mathematical practices as socially learned and transmitted to the members of sociocultural groups to mathematical practices viewed as made up of abstract symbolic systems with an internal logic giving a symbolic system to its mathematical structure (Read, 2004).

If the former is considered, then it is the process by which transmission takes place from one member to another, which is central to elucidating the role of culture in the development of mathematical knowledge. If the latter is considered, then culture plays a far reaching and constructive role with respect to mathematical practices that cannot be induced simply through observation of these practices (D'Ambrosio, 1993; Read, 2004). In this regard, if mathematical knowledge developed by the members of a cultural group consists of abstract symbol systems whose form is the consequence of an internal logic, then students may learn specific instances of the usage of that symbol system as well as derive from those instances a cognitive based understanding of the internal logic of the mathematical symbology system (Rosa & Orey, 2003).

In other words, emic knowledge is acquired in accordance to an *insiders'* point of view. In so doing, we argue here that *emic* mathematical models are grounded in mathematical ideas, procedures, and practices that matter to members of distinct cultural groups that are being modeled. On the other hand, many mathematical models are *etic* because they are built on an outsider's view of these members' work. In this regard, etic mathematical models represent how the modeler thinks this world works while emic mathematical models represent how members who live in these cultural groups think their world really is. It is necessary to highlight this while emic knowledge plays an important role in ethnomodeling research; etic knowledge should be taken into consideration when conducting ethnomodeling research (Agar, 2005; Rosa & Orey, 2012a).

In contrast, if mathematical knowledge consists of a set of socially learned and transmitted mathematical ideas, procedures, and practices, then the cognitive aspects of its development plays an important role when constructing ethnomodels of mathematical practices of sociocultural systems. The cognitive aspect needed in this framework is also a decision process by which the members of cultural groups either accept or reject an ethnomodel as part of their own repertoire of mathematical knowledge (Rosa & Orey, 2003). We believe that the conjunction of these two scenarios appears to be adequate to the depth needed to encompass the full range of cultural mathematical phenomena because mathematical practices are socially learned and transmitted to the members of cultural groups through generations.

According to this context, ethnomodels are described as cultural artifacts that are pedagogical tools used to facilitate the understanding and comprehension of systems taken from the reality of the members of distinct cultural groups. In this regard, ethnomodels may be considered as external representations that are precise and consistent with the scientific and mathematical knowledge that is socially constructed, developed, and shared by members of specific cultural groups. The main objective for the elaboration of ethnomodels is to *translate* emic constructs such as mathematical ideas, procedures, and practices to attempt to establish relations between local conceptual knowledge and the mathematics embedded in these constructs (Rosa & Orey, 2013a).

There are two ways in which we make sense of mathematical phenomena. First, there is a level of cognition shared by members of distinct cultural groups, to varying degrees, with the members of their own group. This shared cognition includes cognitive models the groups may elaborate at a nonconscious level comprising an internal organization of external mathematical phenomena as well as the basis upon which mathematical procedures and practices take place. Second, there are culturally constructed representations of external mathematical phenomena that provide its internal organization (Read, 2004). However, the "form of the representation arises through formulating an abstract, conceptual structure that provides form and organization for external phenomena in a manner that need not be consistent with the form and patterning of those phenomena as external phenomena (Read, 2004, p. 167), which relates to ethnomodeling research.

The implications for mathematical modeling are that models of a cultural construct may be considered as a symbol system organized by an internal logic of the members of distinct cultural groups. Ethnomodels are considered cultural constructs because one of the main objectives of its elaboration is to comprehend the way of thinking of these members as well as to understand how they organize and model their mathematical ideas and procedures from their own point of view in order to mathematize their own reality (Rosa & Orey, 2011).

On the other hand, a model built without a firsthand sense of the world being modeled should be viewed with suspicion. Researchers, investigators, and educators, if not hindered by their prior ideology, should come out with an informed sense of the distinctions that make a difference from the point of view of the mathematical knowledge of the people being modeled. In so doing, they should be able to inform outsiders (etic and global) what matters to insiders (emic and local) (Agar, 2005; Rosa & Orey, 2010b).

EMBODIED AND SITUATED COGNITION IN ETHNOMODELING RESEARCH

Culture is a lens, shaping reality, as well as a blueprint, specifying a plan of action or expectations. At the same time, since there are aspects of a culture that is unique to a specific group of people, these same people grow, learn and act everyday in unique contexts that are unique to the specific economic, social, cultural, political, and environmental ecology in which they live. In this context, language is often the ultimate cognitive artifact (D'Ambrosio, 1990) we have of this unique set of circumstances. These environments allow sociocultural groups to extend their cognitive processes beyond the brain, the body and the immediate environment. This means that cognition is not only embodied and situated, but also distributed because the members of these groups create, process, accumulate, and diffuse mathematical information conjointly.

As alluded to above, cognition is embodied and situated in environments and distributed among agents, symbols, cultural artifacts, and external structures through generations (D'Ambrosio, 1993). These sociocultural group members imply that their actions have social and cultural meaning that requires background knowledge or *know how* in certain situations such as the development of mathematical ideas, procedures, and practices. Situatedness "refers to having one's behavior strongly affected by the environment.... Embodiment is a type of situatedness; it refers to having a physical body [every material object] and thus interacting with the environment through the constraints of that body" (Matarić, 2001, p. 82). In this direction, situated cognition

> takes place in the context of task-relevant inputs and outputs. That is, while a cognitive process is carried out, perceptual information continues to come in that affects processing, and motor activity is executed that affects the environment in task-relevant ways. (Wilson, 2002, p. 626)

This means that, "embodiment in the field of cognitive science refers to understanding the role of an agent's own body in its everyday, situated activity" (Gibbs, 2006, p. 1). Generally speaking, embodiment can be viewed as the mechanism that makes it possible to be situated, and consequently it can be argued that embodiment might be a *stronger* position than situatedness (Lindblom, 2007), which implies that both concepts are of crucial importance in the alternative *embodied*, *situated*, and *distributed* approaches to the study of ethnomodeling.

Situatedness and embodiment are also considered as approaches that may be applied in the conduction of an ethnomodeling research because:

a. *Cognition is situated.* It takes place in the context of a real world environment, involving perceptions and actions (Wilson, 2002) of the members of distinct sociocultural groups, which are performed in order to solve problems faced daily. In the context of ethnomodeling research, situatedness is related to mathematical ideas, procedures and practices that are situated in the sociocultural context of the environments of these members. In other words, cognition is contextualized according to the relevance of the inputs and outputs of developed activities (Lindblom, 2007) such as mathematical practices.
b. *Cognition is embodied.* Embodiment refers to the "experiences that arise from the living body in its interactions with a material/physical as well as a social and cultural world" (Lindblom, 2007, p. 14). In the context of ethnomodeling research, it is related to mathematical ideas and procedures that are embodied in the cultural artifacts and practices developed by the members of distinct sociocultural groups. These embodied activities possess cognitive and/or epistemic meaning because they may be considered as part of problem-solving processes that function as external scaffolds for the development of higher-level cognition (Anderson, 2003) of these members.

Cognition should be considered an activity structured by the body and its situatedness in its environment as embodied action. The embodied view of cognitive science stresses physical, temporal, and functional situatedness, and enforces interaction between the members of distinct sociocultural groups and their environments. Such a holistic view prevents inappropriate simplifications and unrealistic assumptions of mathematical ideas, procedures, and practices because it enforces dealing with unexpected contingencies, provides specificity, and incorporates energetic and resource considerations (Matarić, 1997) that allows for direct cultural interaction, which is undeniably crucial for ethnomodeling research.

We characterize the role and relevance of embodiment in social interaction and cognition in order to develop a thorough and integrated understanding of mathematical knowledge developed by the members of distinct cultural groups through ethnomodeling that supports and explains the relationships that actually exist, which originate from the embodied nature of social interaction and cognition. In this context, embodiment and situatedness relate to ethnomodeling research because of their possibilities for actions that allow the interaction between the activities such as mathematical practices that the members of distinct cultural groups perform in accordance to the surrounding environments. Embodied actions are situated in sociocultural contexts in which the members of these groups apply them in their social interaction as ways of facilitating and coordinating different social, cultural, and cognitive processes (Adolph & Berger, 2006). The nature of social interaction is relational and meaning and intentions are emergent phenomena of these social interactions.

Embodiment may provide concrete evidences of the development of mathematical knowledge when the members of distinct cultural groups solve problems faced in their daily life, which is situated in distinct environments such as physical, political, economic, social, and cultural. In other words, embodied practices developed in sociocultural contexts provide sufficient scaffolding for the understanding of these members as a way of acting and creating meaning together for those practices (Gallagher, 2007). In this regard, we argue here that embodiment is the part and parcel of social interaction and cognition in the most general and specific ways, and in which dynamically embodied actions themselves have meaning and agency for the construction of mathematical knowledge through the application of ethnomodeling activities in classrooms.

CULTURALLY UNIVERSAL (ETICS) VERSUS CULTURALLY SPECIFIC (ETIC) APPROACHES IN ETHNOMODELING RESEARCH

The discussion of current understanding of universality is complex. Often, mathematics is regarded as a neutral and culturally free research area that is not connected to social and cultural values (D'Ambrosio, 1990). Mathematics is taught in schools as a culturally free discipline that involves learning supposedly accepted universal ideas, procedures, concepts, practices, and contents (Rosa & Orey, 2011).

Ethnomodeling research has been developed "to confront the taboos that mathematics is a field of study that is universal and acculturated" (Rosa & Orey, 2006, p. 20). This, we believe, avoids the pervasive view of mathematics as Eurocentric and value-free misrepresents the evolution of modern mathematics (Joseph, 2000). This perception is also reinforced by students' experiences of the way mathematics is taught in schools. In this context, teachers' view of mathematics is transmitted to their students and this helps to shape their views about the nature of mathematics (Brown et al., 1990).

Even though the universality of mathematical truths is not in question, it is only in the last three decades that the view of mathematics as culture free has been challenged (Rosa & Orey, 2006). This means that "there is no sense in regarding mathematics learning as abstract and culture free" (Bishop et al., 1993, p. 1) because the learning process cannot be abstract and context free because it cannot be free of societal influence. Studies conducted by Bandeira and Lucena (2004), Chieus (2004), and Rosa and Orey (2009) that examined mathematics in a variety of cultural contexts confirms this assertion.

It is worth noting that the contextualization of mathematics has been described as the identification of mathematical practices developed by the members of distinct cultural groups (Nasir & Cobb, 2007; Orey, 2000). If mathematics can be considered as a cultural construct, then it is equally a product of cultural development (Rios, 2000; Rosa & Orey, 2007). Mathematics as a cultural construct contradicts the claims that modern mathematics is universal, objective, and culturally neutral. Since mathematical knowledge results from social interactions in which relevant ideas, facts, concepts, principles, and skills are acquired as a result of cultural context, then mathematics is not a universal formal domain of knowledge (Dossey, 1992; Orey, 2000).

In this context, there are at very least six universal mathematical activities such as counting, measuring, designing, locating, explaining, and playing (gambling, guessing, and explaining) that are thought to be practiced by the members of every cultural group. These activities are also widely developed across cultures in order to provide the fundamental facets used to probe traditional daily living and scientific and mathematical practices and are inseparably intertwined with other aspects and activity of any sociocultural group (Bishop, 1991). Through a study of these applications, it is possible to come to an improved understanding of the knowledge of the members of these groups and their experiences in developing and using mathematical thinking practices.

Without a doubt, there are some cultural differences found within these activities. Even though they are considered universal, it is important to recognize that they are merely universal to those individuals who share the same cultural characteristics, historical perspectives, and linguistic background. This means that academic mathematics may look the same in many societies because there is a competitive social, economic, environmental, and political ethics that demands a competitive mathematical development (Rosa, 2010). In other words, dominant cultures have imposed their mathematical ideas on other cultural groups since the beginning of the development of their mathematical knowledge. On the other hand, it is naive to believe that mathematical ideas and concepts do not reflect the cultural values and lifestyles of the members of any given sociocultural group.

It is the unique context and assemblage of culturally constructed symbolisms and procedures that facilitates the manipulation of representations of mathematical knowledge. Students develop representations and procedures in their cognitive systems, which also presents a process that occurs in the context of socially constructed activities (Rosa & Orey, 2008). Mathematical skills that students learn in schools are not logically constructed based on abstract cognitive structures but rather forged out of a combination of previously acquired knowledge, skills, and new cultural inputs. Therefore, mathematics arose out of the needs of organized society, which cannot be divorced from the activities and practices developed by the members of sociocultural groups in a globalized society (Rosa & Orey, 2011).

According to this context, research in mathematics education, more specifically, ethnomodeling can be done from three basic viewpoints, which are phrased in terms of the *culturally-universal (etic)*, *culturally-specific (emic)* and *dialectical* (*glocal*) perspectives.

1. *Culturally-universal perspective* refers to a mathematical phenomenon that is constant throughout the world, which does not vary across cultures. Some mathematical concepts are generalizable across cultural groups and the general idea of mathematical practices is considered a universal phenomenon (Bonjour, 1980; Goldman, 1988; Kline, 1953). In this perspective, an etic approach understands the mathematical phenomenon cross culturally rather than culturally specific meanings (Rosa & Orey, 2011). However, it is naive to state that distinct cultural groups do not share universal mathematics characteristics. For example, many of the everyday activities of members of cultural groups involve a substantial amount of mathematical application (Bishop, 1991).
2. *Culturally-specific perspective* focuses on studying the sociocultural aspects of mathematical phenomena from within a specific cultural

context in order to understanding it as the members of distinct sociocultural groups from within that culture understand it (Gudykunst, 1997). In keeping with the emic approach, a demand for local or culturally specific approaches has emerged besides the Western and Euro-American approach. The examples of local or culturally specific topics and research instruments encouraged some researchers to support the emic perspective. Many theories and methods seem to be susceptible to cultural differences and to demand a cultural contextualization (D'Ambrosio, 1990; Knijnik, 1996; Rosa & Orey, 2010a; Scandiuzzi, 2002).

3. *Dialectical perspective* makes use of both emic and etic knowledge and understandings through the processes of dialogue. While the traditional concepts of emic and etic are important points of view for understanding and comprehending cultural influences on ethnomathematics and modeling research, this third perspective is very important for the development of ethnomodeling research. In this perspective, etic claims that the knowledge of any given cultural group have no necessary priority over its competing emic claims because there is a relationship of interdependency between emic and etic perspectives. This perspective stresses the development of investigations through a mutual exchange between the emic and etic, which could be considered as a qualitative transformation in the research process (Bassanezi, 2002; D'Ambrosio, 1993; Eglash et al., 2006; Rosa & Orey, 2012a).

One of the primary issues raised in mathematics education concerns itself with the position of researchers, investigators, and educators in relation to different types of universals such as the etic approach (culturally universal and global), the emic approach (culturally specific and local), and dialectical approach (cultural dynamism and glocal) perspectives into the mathematics curriculum. Most researchers, investigators, educators, and teachers may operate from the etic position because they believe that mathematical ideas, concepts, procedures, and practices occur in the same way in every culture. Thus, they base their beliefs on Western ideas in which every cultural group construct, develop, acquire, accumulate, transmit, and diffuse the same kind of scientific and mathematical knowledge.

Scholars, researchers, educators, and teachers consciously or unconsciously transmit the views, values, beliefs, attitudes, feelings, and scientific and mathematical ideas of their own culture or the society being studied into universal principles assuming that cultures are homogeneous (D'Ambrosio, 2002). For example, the results of the study conducted by Rosa (2010) about the perceptions of school leaders in relation to English language learners (ELL) students revealed that 17 (65.38%) out of 26 school leaders are limited by their own cultural orientations in relation to school issues, which may not contribute to the academic success of these students who are guided by another cultural orientation.

Although diversity among and within cultural groups have been identified, due to the lack of cultural specific theories in local cultures, Western born-made theories are overemphasized in mathematical classrooms (Rosa, 2010). In the educational context, minimal modifications in the pedagogical practices of mathematics are required because scientific and mathematical knowledge are considered universal and equally applicable across cultures. Thus, if the assumption that the origin, process, and manifestation of scientific and mathematical knowledge are similar across cultures, then universal guidelines and strategies for the pedagogical work would appear to be appropriate in application to all cultural groups.

It is paramount that researchers, investigators, and educators take in consideration that lifestyles, cultural values, and worldviews influence the development of mathematical knowledge because its development arises from cultural contexts (D'Ambrosio, 1993). This is one of the most important issues currently confronting researchers, investigators, and educators. Researchers, investigators, educators, and teachers who believe that cultural background and life experiences influence the development of scientific and mathematical knowledge propose the use of culturally specific strategies in the pedagogical work of teaching and learning mathematics. Thus, such professionals are pointing out that worldwide current guidelines and standards for mathematical instruction are culturally bound (Rosa, 2010).

Considering the mathematics education research field, should researchers, investigators, and educators be focused on culturally universal or culturally specific knowledge? Some of these professionals believe in cultural universality, which focuses on similarities and minimizes cultural factors while others take on techniques and beliefs of cultural specificity, which focus on cultural differences. According to this perspective, researchers, investigators, and educators do not agree on the nature of mathematics (Rosa, 2010). One of the primary issues is whether mathematics is *external* or *internal* to individuals. This argument is pertinent to the relationship of culture and mathematics in that internalists see connections between mathematics and culture while externalists see mathematics as culture free (Dossey, 1992).

On the other hand, many scholars believe mathematics activity is cultural (D'Ambrosio, 1990; Eglash, 1997; Rosa & Orey, 2008). Internalists such as Bishop (1988) and D'Ambrosio (1985) believe that mathematics is a cultural product, developed as a result of various activities such as counting, locating, measuring, designing, and playing. Other mathematicians such as Kline (1980) are externalists because they believe mathematics activity is culture free. Thus, they do not believe in the connection between mathematics and culture. The results in a study conducted by Rosa (2010) revealed that 16 (61.54%) out of 26 school leaders in nine high schools in a school district in Sacramento, California, possess an externalist view of mathematics, which means that they perceive it as being *culture free*.

The question is, then, whether researchers, investigators, and educators have to understand local or cultural specificity against the background of universal and generic theories and methods (etic approach), or whether this behavior can only be understood within its cultural context, and therefore requires culturally specific theories and concepts (emic approach). In our point of view, a better approach to these opposing views may be to try to understand the universality of mathematical ideas, procedures, and practices relevant to researchers, investigators, educators, and the members of cultural groups. This approach takes into consideration the relationship between cultural norms, values, attitudes, and the manifestation of mathematical ideas and procedures (Rosa & Orey, 2010b). Moreover, the practices found in different political, economic, environmental, social, and cultural contexts are equally important.

Researchers, investigators, and educators must also be aware of their own worldviews, and the danger of it influencing their findings. As they become more mindful of how their worldviews and values shape their perceptions, they can become more open to apply aspects of ethnomodeling in their pedagogical practices. This may lead them to a clear decision on the two approaches. These professionals may also use one paradigm or multiple paradigms as a "dialectical perspective" (Creswell & Plano Clark, 2007, p. 27) in order to best fit their worldview. Different paradigms give rise to contradictory ideas and contested arguments (Greene & Caracelli, 2003). These "contradictions, tensions, and oppositions reflect different ways of knowing about and valuing the social world" (Creswell & Plano Clark, 2007, p. 27).

From the non-universalistic viewpoint, distinctions can be made specifying a hypothetical construct as culturally specific and local, which can be distinguished from the culturally-universal and global (Lonner & Berry, 1986). In a multicultural comparison, the etic refers to a mathematical phenomenon that has a common meaning across cultures, often referenced as core mathematical meanings. Conversely, emic refers to different phenomena across cultures, where each emic aspect is related to the shared etic core (Berry, 1969; Brislin, 1983, Rosa & Orey, 2012a).

In this context, the etic approach may be defined as "the collective programming of the mind which distinguishes the members of one group or category of people from another" (Hofstede, 1997, p. 5). The focus of this definition may be considered as the comparison of one cultural group with another. Researchers, investigators, and educators who follow an etic approach in research generally look for universal or culture-free scientific and mathematical concepts and theories (Rosa, 2010). They may be searching for variables and constructs common to all cultures that can be directly compared in order to discover how scientific and mathematical knowledge of the member of those cultural groups are different from or similar to each other.

The emic mathematical practices may be defined through the use of a:

> lens through which all phenomena are seen. It determines how these phenomena are apprehended and assimilated. Second, culture is the blueprint of human activity. It determines the coordinates of social action and productive activity, specifying the behaviors and objects that issue from both. (McCracken, 1988, p. 73)

Emic approaches do not intend to directly compare the mathematical knowledge developed by the members of two or more differing cultural groups but promotes a complete understanding of mathematical ideas and procedures practiced by members of a cultural group under study through thick description (Geertz, 1973). The methods utilized in conducting emic research do not provide culture-free measures that can be directly compared; instead, they provide culture-rich information about these practices. In this approach, information and observations are constructed to reflect the target culture's own scientific and mathematical knowledge, language, and belief system.

The emic approach contrasts with the etic in that it refers to information collected in terms of the conceptual system and categories of the researchers, investigators, and other outsiders. To collect emic data, it is usually necessary to use the local language or dialect and mathematical ideas and procedures in order to gather information in a very open-ended and nondirective way. In the etic approach, the observations and data are constructed in the researcher's system of categories and definitions (Rosa & Orey, 2010a).

Studying sociocultural groups according to preestablished etic procedures may impede the discovery of cultural diversity, whereas emic analysis broadens this view (Headland et al., 1990). The emic approach seeks to understand a particular mathematical phenomenon from the point of view of its adherents while the etic approach does the same but by means of analytical tools and concepts drawn from outside.

The choice of emic versus etic approaches depend on several important factors including the nature of the research question itself, the researcher's resources and training, and the purpose of the study. Thus, it is possible to conclude that both the etic and the emic perspectives refer to similar constructs but from different points of view related to mathematical knowledge developed between-cultures versus the ones developed within-cultures (Luna, 2001, Rosa & Orey, 2010a).

THE DIALECTICAL APPROACH IN ETHNOMODELING RESEARCH

Currently, the emic-etic debate continues to be one of the most intriguing questions in ethnomodeling research since investigators and researchers continue to deal with questions such as:

1. *Are there mathematical patterns that are identifiable and similar across cultures?*
2. *Is it better to focus on these patterns particularly arising from the culture under investigation?*

While emic and etic are often thought to create a conflicting dichotomy, they were originally conceptualized as complementary viewpoints (Pike, 1967). Rather than posing a dilemma, the use of both approaches can deepen our understanding of important issues in cultural psychology (Berry, 1999).

Usually, in ethnomodeling research, an emic analysis focuses on a single culture and employs descriptive and qualitative methods to study a mathematical idea, concept, procedure, or practice of interest. An emic approach focuses on the study within the cultural group context in which the researcher tries to develop research criteria relative to the internal characteristics or logic of the system of the cultural system. Meaning is gained relative to the context and therefore not transferable to other contextual settings (Rosa & Orey, 2013a). For example, it is not intended to compare the observed mathematical patterns in one setting with mathematical patters in other settings. On the other hand, an etic analysis would be comparative, examining many different cultures by using standardized methods.

In the ethnomodeling research, an etic construct is a mathematical-theoretical idea that is assumed to apply in all cultural groups while an emic construct is one that applies only to a specific cultural group (Rosa & Orey, 2012a). Cultural bias in ethnomodeling research may occur if investigators and researchers assume that an emic mathematical construct is actually etic. This results in an imposed etic perspective in which a culture-specific mathematical idea or procedure is wrongly imposed to the members of another cultural group.

The primary goal of the emic approach is a descriptive idiographic orientation of the mathematical phenomena because it puts emphasis on the uniqueness of each mathematical idea, concept, procedure, or practice (Rosa & Orey, 2011). In contrast, the etic approach tries to identify lawful relationships and causal explanations valid across different cultures. Thus, if researchers and educators wish to make statements about universal or etic aspects of mathematical knowledge, these statements need to be phrased in abstract ways. Conversely, if they wish to highlight the meaning of these generalizations in specific or emic ways, then they need to refer to more precisely specified mathematical events.

On a conceptual level, etic statements about the role of mathematical knowledge can be deduced based on empirical data evaluating important attributes of mathematical patterns across cultures. However, emic examples of how such generalizations are expressed in different ways indifferent to national cultures are also developed, accumulated, and transmitted through generations (Rosa & Orey, 2010a). In this regard, the more detailed the description of a mathematical idea, concept, procedure, or practice, the greater becomes the likelihood of finding significant variations. For example, if researchers and educators look at a mathematical a specific mathematical idea in a great degree of detail, it may turn out that many cultural groups give unique meanings and or enactment to general principles.

We also agree with Pike's (1954) point view that the etics perspective may be a way of getting at emics of mathematical knowledge developed by the members of cultural groups. The etic perspective may be useful for penetrating, discovering, and elucidating emic systems present in the mathematical practices of these groups. The rationale behind the emic-etic dilemma is the argument that mathematical phenomena in their full complexity can only be understood within the context of the culture in which they occur. The emic approach tries to investigate the mathematical phenomena and their interrelationships and structures through the eyes of the people native to a particular cultural group.

In the dialectical perspective, the etic perspective claims to acknowledge of any given cultural group have no necessary priority over its competing emic claims. According to this point of view, Eglash et al (2006) stated that there is a necessity to depend "on acts of "translation" between emic and etic perspectives" (p. 347). In this regard, the cultural specificity may be better understood with the background of communality and the universality of theories and methods and vice versa. In our point of view, it is necessary that the insights that have been acquired through subjective and culturally contextualized methods be verified with methods independent of the subjectivity of the observer and researcher in order to achieve a scientific character.

AN EXAMPLE OF A DIALECTAL APPROACH IN ETHNOMODELING RESEARCH

Researchers and investigators have revealed sophisticated mathematical ideas and practices that include geometric principles in craft work, architecture concepts, and practices in the activities and artifacts of many indigenous, local, and vernacular cultures (Ascher, 2002; Eglash, 1999; Gerdes, 1991; Orey, 2000; Urton, 1997; Rosa & Orey, 2009). These concepts are related to numeric relations found in measuring, calculation, games, divination, navigation, astronomy, modeling, and a wide variety of other mathematical procedures and cultural artifacts. It is paramount that research and investigations provide strong challenge to the primitivist view that indigenous and local societies and cultural groups had only simplistic technologies as well as scientific and mathematical knowledge (Eglash at al., 2006). It is crucial that ongoing research and investigation show sophisticated scientific and mathematical practices, not just trivial examples by directly challenging the epistemological stereotypes most damaging to minority groups (D'Ambrosio, 1990). Another controversy stems from the applications of ethnomathematics of local cultural groups to contemporary mathematics education through the application of ethnomodeling.

In research, ethnomathematicians often use the term *translation* to describe the process of modeling local cultural systems with a Western academic mathematical representation (Eglash et al., 2006; Rosa & Orey, 2006; Rosa & Orey, 2010a). However, as with all translation, the success is always partial. Intentionality is one of the areas in which the process is particularly problematic. Often indigenous designs are merely analyzed from a Western view such as the applications of the symmetry classifications from crystallography to local textile patterns. In accordance to this context, ethnomathematics also makes use of modeling by attempting to use it in order to establish relations between the local conceptual framework and the mathematics embedded in relation to local designs. In this regard, the mathematics knowledge can be seen as arising from emic rather than etic origins (Eglash et al., 2006).

In some cases, the translation to Western mathematics is direct and simple such as counting systems and calendars. In other cases, the mathematics is embedded in a process such as iteration in bead work, and in Eulerian paths in sand drawings. In this regard, the act of translation is named ethnomodeling. It is easier to use strictly and explicit numeric systems such as counting than to look at embedded mathematics such as architecture and crafts because it requires ethnomodeling (Eglash at al., 2006).

In this context, these insights must be verified with methods independent of the subjectivity of the observer and researcher in order to achieve a scientific character. In so doing, it is important to analyze the insights that have been acquired through subjective and culturally contextualized methods. The rationale behind the emic-etic dilemma is the argument that mathematical phenomena in their full complexity can only be understood within the context of the culture in which they occur.

Ethnomodeling Process

An ethnomodel that offers an ethnomodeling methodology example was elaborated by a group of Brazilian students who studied wine production. The motivation of their study was to find the volume of wine barrels by applying the techniques learned by ancestors of the wine producers who came to Southern Brazil as Italian immigrants in the early twentieth century (Bassanezi, 2002). Since then the production of wine has become an essential agricultural activity for the economy of that Brazilian region.

In order to conduct their research, initially, students visited wineries to conduct interviews with the wine producers. Subsequently, they collected data that were supplemented by the literature review on the chosen theme. The ethnological and historical research of the construction of wine barrels' theme was the first stage of the ethnomodeling process. In the ethnological study, students identified characteristics of this particular cultural group so that they were able to understand some of the cultural elements that shape their mathematical thinking (Bassanezi, 2002). In this context, students found out that, in addition to produce wine, the producers also construct their own wooden wine barrels by using geometric schemes inherited from their ancestors in Italy.

Dialectical Wine Barrels Ethnomodels

During their research, students also found that in order to construct these barrels with preestablished volume, it is necessary for the wine producers to cut wooden staves so that they can to fit perfectly together. This process drew the attention of the students who were interested in knowing what kind of inherited mathematical ideas the wine producers were using in their geometric schemes (Bassanezi, 2002). For example, Figure 1.1 shows a geometric scheme made by the wine producers in the construction of wine barrels.

In the scheme shown in Figure 1.1, L is the maximum width of the stave, λ is the width to be determined and β is the fitting angle between the staves, which depend on the initial width of the stave L and the volume required for the wine barrel (Bassanezi, 2002). In Figure 1.2, the larger circle with

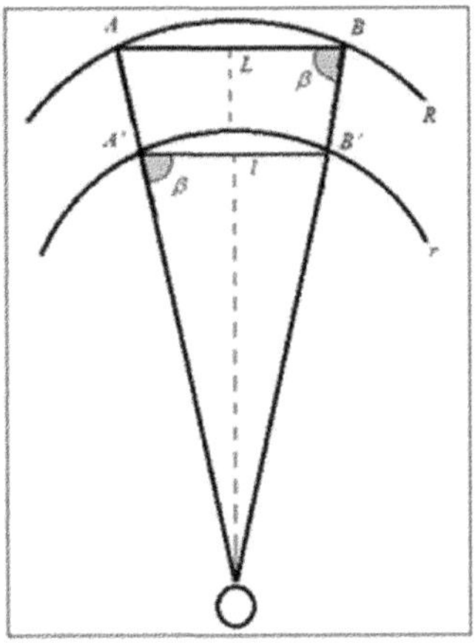

Source: Bassanezi (2002, p. 47).

Figure 1.1. Geometric scheme made by wine producers in the construction of wine barrels.

radius *R* represents the base of the barrel, the smaller circle with radius *r* represents its cover, and *H* represents the height of the barrel.

The wine producers construct barrels shaped like a truncated cone by interlocking wooden staves whose dimensions are 2.5 cm in length and width ranging from 5cm to 10cm (Bassanezi, 2002). Figure 1.2 shows a wine barrel shaped like a truncated cone.

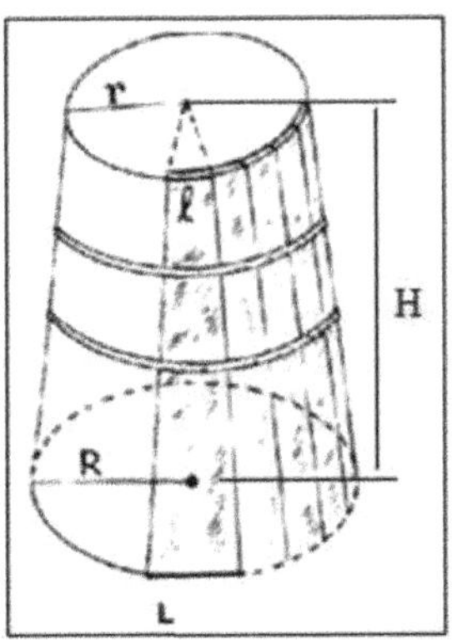

Source: Bassanezi (2002, p. 48).

Figure 1.2. Wine barrel shaped like a truncated cone.

In order to determine the volume of the wine barrel, wine producers approximate the volume of the barrel by applying a procedure named *average cylinder* (Bassanezi, 2002) which is given by formula I:

$$V \cong \pi \bullet r_m^2 \bullet H$$

They also apply the *average radius* procedure, which is given by formula II:

$$r_m = \frac{r + R}{2}$$

By replacing formula II into formula I, formula III is given by:

$$V \cong \pi \bullet \left(\frac{r + R}{2}\right)^2 \bullet H$$

In this process, it is possible to observe that the scheme used in Figure 1.1 is an orthogonal projection of one of the wine barrel wooden staves as shown in Figure 1.2. Figure 1.3 shows an orthogonal projection of the wine barrel wooden slate.

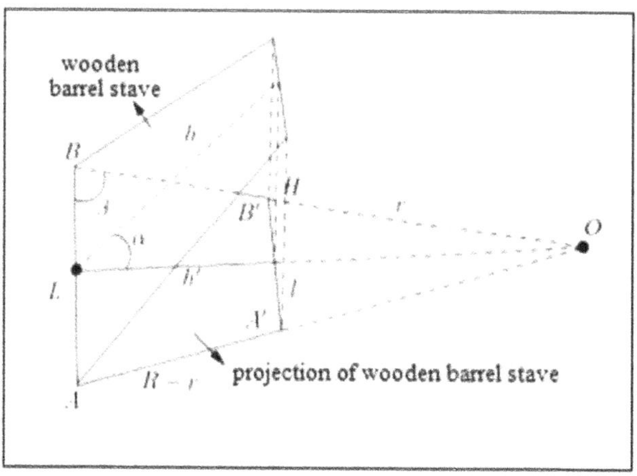

Source: Bassanezi (2002, p. 49).

Figure 1.3. Orthogonal projection of a wine barrel wooden slate.

Figure 1.3 also shows that the fitting angle β between the two wooden staves is obtained by considering that (Bassanezi, 2002):

- R is the radius of the base of the wine barrel.
- L is the width of the wooden stave of the wine barrel in its base.
- All juxtaposed wooden staves form a circumference at the base of the wine barrel.

According to etic approach by developing the mathematical modeling process used in academic mathematics, the volume of the truncated cone is given by the formula:

$$V = \frac{1}{3} \bullet \pi \bullet H \bullet \left(R^2 + rR + r^2\right)$$

On the other hand, in the emic approach by developing the ethnomodeling process used by the wine producers; the volume of the wine barrel is given by the formula:

$$V \cong \pi \bullet \left(\frac{r+R}{2}\right)^2 \bullet H$$

The application of both emic and etic ethnomodels provides an accurate approximation of the volume of the barrel.

This modeling process was investigated from an ethnomathematics perspective as the cultivation of vines and production of wine barrels are strongly linked to the history and culture of people in the Southern region of Brazil. The construction of the wine barrel process is an excellent example that typifies the connection between ethnomathematics and mathematical modeling (D'Ambrosio, 2002) through ethnomodeling. Thus, this method presents an approximated calculation of the area of the volume of the wine barrel that satisfies the needs of this specific cultural group.

Some Considerations About the Ethnomodeling of the Wine Barrel Construction

An emic observation of this mathematical practice sought to understand it for constructing wine barrels from the perspective of internal dynamics and relationships as influenced within the culture of wine producers. On the other hand, an etic approach offers a cross-cultural contrast and comparative perspective by using aspects of academic mathematics to translate this phenomenon in order to amplify the understanding of those from a different cultural background. This approach is necessary to comprehend and explain this mathematical practice as a whole from the point of view of that from the outside.

In this context, the emic viewpoint clarifies intrinsic cultural distinctions while the etic perspective seeks objectivity as an outside observer across cultures. This is the dialectical approach, which concerns the stability of relationships between these two different cultural approaches. In our point of view both approaches are essential to understand human behaviors, especially, social and cultural behaviors that help to shape mathematical ideas, procedures, and practices developed by the members of distinct cultural groups.

For example, the embodied and situated cognition enforced interaction between members of this particular and distinct sociocultural group with their environment. Such a holistic view allowed for realistic assumptions of mathematical ideas, procedures, and practices because it provided specificity and incorporated energetic and resource considerations that allowed for direct cultural interaction.

In this regard, one of the latest trends in mathematics education points out to the need to integrate the teaching of this science with other knowledge areas in an interdisciplinary fashion at all levels of education. In order for this process to be successful as well as mathematics to be valued as a discipline whose contents can be considered as human creations, it is necessary to understand and modify the environment we live. In this regard, we can use ethnomodeling in order to link theory into practice by the inclusion of the dialectical approach into the mathematics curriculum.

THE DIALECTAL APPROACH IN THE ETHNOMODELING CURRICULUM

Mathematical local knowledge of the members of cultural groups that is combined with Western mathematical knowledge systems may result in an emic-etic approach (dialectical perspective) to mathematics education. An emic analysis of a mathematical phenomenon is based on internal structural or functional elements of a particular cultural group. An etic analysis is based on predetermined general concepts external to that cultural group (Lovelace, 1984). The emic perspective provides internal conceptions and perceptions of mathematical ideas and concepts. The etic perspective provides a framework for determining the effects of those beliefs on the development of the mathematical knowledge. In this regard, the acquisition of mathematical knowledge is based on the applications of current mathematics curriculum, which is assessed based on multiple instructional methodologies across various cultures.

The reasons are complex and difficult; one of the reasons for the failing of many educational systems is that curricula developers have ignored the emic perspectives in the school curricula. An emic-etic perspective includes the recognition of other epistemologies, and of the holistic and integrated nature of the mathematical knowledge of the member of any given cultural group. In this regard, an ethnomodeling curriculum provides an ideological basis for learning with and from the cultural and linguistic backgrounds of the members of distinct cultural groups (Rosa & Orey, 2013a).

An ethnomodeling curriculum combining the key elements of local knowledge with an emic-etic perspective is likely to produce students who can manage knowledge and information systems taken from their own reality. In this context, ethnomodeling is considered as part of the critical mathematics education that is a process of learning in which teachers encourage critical examination of multiple sources of knowledge and theories in diverse learning styles with the intention that the acquired knowledge is centered, located, oriented, and grounded on the students' cultural backgrounds, which could be applied, translated appropriately by the students and thus well equipped to be productive locally and globally citizens. Ethnomodeling is a pedagogical approach to reach this goal (Rosa & Orey, 2010a).

In this regard, the development of a confluent curriculum that integrates the "affective domain (emotions, attitudes, values) with the cognitive domain (intellectual knowledge and abilities)" is necessary (McNeil, 1983, p. 11). In the elements of an ethnomodeling curriculum essential in the process of designing an emic-etic pedagogical approach are integration, participation, relevance, and self as object of learning. These elements and the nature of the previous mathematical knowledge of the students lend themselves to the principle of sequencing in curriculum development. Starting with the students' mathematical previous knowledge, educators can move from the familiar to the unfamiliar and from the concrete to the abstract in the process of promoting the acquisition of their mathematical knowledge (Rosa & Orey, 2013a).

In this context, an emic-etic perspective into the ethnomodeling curriculum provides the underlying philosophy of knowledge generation and exchange within and between all subsystems of mathematics education. Key elements of an ethnomodeling curriculum approach ensure the balanced integration of the affective domain of educational objectives essential to recognition and utilization of the students' previous knowledge.

FINAL CONSIDERATIONS

Local or indigenous knowledge, like language, can be used to make a living and solve problems in a particular community or environment. It evolves in situ and is dynamic and creative, and it is constantly growing and adapting to meet new conditions. The expression *local knowledge* refers to the knowledge of any group of people who have lived in a particular and limited area or small region for a long period of time, which allowed them to develop a body of scientific and mathematical knowledge through generations of living in close contact with their own social, cultural, and natural environment (Johnson, 1992). It is embedded in a specific and dynamic system in which spirituality; kinship, local politics, scientific ideas, mathematical practices are tied together and influence one another. Researchers must take this into account when examining a particular part of the local knowledge system. Local knowledge has many positive aspects and incorporating it into the ethnomodeling curriculum contributes to local empowerment and provides valuable input for alternative pedagogical strategies and methodologies. Local knowledge researchers should also play a part in stemming the loss of this knowledge by helping local people record and use their knowledge.

Today, many local mathematical knowledge systems are at risk of becoming extinct because of rapidly changing natural environments and fast pacing economical, social, environmental, political, and cultural changes on a global scale. In this context, many local mathematical practices will disappear only because of the intrusion of foreign technologies or development concepts that promise short-term gains or solutions to problems faced by cultural groups without being capable of sustaining them. We, of course have no doubt that this is happening with the local mathematical knowledge. The tragedy of the impending disappearance of local knowledge is most obvious to those who have developed it and make a living through it. But the implication for others can be detrimental as well, when skills, technologies, cultural artifacts, problem solving strategies, and expertise are lost.

Defined in that manner, the usefulness of emic and etic distinctions are evident. Answers to the most fundamental research questions including the origins of humanity, the characteristics of human nature, and the form and function of human social systems is part of the worldview of every culture. Like all human beings, researchers, educators, and teachers have been enculturated to some particular cultural worldview, and they therefore need a means of distinguishing between the answers they derive as enculturated individuals and the answers they derive as anthropological observers. Defining emics and etics in epistemological terms provides a reliable means of making that distinction.

From an ethnomodeling perspective, the two approaches, emic and etic, can be considered as two sides of the same coin. In this context, culture is considered as a lens that shapes reality as well as a blueprint that specifies a plan of action (Luna, 2001) that is necessary for the members of distinct cultural groups to solve problems they face daily. In this regard, it is necessary that investigators and researchers develop ethnomodeling research conducted by applying both approaches in order to gain a more complete understanding of the mathematical knowledge developed by the members of cultural groups of interest.

Finally, the goal of conducting ethnomodeling research must be the acquisition of both emic and etic knowledge. Emic knowledge is essential for an intuitive and empathic understanding of mathematical ideas of a culture and it is essential for conducting effective ethnographic fieldwork. Furthermore, emic knowledge is often a valuable source of inspiration for etic hypotheses. Etic knowledge, on the other hand, is essential for cross-cultural comparison, the essential components of ethnology, because such comparison necessarily demands standard units and categories.

NOTE

1. The *Underground Railroad* was an organized system to assist runaway slaves to escape to Canada during the slavery period in the United States (Rosa & Orey, 2012b)

REFERENCES

Adolph, K. E., & Berger, S. E. (2006). Motor development. In: D. Kuhn, R. S. Siegler, W. Damon, & R. M. Lerner (Eds.), *Handbook of child psychology: cognition, perception, and language* (pp. 161–213). John Wiley & Sons.

Agar, M. (2005). Agents in living color: Towards emic agent-based models. *Journal of Artificial Societies and Social Simulation, 8*(1). http://jasss.soc.surrey.ac.uk/8/1/4.html

Anderson, M. L. (2003). Embodied cognition: a field guide. *Artificial Intelligence, 149*, 91–130.

Ascher, M., & Ascher, R. (1981). *Mathematics of the Incas: Code of the Quipu*. Dover.

Ascher, M. (2002). *Mathematics elsewhere: an exploration of ideas across cultures*. Princeton University Press.

Babbitt, B., Lyles, D., & Eglash, R. (2012). From ethnomathematics to ethnocomputing: Indigenous algorithms in traditional context and contemporary simulation. In S. Mukhopadhyay & W. M. Roth (Eds.), *Alternative forms of knowing in mathematics: Celebrations of diversity of mathematical practices* (pp. 205–220). Sense.

Bandeira, F. A., & Lucena, I. C. R. (2004). *Etnomatemática e práticas sociais* [Ethnomathematics and social practices]. Coleção Introdução à Etnomatemática [Introduction to ethnomathematics collection]. UFRN.

Bassanezi, R. C. (2002). *Ensino-aprendizagem com modelagem matemática* [Teaching and learning with mathematical modeling]. Editora Contexto.

Berry, J. W. (1969). On cross-cultural comparability. *International Journal of Psychology, 4*, 119–128.

Berry, J. W. (1999). Emics and etics: a symbiotic conception. *Culture & Psychology, 5*, 165–171.

Bishop, A. (1988). Mathematics education in its cultural context. *Educational Studies in Mathematics, 19*, 179–191.

Bishop, A. (1991). *Mathematical enculturation: A cultural perspective on mathematics education*. Kluwer.

Bishop, A. I., Hart, K., Lerman, S., & Nume, T. (1993). Significant influences on children's learning of mathematics: influences from society, science and technology education. *UNESCO Document Series, 47*, 1–25.

Bonjour, L. (1980). Externalist theories of empirical knowledge, *Midwest Studies in Philosophy, 5*, 53–73.

Brislin, R. W. (1983). Cross-cultural psychology. *Annual Review of Psychology, 34*, 363–400.

Brown, S. I., Cooney, T. J., & Jones, D. (1990). Mathematics teacher education. In W. R. Houston (Ed.), *Handbook of research on teacher education* (pp. 639–656). Macmillan.

Caldeira, A. D. (2007). Etnomodelagem e sua relações com a educação matemática na infância [Ethnomodeling and its relationship to mathematics educacion in chidlhood]. In J. C. Barbosa, A. D. Caldeira, & J. L. Araújo (Eds.), *Modelagem matemática na educação matemática brasileira: Pesquisas e práticas educacionais* [Mathematical modeling in the Brazilian mathematics education: Eesearch and educational practices] (pp. 81–97). SBEM.

Chieus, J. G. (2004). Etnomatemática: reflexões sobre a prática docente. In J. P. M. Ribeiro, M. C. S Domite, & R. Ferreira (Eds.), *Etnomatemática: Papel, valor e significado* [Ethnomathematics: Paper, value and meaning] Zouk.

Craig, E. (1998). *Routledge encyclopedia of philosophy: Questions to sociobiology* (Vol. 8.). Routledge.

Creswell, J. W., & Plano Clark, V. L. (2007). *Designing and conducting mixed-methods research*. SAGE.

D'Ambrosio, U. (1985). Ethnomathematics and its place in history and pedagogy of mathematics. *For the Learning of Mathematics, 5*(1), 44–48.

D'Ambrosio, U. (1990). *Etnomatemática* [Ethnomathematics]. Editora Ática.

D'Ambrosio, U. (1993). *Etnomatemática: Um programa* [Ethomathematics: A program]. *A Educação Matemática em Revista, 1*(1), 5–11.

D'Ambrosio, U. (2002). *Etnomatemática: Elo entre as tradições e a modernidade* [Ethnomathematics: A link between traditions and modernity]. Autêntica.

Dossey, J. A. (1992). The nature of mathematics: its role and its influence. In D. A Grouws (Ed.), *Handbook of research on mathematics teaching and learning*. Macmillan.

Eglash, R. (1997). When math worlds collide: Intention and invention in ethnomathematics. *Science, Technology and Human Values, 22*(1), 79–97.

Eglash, R. (1999). *African fractals: Modern computing and indigenous design*. Rutgers University Press.

Eglash, R., Bennett, A., O'Donnell, C., Jennings, S., & Cintorino, M. (2006). Culturally situated designed tools: Ethnocomputing from field site to classroom. *American Anthropologist, 108*(2), 347–362.

Gallagher, S. (2007) Simulation trouble. *Social Neuroscience, 2*, 1–13.

Geertz, C. (1973). Thick description: Toward an interpretative theory of culture. In C. Geertz (Ed.), *The interpretation of culture: Selected essays* (pp. 3–30). Basic Books.

Gerdes, P. (1991). *Lusona: Geometrical recreations of Africa*. E.M. University Press.

Gibbs Jr., R. W. (2006). *Embodiment and cognitive science*. Cambridge University Press.

Goldman, A. (1988). Strong and weak justification. *Philosophical Perspectives, 2*, 51–69.

Greene, J. C., & Caracelli, V. J. (2003). Making paradigmatic sense of mixed-methods practices. In: A. Tashakkori & C. Teddlie (Eds.), *Handbook of mixed-methods in social and behavioral research* (pp. 91–110). SAGE.

Gudykunst, W. B. (1997). Cultural variability in communication. *Communication Research, 24*(4), 327–348

Harris, M. (1980). The epistemology of cultural materialism. In M. Harris (Ed.), *Cultural materialism: The struggle for a science of culture* (pp. 29–45). Random House.

Headland, T. N., Pike, K. L., & Harris, M. (1990). *Emics and etics: The insider/outsider debate*. SAGE.

Hofstede, G. (1997). *Cultures and organizations: Software of the mind*. McGraw-Hill.

Johnson, M. (1992). *Lore: Capturing traditional environmental knowledge*. Dene Cultural Institute and International Development Research Center.

Joseph, G. G. (2000). *The crest of the peacock: Non-European roots of mathematics*. Penguin Books.

Kline, M. (1953). *Mathematics in Western culture*. Oxford University Press.

Kline, M. (1980). *Mathematics: The loss of certainty*. Oxford University Press.

Knijnik, G. (1996). *Exclusão e resistência: Educação matemática e legitimidade cultural*. Artes Médicas.

Lett, J. (1990). Emics and etics: Notes on the epistemology of anthropology. In T. N. Headland & K. L. Pike (Eds.), *Emics and etics: The insider/outsider debate* (Frontiers of anthropology, Vol. 7). SAGE.

Lett, J. (1996). Emic-etic distinctions. In D. Levinson & M. Ember (Eds.), *Encyclopedia of cultural anthropology* (pp. 383–383). Henry Holt.

Li, J., & Karakowsky, L. (2002). Cultural malleability in an East Asian context: An illustration of the relationship between government policy, national culture, and firm behavior. *Administration and Society, 34*(2), 176–201.

Lindblom, J. (2007). *Minding the body: Interacting socially through embodied action* (Doctorate dissertation). Department of Computer and Information Science. Institute of Technology. Linköping University, Sweden.

Lonner, W. L., & Berry, J. W. (1986). *Field methods in cross-cultural research*. SAGE.

Lovelace, G. (1984). Cultural beliefs and the management of agro-ecosystems. In T. Rambo & P. E. Sajise (Eds.), *An introduction to human ecology research on agricultural systems in South East Asia* (pp. 194–205). East-West Centre.

Luna, D. (2001). An integrative framework for cross-cultural consumer behavior. *International Marketing Review, 18*(1), 45–69.

Matarić, M. J. (1997). Studying the role of embodiment in cognition. *Cybernetics and Systems, 28*(6), 457–470.

Matarić, M. J. (2001) Learning in behaviour-based multi-robot systems: Policies, models, and other agents. *Cognitive Systems Research, 2*, 81–93.

McCracken, G. (1988). *Culture and consumption: New approaches to the symbolic character of consumer goods and activities*. Indiana University Press.

McNeil, J. D. (1983). *Curriculum: Comprehensive introduction*. Little, Brown and Company.

Morris, M. W., Leung, K., Ames, D., & Lickel, B. (1999). Views from inside and outside: integrating emic and etic insights about culture and justice judgment. *Academy of Management Review, 24*(4), 781–796.

Nasir, N. S., & Cobb, P. (2007). *Equity in students' access to significant mathematical ideas*. Teachers College Press.

Orey, D. C. (2000). The ethnomathematics of the Sioux tipi and cone. In H. Selin (Ed.), *Mathematics across culture: The history of non-Western mathematics* (pp. 239–252). Kluwer.

Pike, K. L. (1954). *Emic and etic standpoints for the description of behavior*. Summer Institute of Linguistics.

Pike, K. L. (1967). *Language in relation to a unified theory of the structure of human behaviour*. Mouton.

Pollak, R., & Watkins, S. (1993). Cultural and economic approaches to fertility: Proper marriage or mésalliance? *Population and Development Review, 19*, 467–496.

Read, D. W. (2004). Mathematical modeling issues in analytical representations of human societies. *Cybernetics and Systems: An International Journal, 35*(2–3), 163–172.

Rios, D. P. (2000). Primero etnogeometría para seguir con etnomatemática. In M. C. Domite (Ed.), *Anais do Primeiro Congresso Brasileiro de Etnomatemática* [Proceedings of the First Brazilian Congress of Ethnomathematics] (pp. 367–375). FE-USP.

Rosa, M. (2010). *A mixed-methods study to understand the perceptions of high school leader about English language learners (ELL): The case of mathematics* (Doctorate dissertation) College of Education, California State University, Sacramento.

Rosa, M., & Orey, D. C. (2003). Vinho e queijo: Etnomatemática e modelagem! [Wine and cheese: Ethnomathematics and modeling!]. *BOLEMA, 16*(20), 1–16.

Rosa, M., & Orey, D. C. (2006). Abordagens atuais do programa etnomatemática: Delinenando-se um caminho para a ação pedagógica [Current approaches in ethnomathematics as a program: Delineating a path toward pedagogical action]. *BOLEMA, 19*(26), 19–48.

Rosa, M., & Orey, D. C. (2007). Cultural assertions and challenges towards pedagogical action of an ethnomathematics program. *For the Learning of Mathematics, 27*(1), 10–16.

Rosa, M., & Orey, D. C. (2008). Ethnomathematics and cultural representations: Teaching in highly diverse contexts. *ACTA SCIENTIA, 10*(1), 27–46.

Rosa, M., & Orey, D. (2009). Symmetrical freedom quilts: The ethnomathematics of ways of communication, liberation, and art. *Revista Latinoamericana de Etnomatemática, 2*(2), 52–75.

Rosa, M., & Orey, D. C. (2010a). Ethnomodeling: A pedagogical action for uncovering ethnomathematical practices. *Journal of Mathematical Modelling and Application, 1*(3), 58–67.

Rosa, M., & Orey, D. C. (2010b). Ethnomodeling: An ethnomathematical holistic tool. *Academic Exchange Quarterly, 3*, 14–23.

Rosa, M., & Orey, D. C. (2011). Ethnomathematics: The cultural aspects of mathematics. *Revista Latinoamericana de Etnomatemática, 4*(2), 32–54.

Rosa, M., & Orey, D. C. (2012a). O campo de pesquisa em etnomodelagem: As abordagens êmica, ética e dialética [The field of research in ethnomodeling: Emic, etic, and dialectical approaches]. *Educação e Pesquisa, 38*(4), 865–879.

Rosa, M., & Orey, D. C. (2012b). An ethnomathematical study of the symmetrical freedom quilts. *Symmetry Culture and Science, 23*(1), 191–220.

Rosa, M., & Orey, D. C. (2013a). Ethnomodelling as a research lens on ethnomathematics and modelling. In G. A. Stillman & J. Brown (Eds.). *Teaching mathematical modelling: Connecting to research and practice* (International perspectives on the teaching and learning of mathematical modelling). (pp. 117–127). Springer Science+Business Media Dordrecht.

Rosa, M., & Orey, D. C. (2013b). Ethnomodelling as a methodology for ethnomathematics. In G. A. Stillman & J. Brown (Eds.), *Teaching mathematical modelling: Connecting to research and practice* (International perspectives on the teaching and learning of mathematical modelling) (pp. 77–88). Springer Science+Business Media Dordrecht.

Scandiuzzi, P. P. (2002). Água e óleo: modelagem e etnomatemática? [Water and oil: Modeling and ethnomathematics?]. *BOLEMA, 15*(17), 55-58.

Sue, D. W., & Sue, D. (2003). *Counseling the culturally diverse*: theory and practice. John Wiley & Sons.

Wilson, M. (2002). Six views on embodied cognition. *Psychonomic Bulletin & Review, 9*(4), 625–636.

Urton, G. (1997). *The social life of numbers: A Quechua ontology of numbers and philosophy of arithmetic.* University of Texas Press.

CHAPTER 2

(HOW) AM I DOING MATHEMATICS?

Nirmala Naresh

ABSTRACT

People in several walks of life perform mathematical activities out of school, at home, and at work. However, those who perform these activities usually do not make the mathematics used in their everyday activities or at their workplaces explicit. Ethnomathematics researchers have called for theoretical and empirical research to further investigate the nature of mathematical knowledge used in everyday situations and to examine how it is similar to and different from mathematics learned in school. I drew upon the research fields of ethnomathematics and its subset workplace mathematics, to investigate the workplace mathematical activities of a group of bus conductors in Chennai, India. In this report, the focus is on Mr. Muthu, a bus conductor employed by the government agency, Metropolitan Transport Corporation (MTC). I provide an overview of Muthu's workplace mathematical activities; situate his workplace mathematical activities in the broader context of "bus conducting"; and describe his views of mathematics and its learning and discuss if and how his views (of mathematics) influenced his perceptions of mathematics used at this workplace.

Evidence-Based Inquiries in Ethno-STEM Research: Investigations in Knowledge Systems Across Disciplines and Transcultural Settings, pp. 35–55

INTRODUCTION

"I struggled doing mathematics. It was not my favorite subject. At school, I had trouble understanding ideas. [It] was very confusing—How could I be doing mathematics here [at work]?" quipped Muthu, a bus conductor, to my narrative of his workplace mathematical activities. The descriptions and related interpretations were rooted in an investigation of Muthu's work practices and were part of the findings from a larger study that investigated the mathematical activities of a group of bus conductors in Chennai, India. To understand the context of this challenge and my descriptions of Muthu's activities that triggered the challenge, I present here Muthu's story.

Muthu is a bus conductor employed by the Metropolitan Transport Corporation (MTC). Chennai is a Metropolitan city in Tamil Nadu, a state in the southern India. The major form of transport for people in Chennai is the public transportation system of buses provided by the state government agency, the Metropolitan Transport Corporation. Established in 1972, the MTC operates along 783 routes and serves 5 million passengers on any given day (MTC, 2013). The average daily collection per day is estimated to be 28 million rupees (MTC). The MTC has a fleet of 3,637 buses and offers 3,365 services (i.e., regular, express, limited stops, point-to-point). Currently over 18,000 drivers and conductors are employed by this organization.

Every MTC bus has a driver and a conductor. The driver is basically responsible for operating the bus and the conductor is in charge of issuing tickets to the commuters who travel in the bus. The duties of a bus conductor include: (a) issuing a ticket (a token of travel) to commuters based on their entry and exit points, (b) maintaining a record of overall daily earnings and ticket sales using official records such as a waybill (itemized summary of overall earnings for the day) and a traffic return (itemized record of overall ticket sales) for each bus trip, (c) determining the *batta* (an allowance based on the overall ticket collection amount), and (d) submitting the daily earnings to the accountant. Conductors do not use any technological gadgets (i.e., electronic ticketing machines, calculators) to complete situational mathematical tasks specific to their work-related activities.

A larger research study, workplace mathematics of bus conductors in India (Naresh, 2008), was carried out to observe, understand, describe, and analyze the mental mathematical practices of bus conductors and thereby contribute to the growing body of workplace mathematics research in India. In that study, I extended the study of workplace mathematics to a new cultural setting and a different cast of characters: MTC employees in Chennai, India. In this paper, I highlight salient aspects of one bus conductor's (Muthu's) mathematical activities by situating his particular acts in the broader context of "Chennai bus conducting." The following goals are central to this paper: (a) to highlight Muthu's efficient use of mental computation at work (b) to investigate the role of context and related parameters on his mental mathematics activities and (c) to discuss if and how his perceptions of mathematics impacted his views of mathematics used at work.

RESEARCH BACKGROUND

Everyday Mathematics and Workplace Mathematics

Mathematics knowledge has been traditionally considered as absolute and infallible (Ernest, 1991). This absolutist view of mathematics has been challenged since the beginning of the twentieth century and a new wave of "fallibilist" philosophies of mathematics has evolved. Researchers have recognized that the study of mathematical knowledge has a value dimension and hence it is a social and a cultural phenomenon as well (Bishop, 1988). This evolution has greatly influenced mathematics education research in the past few decades and major shifts have occurred in relation to the teaching and learning of mathematics (Presmeg, 2007).

Ubiratan D'Ambrosio launched his *ethnomathematical program* in the 1970s "as a methodology to track and analyse the processes of generation, transmission, diffusion and institutionalisation of mathematical knowledge in diverse cultural systems" (D'Ambrosio, 1990, p. 78). Ascher (2002) argues that mathematical thinking and learning take place in any culture and it is important to study the mathematics of different groups of people from all over the world. The author explains that such investigations broaden and modify the history of mathematics to one that has a multicultural, global perspective, and emphasize the interplay between mathematics and culture (Ascher, 2002). In the last two decades ethnomathematics researchers have analysed and documented the mathematics practices of adults as well as children that take place outside the school settings (e.g., Carraher et al., 1987; Gerdes, 1996; Jurdak & Shahin, 1999, 2002; Saxe, 1991). This association has given rise to the recognition of different forms of mathematics such as indigenous mathematics, everyday mathematics, folk mathematics, and workplace mathematics.

Workplace mathematics, a subset of the ethnomathematics research field is concerned with research conducted in work settings. This line of research gives some insight into how people conceptualize the role of mathematics in their work. Much of the early research on workplace mathematics described mathematical activities of workers particularly focusing on the developed countries. However, the accumulation of such studies in developing countries is now becoming substantial (e.g., Gahamanyi et al., 2009;

Millroy, 1992; Matos & Santos, 2008). More recent research in workplace mathematics has attempted to uncover the mathematical practices of specific groups (e.g., taxi drivers, nurses, automobile workers, carpet layers) and identify and establish abstractions that underlie these practices (Eglash et al., 2006; Gerdes, 2010; Magajna & Monaghan, 2003; Naresh & Chahine, 2013; Pozzi et al., 1998). Such studies assume significance because they "address lacunae in literature on the history of mathematical thought through the ages" (Presmeg, 2007, p. 449). The larger research study aimed to fill part of the gap by investigating the workplace mathematics of Chennai's bus conductors.

The Role of Mental Computation in Everyday Practices

Mental computation is a tool that people use when making quantitative estimates without having recourse to written algorithms. Efficient mental calculators used a variety of strategies (Chahine & Naresh, 2013; Hope & Sherril, 1987; Jurdak & Shahin, 1999) and these strategies differed across everyday activities and academic activities. Researchers argue that adults and children compute more efficiently mentally when problems are embedded in a specific context than when posed as a plain computational problem and have called for further investigations of everyday practices that involve mental mathematics (Carraher et al., 1987; Saxe, 1991). One of the goals of the larger research study was to highlight salient features of bus conductors' mental mathematical activities and thus it adds to existing research on mental computation.

One can adopt two different approaches with respect to mental computation, namely, traditional and contemporary views. According to the traditional view, mental computation is considered a skill that only a few individuals possess and that is available to them. The contemporary view associates mental computation with higher order thinking processes. According to this view, mental computation is considered a vehicle for promoting thinking, conjecturing, and generalizing based on conceptual understanding rather than as a set of skills (Trafton, 1986). A traditional information-processing perspective has been significantly used in research involving mental computation. This perspective involves two assumptions, decomposition and decontextualization (Silver, 1994). Decomposition implies that competence or skill can be broken down into individual elements of skill or knowledge. Decontextualization implies that knowledge exists in the mind of the individual and is independent of the situation in which it is used. This theory has been criticized mainly because research evidence (e.g., Carraher et al. 1987) suggests that, contrary to the view that an individual possesses a set of skills and competencies that can be decontextualized, research participants were able to apply these in certain everyday situations and not in certain school settings. Thus, researchers have argued that it would be worth considering researching mental computation from an alternate perspective, one that would take into consideration the influence of context and context related parameters on an individual's use of mental computation (Reys & Barger, 1994; Silver, 1994). This is the approach that I adopted to describe bus conductors' workplace mathematics.

THEORETICAL FRAMEWORK

The Reformulation, Translation, and Compensation (RTC) Model

The RTC model (Reys et al., 1982) serves as a useful framework to describe mental computational strategies in terms of the following processes: (a) reformulation (i.e., the alteration of the numerical data in the problem), (b) translation (i.e., modifying the mathematical structure of the problem), and (c) compensation (i.e., making adjustments to the problem to compensate for the reformulation and transformation techniques used). Shimizu and Ishida (1994, p. 174) provide the following example to illustrate the use of RTC strategies to mentally estimate the value of 54215 * 0.3:

> (1) 54,215 * 0.3 = 5, 421.5 * 3 (using arithmetic properties), (2) rounding 54,215 to 54,000 (reformulation process), (3) 5,400 * 3 = 5.4 * 3 * 1000 (using arithmetic properties), (4) 5 * 3 = 15, 6 * 3 = 18 (averaging or transformation strategy), (5) the average of 15 and 18 is 16 (compensation), and (6) the final answer is 16,000.

In this chapter, I use the RTC model to elucidate Muthu's mental mathematical strategies to situational computational tasks and to provide a rationale for Muthu's mental mathematical actions.

RESEARCH METHODS

The overall research study was qualitative in nature and an instrumental case study (Stake, 2000) methodology was employed to explore the research goals. After gaining entry into the MTC, a purposive sampling strategy was used to select the five conductor-participants, who were carefully and appropriately chosen based on their years of service with the

MTC, educational qualifications, service records, and their willingness to participate in the study. In this chapter, I focus on one bus conductor's (Muthu's) case study to explore in depth the mathematical activities at this workplace. Data sources included: (a) a conductor's manual (CM), (b) official records (OR) specific to Muthu's bus trips (i.e., work schedules, the bus route information manual, waybill, traffic return; these documents outlined the duties of the bus conductors and helped me better understand the MTC bus schedules and bus routes), (c) the MTC website (www.mtcbus.org), (d) field observations and related notes (FO/FN), (e) descriptive and reflective summaries of field observations (SFO), (f) summaries of informal conversations with Muthu (IC), and (h) transcriptions of formal and semi-structured interviews (FI/SSI).

I accompanied and observed Muthu two times a week during his work shifts before, during, and after the bus trips, observing a total of 10 shifts. During these sessions, I observed the physical setting (i.e., conductors' lounge, the bus depot, the bus), Muthu's activities and interactions with the passengers, and the duration of such activities. Based on the first few observations, I singled out several episodes that were most helpful in examining Muthu's use of mental computation at work. I scrutinized his work-related mathematical activities and identified and described solution strategies that were used to complete these activities. Observation sessions always included short informal and semistructured interviews, which took place during break times. Informal interviews were designed to elicit information about the Muthu's perceptions of mathematics, his views about the role of mathematics at work, and his opinions about formal school-taught mathematics. Semistructured interviews dealt with questions related to my on-site observations and to Muthu's work-related documents.

In order to investigate the influence of context-related parameters on Muthu's mental computational activities, I asked Muthu to solve two types of computational tasks outside work settings, contextual (CT) and plain (PT) (Naresh, 2008). I administered the tasks orally in an informal setting and Muthu was given 10–15 seconds to complete each task. The CTs were administered first; the PTs were administered a week later in the same setting. The PTs involved the same mathematical computation as the CTs, but were completely stripped of the context. I compared Muthu's performances on CTs and PTs to investigate the influence of context on his computational skills. During the semistructures interviews, I used a stimulated-recall method to help Muthu reflect on his mental computational acts and explain his decision-making processes. During these sessions, I probed Muthu to justify his choice of solution strategies, use of alternate strategies, and use of school-taught strategies.

I employed three levels of data analysis for this study. Level one of data analysis took place simultaneously with data collection process. To identify mathematical aspects of Muthu's everyday practice, I carefully scrutinized information from several sources such as field notes, the conductors' manual, and summary of informal sessions. At the end of each field visit, I reviewed my field notes and personal reflections and identified features of Muthu's workplace that needed further scrutiny in the subsequent sessions.

Level two of data analysis resulted in the creation of a case study database. This folder included information about Muthu's background information, a narrative that included information about Muthu's workplace activities, his work-related goals and activities, his mathematical activities, a summary of informal chat sessions, transcripts of formal interviews, and Muthu's responses to the two types of structured tasks. This case study database served two purposes. First, it acted as a repository from which I was able to view and review the evidence as and when needed. Second, creation of this database also increased the reliability of the case study.

For the third level of data analysis, I used the RTC model to further scrutinize the data from the case study database. I used this framework to describe Muthu's solution strategies to work-specific computational tasks, contextual and plain word problems. I used Muthu's descriptions of solutions strategies to identify and attest components of RTC model evident in those strategies.

MUTHU'S STORY

Muthu was a junior conductor who had nine years of service with the MTC. He was highly recommended for this study because of his excellent service records and he had a polite demeanor. Muthu has served on many bus routes operated by the MTC. He has served on many types of bus services including the ordinary service (distance covered is less than 20 km, ticket fares are the least, picks up passengers at every bus stop along that route, buses are not air-conditioned), limited stop service (distance covered is between 20 km to 30 km; fares are slightly higher than the ordinary service; picks up passengers at selective bus stops along that route; some of these buses are air-conditioned) and the express bus service (distance covered is more than 30 km; fares are higher than the limited stop service, picks up passengers at selective bus stops along that route, some of these buses are air-conditioned). Muthu frequently served on the coveted express bus service (bus number 170 T), which is one of the longest routes operated by the MTC. According to his immediate supervisor, Muthu was a sincere, dedicated, and a hardworking employee. The daily revenue earned from his bus services were almost above the average daily earnings of the MTC.

Muthu's Reflections: School Mathematics Learning Experiences

Muthu was born and brought up in the suburbs of the city of Chennai. He completed his school education in an all men's school in Chennai. He attended a men's college in Chennai and graduated with a bachelor's degree in economics. He called himself an average student in mathematics and said noted that he had not been particularly fond of school mathematics. When asked to summarize in one word his school mathematics learning experiences, Muthu said that he "struggled." He was barely able to recall any mathematical topics that he had learned in school. He recalled that he got a "mere passing grade" in mathematics throughout his middle school and high school years. When asked to elaborate on this topic, he hesitated and then spoke with a tone of guilt in his voice: "In maths classes, I struggled. The teacher never helped me nor answered any questions. My high school teacher never said or explained anything to me." However, he felt guilty for blaming the teachers and attributed part of his struggle to his indifference towards the subject: "I wasted my time in classes. Maths was not my favorite subject. I liked commerce better than math."

When I questioned Muthu about his views of mathematics he was puzzled. He said that he had never been asked such a question before. H added, "Why do you want to know? Why does it matter?" After I reiterated my research purposes and my interest in his views, he described his view of mathematics as, "It is a subject one learns at school and college. There is nothing else to think about it. One has to do mathematics." He believed that mathematics was important because it held the "key" to higher studies. He added, "One has to learn complex stuff like algebra and other formula-based math to succeed. Sometimes you have to do even construction with compass and set squares."

Muthu regretted his poor performance in mathematics at school and wished he had a better teacher, worked harder, performed well in exams, and pursued higher studies. He believed that a good performance in mathematics would have given him a "free pass" to higher education and better job prospects. Muthu's average performance in school mathematics and his unpleasant school mathematics learning experiences did not seem to have had any negative impact on his views of mathematics. During our conversations, he mentioned several times that it is very important for someone to concentrate on mathematics because it is needed for higher studies and for landing good jobs. To that end, he proudly mentioned that he had arranged for extra "math tuition lessons" for his children so they can excel in school mathematics.

Muthu's Reflections: Perceptions of Mathematics at Work

When asked to comment on the use of mathematics at work, Muthu immediately said, "I don't think about mathematics at work." His justified his comment by stating, "every time I board the bus, I am completely absorbed in work and never thought about anything else." During our interview sessions, I probed him to think deeper about specific aspects of his work that required the use of mathematics.

Researcher (R): Do you think you do math?

Muthu: No! I don't think like that. Sometimes I give change to passengers and so on. It is just my job. I will only call it job activity. Not mathematical activity.

R: Why?

Muthu: That is what you do for the job. See you start with taking the ticket bundles, then we start the first bus trip, and then we do it internally. But on the outside, we are completing my duties like issuing tickets and so on. Calculate 100s, 50s, 10s and so on. I will only say that I am working.

R: How are you enacting the ticket transactions? Are you doing any calculations internally or externally?

Muthu: I give tickets and fill out forms. I am thinking hard in my mind—quickly too— about a lot of things, like calculating the ticket fare using the passenger information.

R: How do you do that?

Muthu: See ticket fares are based on entry and exit points. So I have to think about it internally and determining the ticket fare too—it happens simultaneously.

R: So, what do you call this process? Mathematical thinking ...

Muthu I don't know if I am doing mathematics for my job. If I am a teacher, and I teach students to do problems, then I can say that I use mathematics at school. Remember the algebra and compass stuff ... I do well back then—So how can I say that I am doing mathematics [now]?

Although Muthu acknowledged that he did engage in mental arithmetic to complete work-related activities, he strictly termed those engagements as the "requirements of his job." He did not think that he was "doing mathematics" as the types of engagements were different from those that happened at school in terms of the mathematical content, nature of the tasks (contextual vs "school-like"), and the methods of engagement (quasi-structured at work-setting vs highly structured in a school setting).

A Description of Muthu's Duties

The MTC has set rules that concern a conductor's overall appearance and behavior, his responsibilities during a work-shift, and his obligations to the commuters. These rules, as listed in a conductor's manual are: (a) Display a route information board correctly and clearly on the bus, (b) Check if the bus is clean and tidy before each the start of each single, (c) Maintain a fully stocked first aid box on the bus for passengers' safety at all times, (d), Regulate entry and exit of passengers into their bus at every stop and then signal the driver to start the bus, (e) Move up and down the bus; collect required ticket fare; and issue tickets to every passenger over the age of 3, (f) If needed, collect right luggage fares from passengers and issue appropriate tickets, (g) Maintain an updated record of ticket sales (within and between each bus trip) and present it to checking inspectors upon request, (h) Facilitate alternate travel arrangements for passengers in case of accidents or bus breakdowns and if unable to do so, issue a full refund to the passengers. An MTC bus conductor is an employee of the state government agency. As a public servant, at all times during a work-shift, Muthu is expected to behave professionally, act in the best interest of his commuters, and thereby serve the community.

A Snapshot of Muthu's Activities During a Work-Shift

During a work-shift, Muthu's work-specific engagements unfolded over three different phases: (a) sign-on time (beginning of the shift), (b) spread-over time (during bus trips), and (c) sign-off time (end of shift). At the start of a work-shift, Muthu first reported to his immediate supervisor. During this phase (sign-on time: duration 30–45 minutes), Muthu obtained information about the assigned bus (i.e., bus route, bus number, bus driver), collected the required artefacts (i.e., imprest, ticket bundles, official records), and located the bus. During the spread-over phase (duration: 6–8 hours), Muthu made several trips along his assigned bus route from point A to point B, selling tickets and directing passengers' entry into and exit from the bus. In addition, he kept a record of the number and cost of tickets within and between each bus trip. At the end of the last bus trip on his work shift, Muthu determined his *batta* (daily allowance) and shared half of that amount with his bus driver. During the sign-off time (duration: 30 minutes), Muthu reported to the accounts section and submitted the trip earnings from his work shift to the accountant in charge. Once the accountant signed off on the waybill and Muthu officially signed off from that shift, his job was over for the day.

Mental Computation-Work Setting

At work, Muthu used mental computation (a) to complete ticket transactions, (b) to determine the overall trip earnings, and (c) to determine his everyday allowance (*batta*).

Ticket transactions. A ticket transaction could be initiated by a passenger or by the conductor and typically took 10 to 20 seconds. Muthu noted that, ticket transactions are "simplistic" when passengers promptly provide all relevant information as well as the correct ticket fare. However, in any given trip, such cases are rare. I present one scenario to highlight the complexities associated with a nonsimplistic ticket transaction (FO/FN).

Bus number 170 T is operated as an express bus service and runs between KKD Nagar (origin) and Vandaloor Zoo (destination). The entrance and exit were located in the back and the front of the bus respectively. The described actions took place during the second trip of the work shift. While the work shift began at 5:00 A.M., the trip in action began at 10:05 A.M. The bus was overcrowded, its normal capacity of 70 passengers swollen by another 40 people. Muthu was dressed in his uniform of navy blue pants and a light blue shirt. His nametag was pinned above the pocket on the left side of his shirt and a whistle hung from a cord around his neck. A brown faux leather conductor bag was tightly tucked under his left shoulder. The bag contained currency and coins of different denominations and some official documents. In the palm of his left hand, Muthu held ticket bundles with several different denominations. Muthu skilfully navigated his way through the crowded bus enacting on-the-fly ticket transactions to authenticate each passenger's trip. Many passengers in the bus travelled to destinations that lay along that bus route. A few did not and asked the conductor for additional directions. Some passengers asked for tickets by mentioning their points of exit. Many others did not. Muthu sought the required information as and when needed and then completed the transactions. Some passengers tended the exact amount towards their ticket fares. Many did not. Muthu either asked for additional ticket fare or issued a balance ticket fare as appropriate.

Here is an example of a ticket transaction scenario. In this situation, the passenger requested tickets by mentioning an exit point. The passenger's exit point was along that bus's route (FO/FN).

Three passengers boarded the bus at Vandaloor Zoo. One of the passengers requested the conductor to issue three tickets to SIDCO and handed over a 500-rupee note. (₹ is the symbol for the Indian currency rupee). He also requested Muthu to issue the balance amount [₹455.00] using smaller currency denominations. Muthu issued three tickets and wrote behind one of their tickets 500/3 and signed below. He asked the passenger to collect the balance amount before exiting the bus.

First, Muthu used his prior knowledge about bus stops and fare stages (a fare stage denotes a major bus stop at which ticket fares are subject to change) to determine the fare stage numbers (FSN's) specific to the passenger's entry and exit points. Knowing that the passengers entered at FSN 25 and will exit at FSN 2, Muthu determined the ticket fare for one passenger as ₹15.00 and the total ticket fare as ₹45.00. Later, upon request, Muthu created Table 1 to explain the procedures for determining the FSNs. However, at no point during his bus trip, he consulted such a chart to determine the FSNs. Muthu engaged in the process of determining the FSNs mentally and concurrently during the ticket transaction.

Table 2.1.
Ticket Prices for Bus Number 170 T (Express)

Fare stages	25, 24, 23	22, 21	20	19,18	17,16	15,14	13	12–8	7,6,5	4–1
Ticket price	5	7	8	9	10	11	12	13	14	15

Muthu cited several reasons for not issuing the balance amount right away. The passenger's exit point was closer to the last bus stop and thus he knew that the passenger was not in a hurry. Moreover, he did not have the time to identify and match the balance amount (as per the commuter's request) as there were a number of other passengers to attend to. Later, Muthu identified the passenger and issued the pending balance amount of ₹455.00 using ten 20-rupee notes, 25 ten-rupee notes and a five-rupee coin. Muthu noted that he was happy to oblige the passenger as he wanted to reserve the larger currency denominations in his possession (for easy storage) and dispense the smaller currency denominations. By being flexible, he not only pleased the commuter but also made a part of his job (storing trip earnings) easier. During the bus trips, Muthu summoned several such 'tricks of the trade' which he employed as needed to complete the ticket transaction quickly and efficiently (Naresh, 2008).

Determination of overall trip earnings. Every day at the end of his work shift, Muthu used an official record called a waybill to record the number of tickets sold in each denomination and then used this information to calculate overall revenue for that work shift. Corresponding to each ticket denomination, Muthu first determined the number of tickets sold, using the beginning and ending numbers on his ticket bundles. Next, he determined the collection amount for each ticket denomination. In Table 2.2, I present information adapted from Muthu's waybill contents (Naresh, 2008) from one of his work shifts (OR).

Muthu took less than 15 seconds to determine—mentally—the collection amount for each ticket denomination. He recorded the solutions in his waybill to keep track of his calculations from time to time. An analysis of

Table 2.2.
Mental Strategies Employed in Completing the Waybill

Ticket Denomination	Starting & Ending Numbers On Ticket Bundles	Number of Tickets Sold	Research-Identified Solution Strategy
2.00	442079 - 442100 442200 477800 477900 478000-478010	100-79 = 21; 100; 100; 100; 100; 10; Total: 431.	To find 431x2, double 431. 400 + 400 = 800; 31+31 = 62; Total: 800 + 62 = 862.
3.00	372037 - 372100 - 400900 - 401000 - 401030	100 - 37 = 63; 100; 100; 30; Total: 293.	To find 293 x 3, first find 300 x 3 as 900; (reformulation) Subtract 7 x 3 = 21 from 900; (compensation) 293 x 3 = (300x3) - (7x3) = 900 - 21 = 879.
3.50	968900 - 968961	61	61x3.50 = (60x3.50) + (1x3.50) =(60x3) + 60/2 + 3.50 (Translation to find 60x0.5) = 180+30+3.50 = 213.50.

his mathematical descriptions compels us to infer that he consistently used associative, distributive, and commutative properties of whole numbers, fractions and decimals, and the reformulation, translation and compensation (RTC) strategies to solve mental mathematical tasks (Reys et al., 1982).

Determination of Daily Allowance. In addition to his monthly salaries, MTC conductors and drivers earned a daily allowance (*batta*) for completing a work shift. Conductors calculated this amount using the overall trip earnings from a work shift. The everyday allowance is composed of both a fixed and a variable allowance. The fixed allowance is preset for each bus type; the variable allowance is calculated using the overall trip earnings from a given work shift. I use data from one of Muthu's work shifts to describe his mental computational acts specific to this phase of his work.

The fixed allowance for the bus service L23C was ₹18.50 (OR). For this bus type, the variable allowance was fixed at ₹2.35 for every 100 rupees collected in fares. The waybill contents for this scenario indicated that Muthu remitted ₹2894 to the accountant and claimed ₹89 as the *batta* (which he split evenly with the driver). The waybill records indicated that Muthu claimed Rs. 89 as the *batta* and remitted Rs. 2894 to the accountant. Here is Muthu's description of batta calculation strategies (SSI).

In an ordinary service, for every 100 rupees collected, the variable allowance is ₹2.35. So the total amount collected is ₹2953. I decompose 2953 as 1000+ 1000 + 953. The variable *batta* for this amount is 23.50 + 23. 50 + 23.50 = 69.50. I also sold three token tickets that day; for each ticket I will get 10 paise each. Thus, the total variable *batta* is ₹70.00 after rounding to the nearest rupee. The fixed allowance is 18.50. If this amount is added to Rs. 70.00, I get the *batta* as Rs. 88.50. In order to subtract this amount from 2,953.00, I first subtracted 100 to get 2,853. I then compensated by adding the extra amount that I took away by adding 11.50 to it. Thus, the total amount after deducting *batta* is 2864.50. To this amount, I added the imprest (advance amount of ₹30.00) to get the total amount as 2894.50. This is what I owe the accountant.

Note that Muthu allotted ₹23.50 as the *batta* amount corresponding to ₹Rs. 953. When questioned, he said that it is acceptable to round off ₹953 to ₹1000 for *batta* calculation purposes. However, the waybill entries showed that the conductor had received ₹89.00 as the daily allowance. The accountant indicated that the computer is programmed to round off any paise amount to the nearest rupee.

Mental Computation: Nonwork Setting

I now present findings related to Muthu's use of mental strategies as he solved computational tasks outside work settings. Table 2.3 presents a summary of his responses to the contextual tasks (CTs) and Plain computational tasks (PTs). Details of two tasks are subsequently presented to illustrate the contrasts in his thinking.

Table 2.3.
A Numerical Summary of Muthu's Responses to CTs and PTs

Prob. #	1	2	3	4	5	6	7	8	9	10	11	Total Correct	%
Context	√	√	X	√	√	√	√	√	√	√	√	10	91
Plain	X	X	X	X	X	X	X	X	X	X	√	2	18

Muthu answered all but one of the contextual tasks correctly. The only task he missed was task number three which is stated here:

Contextual task 3: Four members of a family want to travel to destination A. The ticket price to travel to destination A is Rs.5.30. One passenger initially gives Rs.15 towards the total ticket cost. How much more does the passenger need to pay?

Muthu calculated the ticket fare for four passengers as 21.10 using the following strategy: 5*4 =20 (in rupees); 30*4 =120 (in paise), 120 paise is the same as 1.20 rupees, so the total ticket fare is 21.20. Instead of answering the questions, Muthu wrote down this number the answer. During the review session, he chided himself for not reading the question well; then chuckled and noted that the absence of the contextual parameters (i.e., passengers, monetary units, ticket bundles) "got to him" in this scenario. Muthu found all the problems in this category easy because the contexts were "good and familiar." He used strategies similar to the ones that he used at work to solve most of the problems.

Although Muthu was highly successful in solving contextual tasks, he was completely perplexed when he was presented with the plain computational tasks. Contextual tasks helped Muthu use efficient strategies, but this was not the case with plain tasks. To illustrate this point, I present his solution to the following problem.

Contextual task 9: A typist types 25 words per minute. How long will it take the typist to type a document that contains 430 words?

Muthu's solution: I know that I had to do 430/25. When you listen to the problem, I know that in four minutes, 100 words can be types, and for 400 words, it is 4 x4, which is 16; add one more minute to type 25 words. So the answer is 18 minutes if you want full minutes, or17 minutes and few seconds.

However, when the same problem was presented devoid of context, to determine 430/25, Muthu resorted to the long division method without any success. Muthu acknowledged that the context of the problem greatly influenced the choice of his strategies. He said that the plain computational problems reminded him of school mathematics, and he found it difficult to recall appropriate school-taught strategies to solve them. He lamented, "I was not very good at mathematics at school—now I have trouble solving school-type problems." However, after he explained his solution strategies to contextual tasks, he was able to see that the plain tasks could be solved the same way as he solved the computational tasks.

RESEARCH GOALS—A DISCUSSION

Every day, as part of his work activities, Muthu solved several in situ computational tasks associated with the ticket transactions and waybill and *batta* calculations. For a given bus service, a bus trip could be characterized in terms of the number of fare stages between the origin and destination. Muthu's activities between any two successive fare stages are outlined in Figure 2.1.Muthu completed additional activities during the last two fare stages on the last bus trip on his work shifts (Figure 2.2).

Muthu noted that the use of context-related artefacts such as monetary units greatly helped him to complete mental calculations quickly. Official records such as the waybill and the traffic return helped him to keep track

Figure 2.1. Muthu's activities between fare stages.

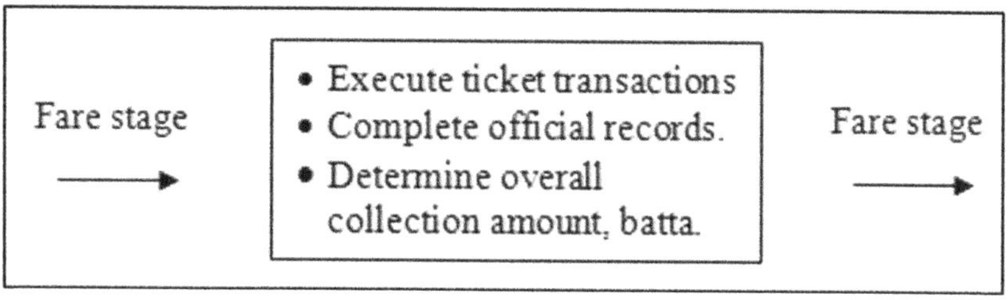

Figure 2.2. Muthu's activities between fare stages (on the last bus trip).

of the number of tickets sold in each denomination and to document his solutions to mental math tasks. Below, using his own words, I highlight context-related parameters that helped him efficiently complete computational tasks at work (SSI).

I know how to solve problems without using money. But when I use money, I calculate very quickly. There is also another benefit. You can check if you are doing correct calculations by looking at the money in hand. I think I use lot of different ideas here, sometimes passengers help, money is a big factor, math that I already knew from school, use of bus tickets, and much more. The math I do at work is possible because of what is available at work. I adapt and use all these factors to be an efficient conductor.

Amid tough working conditions, Muthu used context and related parameters (social interactions with passengers, monetary units, ticket bundles, prior knowledge about MTC bus routes, fare stages, ticket fares) to acquire, adapt, and use efficient mental strategies to complete *in situ* computational tasks. In so doing, Muthu (and his fellow conductors) contributed to a repertoire of knowledge that I term "conductors' mathematics" whose characteristics are shaped by the dynamics, context, and tools specific to their workplace. These findings are consistent with those of other studies where workers used tools and artefacts specific to their workplace to complete the workplace activities (Magajna & Monaghan, 2003; Pozzi et al., 1998).

An analysis of Muthu's use of mental computation both in work-related and non-work-related settings reveal that context and related parameters had a great influence on the choice of his solution strategies. While Muthu was highly successful in using efficient mental strategies to solve problems at work, he was less successful in solving non-work-related problems, especially those that were devoid of context. He totally relied on using traditional algorithms to solve such tasks. Muthu explained that the settings at work were different from those outside of work and hence he was unable to use the same strategies on both occasions. In his view, the in situ mental computational tasks that emerged at work were inextricably tied to his work-related activities, had a clear purpose and hence were meaningful. On the contrary, the non-work-related problems lacked a practical purpose. Muthu's views about mathematics did impact his choice of solutions strategies specific to plain computational tasks. These tasks reminded him of the school mathematical problems and required the use the traditional algorithms, as he believed that "the chance of making mistakes is less. Formulas always work. They are never wrong." However, he was unable acquire and appropriate such algorithms to solve the problems as they were not quickly and readily available. Muthu was unable and perhaps unwilling to see the effectiveness of using simpler mentally manageable solutions to solve the two types of tasks.

A closer scrutiny of Muthu's mental strategies supports conclusions related to the use of mental computation in everyday situations (Chahine & Naresh, 2013). Like his counter-parts engaged in the acts of candy selling (Carraher et al., 1987; 1987; Saxe, 1991), street vending (Jurdak & Shahin, 1999) and newspaper vending (Santos & Matos, 2002), Muthu used situation-efficient strategies to solve computational tasks. On the theoretical level, research evidence supports a model of mental computation as a vehicle for promoting thinking, conjecturing, and generalizing based on situational and conceptual understanding, rather than the traditional view of such computation as a set of skills to be mastered.

People in several walks of life perform mathematical activities out of school, at home, and at work. However, adults who perform these activities usually do not make the mathematics used at their workplace explicit. In particular, Vergnaud (2000) claims "most of the knowledge used at the workplace remains implicit, sometimes unconscious" (p. xvii). This was evident in the bus conductors' workplace. In Muthu's case, his perceptions of mathematics and related struggles with school mathematics may help us describe why (his) workplace mathematical activities were invisible to his eyes. Muthu considered mathematics as a "subject learned at school" and associated it mostly with "procedures and formulae" used in school classrooms. Since he had not fared well in the subject in such settings, he truly believed that he was not capable of doing mathematics outside school settings. Thus, in his eyes, his workplace activities (interpreted as mathematical activities) were completely removed from what he perceived as "genuine mathematical activities." For example, consider a ticket transaction activity—Implicit in each transaction is a mental computational task that needed Muthu to engage in mental computation (i.e., determining the fare stage numbers, ticket fares and balance amount). However, in Muthu's view, associated with each ticket transaction is the need to authenticate a passenger's travel in the bus. Though initially Muthu challenged my interpretations of his workplace activities, later, upon reflection he noted,

You may be right. However, when I work, I am completely focused. I do not think about what mathematics I am using. I see money, I hear words, my hands deal with tickets and money and I calculate in my heart. There is little time to think about anything else. I cannot afford to make mistakes in my calculations because I may end up losing department's money. So, I only concentrate on doing my job quickly and correctly. A conductor's job is stressful, risky, and highly demanding. In an office job, if you make mistakes, you get the time and chance to correct it later. But in our job, we cannot do today's work tomorrow—or this minute's work the next minute.

Every day, Muthu reported to work with two main goals: (a) to generate maximum revenue through ticket sales (economic goal), and (b) to provide the best possible commuter service (service goal). As an employee of the MTC, he was expected to fulfil the economic goal to the employer's satisfaction. On the other hand, as a public servant, he had to fulfil the service goal to the customers' satisfaction. Muthu accepted and enacted this dual role willingly and effortlessly.

By recognizing concomitant perceptions of Muthu's work-related activities, we can rationalize his initial resistance to my interpretations of his workplace mathematical practices. At the same time, an acceptance of such parallel perceptions helps us, ethnomathematics researchers, to valorize Muthu's mathematical practices that are often obscured by the demands of his workplace. In effect, while Muthu's workplace mathematical engagements are invisible to him due to the pressures of his practice, to us, ethnomathematics researchers, they serve as exemplars of ethnomathematical activities.

REFERENCES

Ascher, M. (2002). *Mathematics elsewhere: An exploration of ideas across cultures.* Princeton University Press.

Bishop, A. J. (1988). Mathematics education in its cultural context. *Educational Studies in Mathematics, 19*, 179–191.

Carraher, T. N., Carraher, D. W., & Schliemann, A. D. (1987). Written and oral mathematics. *Journal for Research in Mathematics Education, 18*, 83–97.

Chahine, I. C., & Naresh, N. (2013). Mapping the cognitive competencies of street vendors and bus conductors: A cross-cultural study in workplace mathematics. *Latin American Journal of Ethnomathematics, 6*(3), 7–35.

D'Ambrosio, U. (1990). The role of mathematics education in building a democratic and just society. *For the Learning of Mathematics, 10*(3) 20–23.

Eglash, R., Bennett, A., O'Donnell, C., Jennings, S., & Cintorino, M. (2006). Culturally situated designed tools: Ethnocomputing from field site to classroom. *American Anthropologist, 108*(2), 347-362.

Ernest, P. (1991). *The philosophy of mathematics education.* The Falmer Press.

Gahamanyi, M., Andersson, I., & Bergsten, C. (2009). Using mathematics as a tool in Rwandan workplace settings: The case of taxi drivers. *Proceedings of the Sixth Congress of the European Society for Research in Mathematics Education* (CERME), Lyon, France. http://ife.ens-lyon.fr/publications/edition-electronique/cerme6/wg8-05-gahamanyi.pdf

Gerdes P. (1996). Ethnomathematics and mathematics education, In A. J. Bishop, M. A. Clements, C. Keitel, J. Kilpatrick, & F. K. S. Leung (Eds.), *Second international handbook of mathematical education,* (pp. 909-943). Kluwer.

Gerdes, P. (2010). *Otthava: Making baskets and doing geometry in the Makhuwa culture in the northeast of Mozambique*. Lúrio University, Nampula & Lulu.

Hope, J. A., & Sherril, J. M. (1987). Characteristics of unskilled and skilled mental calculators. *Journal for Research in Mathematics Education, 18*, 98–111.

Jurdak, M., & Shahin, I. (1999). An ethnographic study of the computational strategies of a group of young street vendors in Beirut. *Educational Studies in Mathematics, 40*(2), 155–172.

Jurdak, M., & Shahin, I. C. (2002). Problem solving activity in the workplace and the school: The case of constructing solids. *Educational Studies in Mathematics, 47*, 297–315.

Magajna, Z., & Monaghan, J. (2003). Advanced mathematical thinking in a technological workplace. *Educational Studies in Mathematics, 52*(2), 101–122.

Matos, J, F., & Santos, M. (2008). Activity, artefacts and power: Contribution of activity theory and situated learning to the analysis of artefacts in mathematical thinking in practice. In J. F. Matos, P. Valero, & K. Yasukawa (Eds.), *Proceedings of the Fifth International Mathematics Education and Society Conference* (pp. 1–9). http://mes5.learning.aau.dk/Papers/Matos_Santos.pdf

Metropolitan Transport Corporation. (2013). http://www.mtcbus.org

Millroy, W. L. (1992). An ethnographic study of the mathematical ideas of a group of carpenters. *Journal for Research in Mathematics Education.* Monograph No. 5. National Council of Teachers of Mathematics.

Naresh, N. (2008). *Workplace mathematics of bus conductors in Chennai, India* (Doctoral dissertation). Illinois State University. gradworks.umi.com/3353093.pdf

Naresh, N., & Chahine, I. C. (2013). Reconceptualizing research on workplace mathematics: Negotiations grounded in personal practical experiences. *Journal of Research in Mathematics Education, 3*(2), 316–342.

Pozzi, S., Noss, R., & Hoyles, C. (1998). Tools in practice, mathematics in use. *Educational Studies in Mathematics, 36*, 105–122.

Presmeg, N. C. (2007). The role of culture in teaching and learning mathematics. In F. K. Lester (Ed.), *Second handbook of research on mathematics teaching and learning* (pp. 435–460). Information Age.

Reys, B. J., & Barger, R. H. (1994). Mental computation: Issues from the United States perspective. In R. E. Reys & N. Nohda (Eds.), *Computational alternatives for the twenty-first century* (pp. 31–47). National Council of Teachers of Mathematics.

Reys, R. E., Reys, B. J., Rybolt, J. F., & Wyatt, J. W. (1982). Processes used by good computational estimators. *Journal for Research in Mathematics Education, 13*, 183–201.

Santos, M., & Matos, J. F. (2002). Thinking about mathematical learning with Cabo Verde ardinas. In G. de Abreu, A. J. Bishop, & N. C. Presmeg (Eds.), *Transitions between contexts of mathematical practices* (pp. 81–122). Kluwer.

Saxe, G. (1991). *Culture and cognitive development: Studies in mathematical understanding*. Lawrence Erlbaum Associates.

Shimizu, K., & Ishida, J. (1994). The cognitive processes and use of strategies of good Japanese estimators. In R. E. Reys & N. Nohda (Eds.), *Computational alternatives for the twenty-first century* (pp. 161–178). National Council of Teachers of Mathematics.

Silver, E. A. (1994). Treating estimation and mental computation as situated mathematical processes. In R. E. Reys & N. Nohda (Eds.), *Computational alternatives for the twenty-first century* (pp. 147–161). National Council of Teachers of Mathematics.

Stake, R. E. (2000). Case Studies. In N. K. Denzin & Y. S. Lincoln (Eds.), *Handbook of qualitative research* (2nd ed., pp. 134–164). SAGE.

Trafton, P. R. (1986). Teaching computational estimation: Establishing an estimation mind-set. In H. L. Schoen & M. J. Zweng (Eds.), *Estimation and mental computation* (pp. 16–30). National Council of Teachers of Mathematics.

Vergnaud, G. (2000). Introduction. In A. Bessot & J. Ridgway (Eds.), *Education for mathematics in the workplace* (pp. xvii–xxiv). Kluwer.

CHAPTER 3

UNRAVELING THE MATHEMATICS OF NDEBELE BEADWORK

Transformations on an Indigenous Cartesian Plane

Alesia Mickle Moldavan

ABSTRACT

This chapter is a personal reflection on my participation in the immersion experience within the Ndebele context. Drawing on the history of Ndebele beadwork as well as what beaded designs communicate both symbolically and mathematically, I highlight the potential role that indigenous cultures play in producing, appropriating, and disseminating mathematical ideas.

Evidence-Based Inquiries in Ethno-STEM Research: Investigations in Knowledge Systems Across Disciplines and Transcultural Settings, pp. 57–74

Growing up in a predominately White, middle-class neighborhood, I attended school with peers who shared similar lived experiences and cultural backgrounds to that of my own. My initial understanding of mathematics developed from the myth that mathematics was universally studied and applied in one manner, the way I learned and used mathematics. My naïveté and lack of exposure to studying mathematics from different cultural perspectives perpetuated this belief. Only after I left the confines of my community and pursued a career in mathematics education did I begin to understand the complexity of the world's cultures and the error in describing mathematics as a culture-free discipline.

After I received my teaching degree, I moved to a major metropolitan city and began teaching secondary mathematics at a Title-I high school in a predominately African American school district in the southeastern United States. I found myself surrounded by people of many nationalities, religious beliefs, and cultural backgrounds. Thereby, my students contributed unique cultural perspectives, interests, and prior mathematical experiences that forced me to reconceptualize how I would best cultivate a mathematics learning environment that acknowledged my students' academic assets and interests. I quickly learned that I had to make mathematics culturally relevant for my students to find ownership and significance in their learned mathematics content, a challenge with which I knew I needed assistance.

I enrolled in graduate school with the goal of learning how to better reach my students. After teaching all day, I spent my evenings in class at a large, urban research university. The university's focus exposed me to various cultures and pedagogies that embraced diversity in mathematics education. My coursework better prepared me to teach students from diverse backgrounds in urban environments and introduced me to the study of ethnomathematics—mathematics practiced among various cultural groups (D'Ambrosio, 1985).

I found great interest in researching ethnomathematics and its view of mathematics as a human endeavor (Swetz, 2009). Mathematics from a cultural perspective made sense to me given the ways in which mathematics evolved through history in response to people using it to better understand the world. Furthermore, the idea that mathematics could be described in terms of cultural practices (Mukhopadhyay et al., 2009) encouraged me to study mathematics through cultural activities. By applying Bishop's (1988) six mathematical concepts (i.e., counting, measuring, locating, playing, designing, and explaining), I could observe and appreciate the mathematical practices of various cultures. I could also self-identify with its applications and position mathematics in academic relevance.

To broaden my awareness of mathematics used in indigenous cultures, I participated in a research group under the guidance of Dr. Iman Chahine. We traveled to South Africa, where I had the opportunity to conduct research at the Mapoch Ndebele Village. After participating in their cultural rituals, I worked closely with the Ndebele women to observe their daily tasks of weaving beads to craft material art. This chapter is a personal reflection on my participation in the immersion experience within the Ndebele context. Drawing on the history of Ndebele beadwork as well as what beaded designs communicate both symbolically and mathematically, I highlight the potential role that indigenous cultures play in producing, appropriating, and disseminating mathematical ideas (Chahine & Kinuthia, 2013).

INDIGENOUS ART OF THE MAPOCH NDEBELE VILLAGE

The Mapoch Ndebele Village lies in the northeastern region of South Africa just outside of Pretoria. As descendants of the Nguni tribes, the Ndebele people share similar cultural practices to those found in the Zulu community (Powell, 1995). I had the opportunity to visit the Mapoch Ndebele Village with my research group and follow in the footsteps of the village's men, women, and children. Through first-hand observation, participation, and conversation, I learned that the present-day community continues to practice the traditional customs of the Ndebele (e.g., indigenous art, ceremonies, beliefs). A closer look at the indigenous art, including the village's homesteads (see, e.g., Figure 3.1) and the material artifacts (e.g., mats, basketry, beadwork), reveals insight into the ancestral traditions and history of the Ndebele people.

Walking through the Mapoch Ndebele Village was like strolling through an art gallery. On the outside of each homestead, the Ndebele women painted colorful, geometric murals. Most of the murals' designs consisted of linear lines joined together to form geometric shapes. The homestead's doors and windows often served as axes of symmetry from which the geometric shapes would be strategically placed to create intricate symmetrical designs (see, e.g., Figure 3.2).

I inquired about the tools used to assist with the placement of each brushstroke and was informed that stencils and rulers were not used. Instead, the Ndebele women used freehand measurements and spatial awareness to guide their paintings. As with any paint exposed to nature's elements, the Ndebele women would retrace or completely redesign their murals after rainy seasons. The paintings reflected both the individuality of the dwellers as well as the tribal customs (Plummer & Chahine, 2014).

While the Ndebele women tended to the upkeep of their homesteads' murals, they also spent their days weaving material artifacts, including mats, baskets, and beadwork. To assist with weaving mats, the Ndebele women used a wooden weaving tool to lace wet grasses together with dyed

Figure 3.1. Mural designs in Mapoch Ndebele Village.

Figure 3.2. Painted homestead with geometric patterns in Mapoch Ndebele Village.

string (see, e.g., Figure 3.3). This task was time consuming and would often take two full days to create a body-sized mat.

Figure 3.3. Ndebele woman weaving a grass mat using the wooden weaving tool.

As for weaving baskets, the Ndebele women used their hands to shape basket sizes and weave detailed patterns. The baskets served different purposes, such as food storage, drinking vessels, and ceremonial gifts. Similarly, the weaving of beadwork was also performed by the Ndebele women and used to commemorate traditions and communicate one's life story.

TRACING THE HISTORY OF NDEBELE BEADWORK

Although the earliest Ndebele beadwork can be traced back to the second half of the 19th century, recent discoveries suggest that the weaving of beads may have existed even earlier (Powell, 1995). When speaking with the Ndebele women, I learned that beadwork evolved with the passing of ancient traditions from one generation to the next. Daughters would learn from their mothers who would have learned from their mothers and sisters. With the passing of time, beads were often recycled into new pieces or buried alongside their creators. Even with today's influence on modernization, beadwork is still created to honor tradition and cover Ndebele men, women, and children.

Powell (1995) describes some of the earliest Ndebele beadwork as simplistic with minimally used geometric designs surrounded by white beads. The white beads were thought to be plentiful during trade with the Europeans. A common motif used to accent the white beads resembles that of a H—a line connecting a double bar. This motif appears in different directions and oftentimes has additional lines extending off the bars to resemble the formation of a homestead.

Another commonly used motif in Ndebele beadwork is that of a triangle. von Kapff (2011) explains that the triangle's orientation and configuration to create other images (e.g., axe-heads, the Ndebele flower, razorblades, huts) communicates different meanings. As seen in Figure 3.4, a triangle pointing down represents an unmarried male, while a triangle pointing up represents an unmarried female. Two triangles meeting together to form an hourglass shape symbolizes a married man. On the other hand, two triangles with their bases meeting to form a diamond represents a married woman, a symbol for fertility as well. How the triangles and diamonds are repeated communicates certain messages, such as the masculine "Assesais of Shaka" and the feminine "Shields of Shaka." Although a single motif embedded in a geometric bead pattern can reveal insight in tribal courtship and social relationships, the addition of other motifs and color adds to the complexity of interpreting bead messages.

After the turn of the 20th century, colored beads began to make more of an appearance, which is when Ndebele beadwork became more elaborate and colorful (Powell, 1995). Although there is no fixed code for interpreting each color's meaning, Chahine and Kinuthia (2013) identify the common positive and negative connotations of inspired bead color designs prevalent in the Zulu cultures. White beads are seen as exhibiting only positive connotations that signify spiritual purity, happiness, luck, and children. Black beads indicate positive connotations of marriage and dark skin as well as negative connotations of sorrow and despair. Red signifies strong passion and longing for the heart but can also mean anger and impatience. Blue communicates fidelity and request or dispute and animosity. Green indicates joy and comfort as well as sickness and discard. Yellow represents wealth and garden or thirst and emotional withering. Pink signifies promise and status or poverty and laziness. Nevertheless, Wickler and Seibt (1995) warn that a color's shade or hue can alter the meaning as well as how a tribe defines a color's purpose, such as its use in traditional ceremonial occasions.

Additionally, combinations of colors and symmetrical arrangements may also be used to decode specific messages (Schoeman, 1968). For instance, Chahine and Kinuthia (2013) identify the color combination of blue, white,

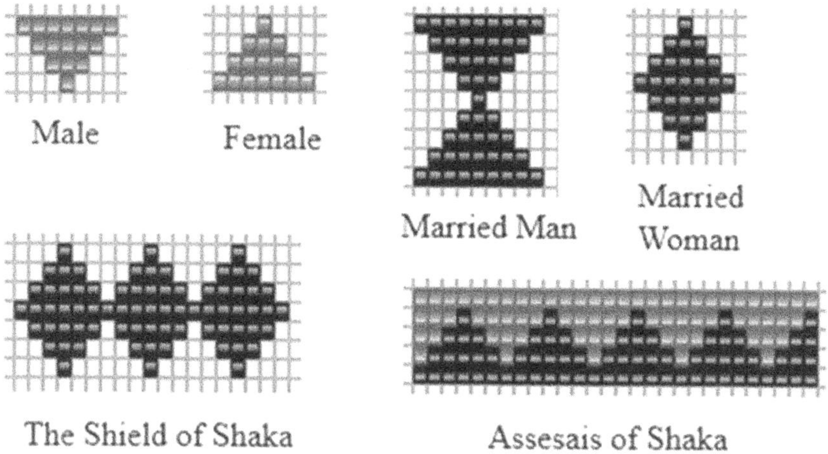

Figure 3.4. Bead symbols and their meanings.

and black as one that communicates a favorable, positive message, whereas the grouping of pink, red, and yellow conveys an unfavorable, negative message. The interweaving of colors' positive and negative connotations can aid in revealing one's "cultural position and personal and social accomplishments" (p. 17).

With increased accessibility to more colors over time, the Ndebele women shared how the beadwork became even more vibrant and eye-catching. Additionally, contemporary influences (e.g., electricity, television, telephone poles, chimneys, airplanes), inspired new interpretations of traditional motifs. Although the Mapoch Ndebele Village had limited access to such amenities, the Ndebele women incorporated such images in their designs to acknowledge today's modern ways. The vibrant colors and modern-day motifs were also used to appeal to tourists and support income for the village.

Most of the Ndebele people I encountered wore beads not only as accessories but also as clothing. Through conversations with the Ndebele women, I learned that the beaded garments represent cultural clothing worn by men, women, and children to honor their ancestors and traditions. The wearing of beads begins soon after birth and continues through childhood and into adulthood. When a child is born, a mother strings white beads around the newborn's waist. Powell (1995) identifies the white string of beads as *umucu* and discusses how the beads signify good luck and induction into the tribe. As children, both boys and girls wear leather and canvas aprons depicting geometric patterns with beaded tassels. Powell

(1995) refers to the aprons as *lighabi*. When the children mature and reach puberty, they attend initiation to prepare for the next chapter of their lives.

Once every four years, male initiation is performed in a secluded place for a period of three months (Powell, 1995). From speaking with the Ndebele women, I learned that the male initiation is a time for men to teach the boys the history, rituals, and traditional practices of the Ndebele culture. Powell (1995) also describes that during this time the boys are circumcised, which signifies the transition into manhood. When the men return home, the village celebrates the completion of initiation with the exchange of beaded gifts and a feast. One Ndebele woman showed me a headband worn for this celebration, which consisted of two narrow strips of beading that was hung from either side of her head. Powell identifies the beadwork as *linga koba* given that it resembles long tears, which is symbolic of the mothers celebrating their sons' transitions into manhood while also expressing sorrow for them no longer being their boys.

During the female initiation, the girls are secluded for a month where they are taught the roles of womanhood in the Ndebele community (Powell, 1995). During conversation, the Ndebele women shared that female initiation serves as a time for girls to work with Ndebele women to practice the Ndebele norms of taking care of husbands and families as well as learning skills, such as cooking, painting, weaving, and beadwork. Upon returning from seclusion, a ceremony is held to celebrate their transition into womanhood. At the ceremony, a mother or grandmother presents the initiated daughter with a stiff beaded apron and beaded hoop jewelry to wear on the daughter's neck, arms, and legs (see, e.g., Figure 3.5). Powell (1995) refers to the stiff beaded apron as the *isiphephetu* and the beaded hoop jewelry as *izigolwani*.

Special beadwork designs are also associated with wedding ceremonies and marriage. During the wedding ceremony, the Ndebele women shared that the bride wears a beaded veil and a beaded cloak made from goatskin. Priebatsch and Knight (1978) identify each as *isiyaya* and *linaga*, respectively. Additionally, the bride wears a long-beaded train, known as a *nyoga*, which extends from the bride's shoulders down to the floor (Priebatsch & Knight, 1978). Throughout the celebration, the bride can be found carrying a wedding stick that is approximately a foot in length and covered in beads (Stalcup, 1999).

During my immersion, I had the opportunity to observe a married Ndebele woman fully dressed in traditional clothing communicating to others her married status (see, e.g., Figure 3.6). She wore copper and brass rings around her neck to signify her faithfulness to her husband. Powell (1995) refers to these rings as *dzilla* and describes how they are gifted from the woman's husband and are not removed until his death. The number of rings also reflects the family's wealth. Additionally, the married Ndebele

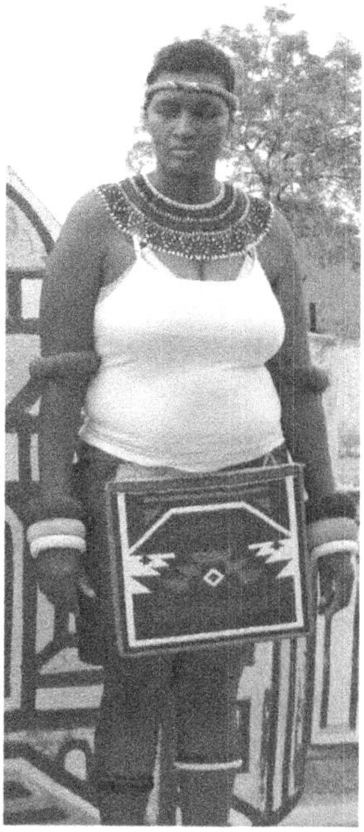

Figure 3.5. An initiated Ndebele woman wearing isiphephetu and izigolwani.

Figure 3.6. A Married Ndebele woman wearing dzilla, izigolwani, and liphotu.

woman wore a beaded panel on goatskin. Powell identifies the beaded panel as *liphotu* and describes how it symbolizes the marriage of husband and wife with hanging tassels to communicate the expectation of children. When a woman has children, she wears a more elaborate apron consisting of five beaded panels. Powell identifies this apron as *ijogolo* and explains how the middle panel depicts the mother while the other panels represents her children. Both aprons are worn interchangeably to symbolize marriage; however, the *liphotu* is worn more often for its less formal appeal.

Beadwork can also be found as adornments on everyday items (e.g., hot pads, pottery, brooms, pursues, headwear, toys). When a family hosted me through the night, I interacted with the family's youngest daughter who shared with me her dolls (see, e.g., Figure 3.7). The dolls were covered in beads and traditional clothing (e.g., hoops, rings, aprons).

Figure 3.7. Ndebele beaded dolls, pottery, and a broom.

I was informed that mothers and sisters would present these dolls to brides and young girls as fertility charms. The next day I had the opportunity to observe and practice my own beadwork with the guidance of the Ndebele women. While the Ndebele women make bead weaving look effortless, I can say firsthand grasping the technique is harder than it looks.

BEAD WEAVING TECHNIQUES

Ndebele women create beadwork known for vibrant colors and geometric designs (Priebatsch & Knight, 1978; Stalcup, 1999). During my immersion, I worked alongside the Ndebele women as they took their needles and thread and began stringing beads. They shared with me several weaving techniques the same way they learned from their mothers and sisters by show-and-tell. The Ndebele women modeled how they counted the threaded beads, measured lengths in terms of beads' dimensions, and explained the complexities of the desired patterns, all mathematical activities as described by Bishop (1988). I retrace my steps and those of other Ndebele women I worked with to describe two weave patterns, the herringbone and the diamond netting weaves.

Also known as the Ndebele weave, the herringbone weave orients beads in an arrangement that resembles the bones of a herring fish. To create this design, a pair of beads rotate 45 degrees in toward the line of reflection passing between the pair to form a V-shaped pattern (see, e.g., Figure 3.8).

Figure 3.8. A template of sample herringbone weave patterns created by the author.

To create a piece using this technique, I was instructed to make a ladder stitch base, which served as the foundation to build the herringbone pattern. I made the ladder stitch base by stringing two beads and placing them side-by-side. Then, I had to pass the thread back through the first bead and then the second bead. To add another bead's length, I had to add a bead and take my thread back through the previous bead and then the last bead added. The process continued until I reached my desired width of the piece, which I was told typically consisted of a row with an even number of beads for pairing purposes.

After establishing the ladder stitch base, I was then instructed by the Ndebele women to start the second row by stringing two beads and passing the thread through the bead stacked below the previously added bead. Then, I weaved my thread back up through the bead next to the one with a bead stacked above it. I continued to add two more beads and complete the same process until I reached the end of the row. To make the turn, I had to pass the thread through the bead stacked below (the end bead), the second bead next to it, and the last added bead. Then the previously implemented process continued until I reached the end of the third row and made another turn. As I learned the pattern, I reflected on the mathematics I used (e.g., counting, mapping, translation, reflection) to conceptualize the pattern and strategically plan for which colored beads came next.

Another commonly used weaving technique used by Ndebele women is that of diamond netting. This technique uses a strand of beads and builds from a previously made string of beads by adding an odd number of beads and then passing back through a single bead. This continues as the Ndebele women add more rows to create a netted look. Two common diamond netting techniques used to create this pattern include the basic and three-bead weaves.

A basic diamond netting technique (see, e.g., Figure 3.9) can be made by stringing a predetermined number of beads to complete a diamond with one bead in between each diamond's vertices.

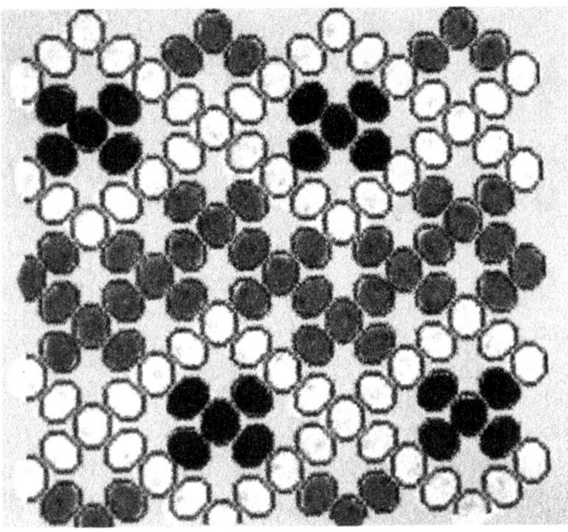

Figure 3.9 A template of a sample basic diamond netting technique created by the author.

If a width of four diamonds is desired, the Ndebele woman starts with 16 beads. For each additional diamond, she adds four beads. After stringing the initial 16 beads, a diamond is closed by pulling the thread back through the eighth bead counted from the back. The Ndebele woman then adds three beads to complete the next diamond. To close the second loop, she threads through the 12th bead from the back. Next, she strings three more beads for the third diamond and then passes through the 16th bead from the back. For the last diamond on the row, she adds three more beads and weaves back through the first bead placed. To complete a diamond on the third row, she adds four beads. Every edge diamond, after the first, has four beads. She continues the row's remaining two diamonds by adding three beads each. A finished example of a piece I made using larger beads with a width of two diamonds can be seen in Figure 3.10. In this piece, yellow triangles are translated and reflected across the center column or axis.

Figure 3.10. Basic diamond netting technique with yellow triangle pattern.

Following the basic diamond netting technique, the three-bead diamond netting technique varies in that it has two beads in between each diamond's vertices (see, e.g., Figure 3.11).

Figure 3.11. Three-bead diamond netting technique found on a broom.

To create a piece using this technique, the Ndebele woman strings a selected pattern of beads. For instance, one yellow bead then two blue beads, and so on (see, e.g., Figure 3.12).

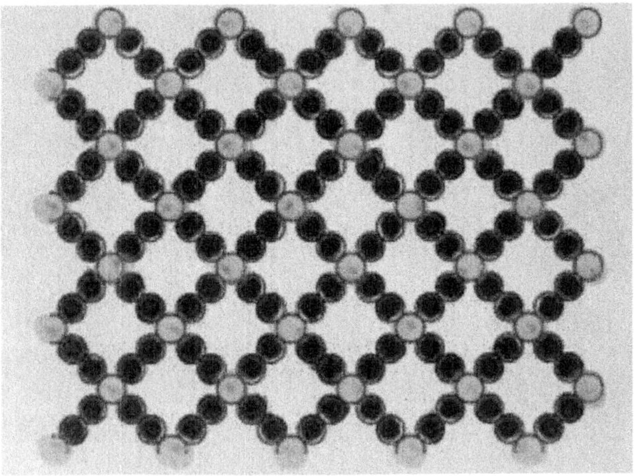

Figure 3.12. A template of a sample yellow and blue patterned three-bead diamond netting technique created by the author.

To begin the second row, she strings two evolutions of the pattern (i.e., one yellow bead then two blue beads and one yellow bead then two blue beads). Then she takes the thread and goes back through the third bead from the end (i.e., the yellow bead). Next, she adds five beads, following the desired pattern, and uses the sixth bead from the previous row to complete the evolution of two patterns. This continues until she reaches the end of the row, where she would prepare for the third row's turn. With more practice, the process becomes repetitive, especially if the Ndebele woman wants to continue a simple pattern throughout the piece. Incorporating different designs, such as diamonds of various sizes and colors (see, e.g., Figure 3.13), can increase the challenge, for the Ndebele woman must strategically plan where the added beads appear in relation to the desired pattern.

TRANSFORMATIONS ON AN INDIGENOUS CARTESIAN PLANE

As I learned the various weaving techniques from the Ndebele women and discussed the complexity of the geometric designs appearing in their pieces, the Ndebele women, although unknowingly, employed principles

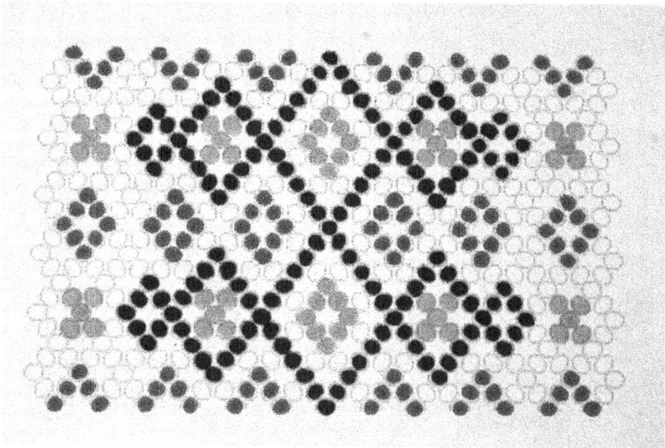

Figure 3.13. A template of various sized and colored triangles using a three-bead diamond netting technique created by the author.

of transformational symmetry to describe and explain their beadwork patterns. They knew where to place motifs using properties of reflection by folding their material artifacts in half or pointing to specific rows and columns that served as axes of symmetry. I observed the Ndebele women keep counts of the beads and note the placement of specific colors in patterns. They described designs in terms of rows and columns and methodologically planned bead positions to establish balance and design.

While I conversed with the Ndebele women, I recalled my experience teaching rigid transformations to my students in geometry. For example, the alignment of rows and columns of beads in a strip pattern design could be represented as points on a Cartesian plane. The depiction of a motif translating, reflecting, and rotating about different points and lines found in a Cartesian plane closely followed transformational symmetry rules. Thinking back to concepts of rigid transformations, also referred to as isometries, I could easily discern the close resemblance between mathematical ideas inherent in Ndebele weaving practices and what we teach in our mathematics classrooms.

To illustrate, I teach my students that a translation (notation $T_{(a,b)}$) shifts a pre-image a fixed distance in a determined direction. For example, Figure 3.14(a) $T_{(2,1)}(\Delta ABC) = \Delta A'B'C'$ depicts where ΔABC is horizontally translated 2 units right and vertically translated 1 unit up. With a reflection (notation r_m), I present a pre-image reflected over a line (e.g., line m) such that the distance between the pre-image and the line of reflection is the same as the distance between the image and the line of reflection. Figure 3.14(b) shows $r_{x=1}(\Delta ABC) = \Delta A'B'C'$ where ΔABC is vertically reflected over the line $x = 1$. In this case, the triangle that begins in the first quadrant now appears in the second quadrant. For a rotation (notation R_θ), I describe how a pre-image's vertices rotates a specified angle about a fixed point.

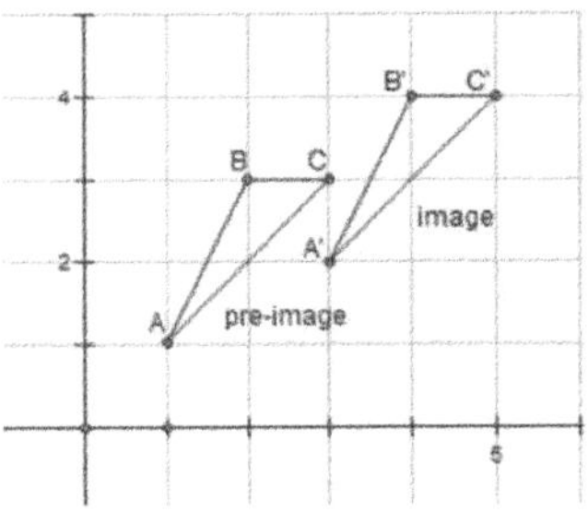

(a) Translation

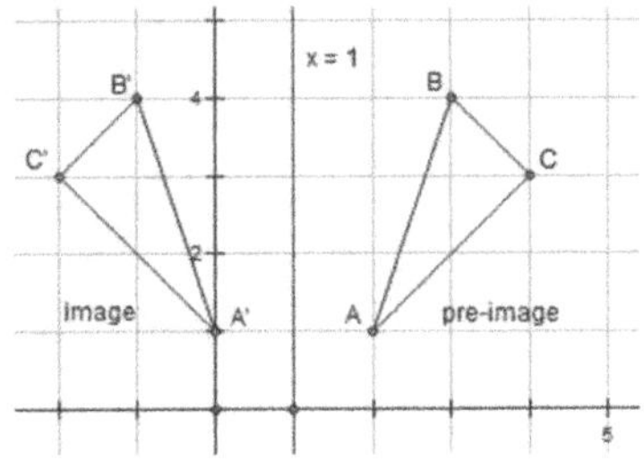

(b) Reflection

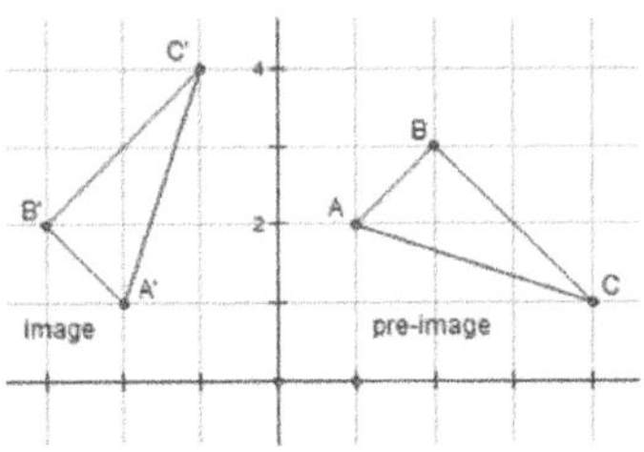

(c) Rotation

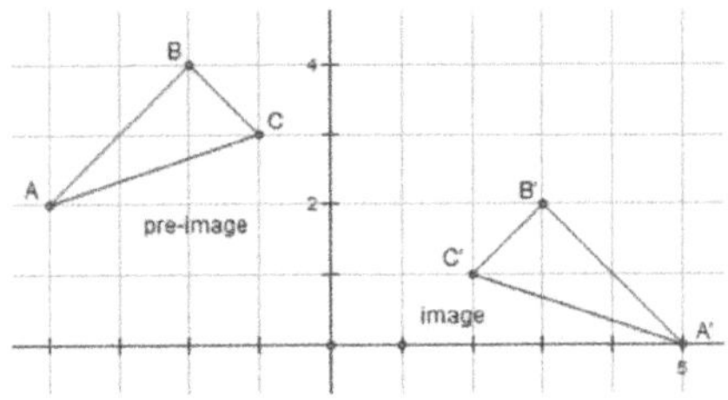

(d) Glide Reflection

Figure 3.14. Triangles depicting (a) translation, (b) reflection, (c) rotation, and (d) glide reflection on a Cartesian plane.

Figure 3.14(c) depicts $R_{90°}(\Delta ABC = \Delta A'B'C')$ where the pre-image is rotated 90 degrees about the origin. Here, the vertices of the pre-image begin in the first quadrant and are rotated to the second quadrant. Lastly, I present a glide reflection as a composition of a translation and a reflection, where order matters. Figure 3.14(d) shows $r_{x=0} \cdot T_{(-1,-2)}(\Delta ABC) = \Delta A'B'C'$ where ΔABC is horizontally translated 1 unit left and vertically translated 2 units down in the second quadrant. Then, the vertices are vertically reflected across the y-axis from the second quadrant to the first quadrant.

Describing symmetries using a Cartesian plane is one way for my students to visualize geometric transformations. Although my students use points and images on a Cartesian plane to communicate their understanding of translations, reflections, rotations, and glide reflections, I realize that my students may benefit from practicing these mathematical concepts from different perspectives. As I spoke with the Ndebele women, they discussed similar principles of transformational symmetry in their beadwork. While the women did not orient their motifs on a Cartesian plane, it was obvious that they used their own indigenous plane made of beads to share similar mathematical concepts.

FINAL THOUGHTS

Throughout my immersion in the Mapoch Ndebele Village, I observed the mathematical brilliance of Ndebele women as they engaged in their bead weaving practices passed down from generation to generation. The various bead patterns and weaving techniques described and modeled by the Ndebele women revealed unique ways of communicating and applying mathematics found in cultural activities. My work with the Ndebele women and my efforts to unravel the mathematics of Ndebele beadwork enabled me to understand that the relevance of mathematics lies in its purpose, including who it serves and how it is used.

Bishop (1988) reminds us that to study the complexity of mathematics as a human creation and cultural product means that mathematics teachers and learners must view mathematics for its varied practices around the world. I believe that engaging in cultural practices first-hand through field experiences can serve as one means to explore mathematics. Although organizing field experiences can be demanding, the reward of opening one's perspective to different mathematical applications can be worthwhile.

After I returned from my travel, I shared my Ndebele experience with my students. I showed them my souvenirs, which included the beaded pieces that I created alongside the Ndebele women. I asked my students to apply their understanding of transformations and symmetry to the bead patterns. Without hesitation, my students described the triangles being reflected and translated across the center column of beads. I asked them to think back to how they were taught about transformations on a Cartesian plane and how viewing artifacts that encompassed such properties made the mathematics relevant. In response, my students found the practice with the figures on a Cartesian plane vital to their success on their year-end test; however, they wished they found better ways to make the mathematical concepts relevant to their personal encounters with mathematics in their daily lives. Moving forward, I continue to challenge myself as well as others to bridge this disconnect by making mathematics culturally relevant to students. In doing so, I hope to teach that a Westernized view of mathematics is only one windowpane of many through which to view the broader picture of mathematics.

REFERENCES

Bishop, A. J. (1988). *Mathematical enculturation: A cultural perspective on mathematics education*. Kluwer.

Chahine, I. C., & Kinuthia, W. (2013). Juxtaposing form, function, and social symbolism: An ethnomathematical analysis of indigenous technologies in the Zulu culture. *Journal of Mathematics and Culture*, *7*(1), 1–30.

D'Ambrosio, U. (1985). Ethnomathematics and its place in the history and pedagogy of mathematics. *For the Learning of Mathematics*, *5*(1), 44–47.

Mukhopadhyay, S., Powell, A. B., & Frankenstein, M. (2009). An ethnomathematical perspective on culturally responsive mathematics education. In B. Geer, S. Mukhopadhyay, A. B. Powell, & S. Nelson-Barber (Eds.), *Culturally responsive mathematics education* (pp. 65–84). Taylor and Francis.

Plummer, A., & Chahine, I. C. (2014). Mathematics lessons from "a place of light." *Mathematics Teacher*, *108*(4), 254–257.

Powell, I. (1995). *Ndebele: A people & their art*. Cross River Press.

Priebatsch, S., & Knight, N. (1978). Traditional Ndebele beadwork. *African Arts*, *11*(2), 24–27.

Schoeman, H. S. (1968). A preliminary report on traditional beadwork in the Mkhwanazi area of the Mtunzini district, Zululand. *African Studies*, *27*(2), 57–81.

Stalcup, A. (1999). *Ndebele beadwork: African artistry*. Powerkids Press.

Swetz, F. J. (2009). Culture and the development of mathematics: An historical perspective. In B. Geer, S. Mukhopadhyay, A. B. Powell, & S. Nelson-Barber (Eds.), *Culturally responsive mathematics education* (pp. 11–41). Taylor and Francis.

von Kapff, U. (2011). *The Zulu: An A-Z of culture and traditions*. Struik Travel & Heritage.

Wickler, W., & Seibt, U. (1995). Syntax and semantics in Zulu bead colour communication system. *Anthropos*, *90*, 391–405.

CHAPTER 4

MATHEMATICAL CARD GAMES

A Study of Cardfight! Vanguard

Alexandra Starke and Jessica Steele

ABSTRACT

The role of mathematics in gaming is evident in the nature of games, and the strategy used in playing them. Most strategy based games use mathematical concepts such as probability, counting principles, and logic. However, specific games emphasize different aspects of mathematics. Topics such as counting, additive properties, combinations and permutations, patterns, classifying, and ordering often arise when playing strategy based games. While players may not set out to specifically use math in order to accomplish goals or win the game, it is interesting to investigate how mathematics plays an underlying role in each of these games. While playing various games, participants are consistently reinforcing their mathematical skills and relying on their mathematical abilities to successfully execute each turn. This investigation focuses on a specific card game, Cardfight! Vanguard, and the mathematical nature of its design and gameplay.

Evidence-Based Inquiries in Ethno-STEM Research: Investigations in Knowledge Systems Across Disciplines and Transcultural Settings, pp. 75–87

CONTEXT OF THE STUDY

A community is defined as "a unified body of individuals: such as a body of persons or nations having a common history or common social, economic, and political interests" ("Community," 2003). Communities provide individuals an opportunity to express themselves freely while experiencing shared passions, interests, and hobbies. When observing a community, it is important to consider the reasons for which they have come together as a cohesive whole. The gaming community has been experiencing significant growth over the past decade. This community includes those interested in video games, card games, board games, and role-playing games as some examples of games relevant to the American culture. While some of these types of games have been played in America in family and social settings for years, games played in tournament format by large groups are relatively new. According to ICv2, an online trade magazine that focuses on "geek" culture, there has been expansive growth in gaming sales from 2014 to 2015. There has been notable growth in all areas of gaming interest, particularly hobby board games and collectable games, however, Griepp (2016) claims that "the fastest growth rate was for hobby card and dice games, which added a digit and grew 75% to $105 million in 2015, up from $60 million in 2014" (p. 1). This growth in sales has created the need for a new outlet for the gaming culture including the establishment of gaming cafes where patrons can borrow various board games to play while enjoying a meal or a drink, and game stores where the number one purpose is not to sell games, but instead to provide a space for people to come together and simply, play.

The culture of the gaming community accepts and welcomes people of all backgrounds, while celebrating and emphasizing individuality. According to Gilsdorf (2012), "when a collection of people follows a similar trend in assigning meanings and beliefs, they have what anthropologists call a culture" (p. 4). The gaming culture places a large emphasis on teamwork, camaraderie, and of course, friendly competition. Tournament style games provide patrons with the opportunity to interact with numerous different people, and sometimes provide gamers with the chance to travel the world. For example, *Magic: The Gathering* hosts a yearly World Championship where certain invitees from different regions, including, the Asian-Pacific region, Japan, Europe, Latin America, and North America, come together to participate in an expansive tournament resulting in a "World Champion." In a more local sense, the gaming culture consists of monthly local events, and weekly game nights. These game nights allow individuals to come together, play for store credit, and interact with about two hundred of their closest friends. The importance of this culture resides in the personal connections and relationships that development from a result of shared interests and experiences.

The role of mathematics in the gaming community is evident in the nature of these games, and the strategy used in playing them. Most strategy based games use mathematical concepts such as probability, counting principles, and logic. However, specific games emphasize different aspects of mathematics. Topics such as counting, additive properties, combinations and permutations, patterns, classifying, and ordering often arise when playing strategy based games. While players may not set out to specifically use math in order to accomplish goals or win the game, it is interesting to investigate how mathematics plays an underlying role in each of these games. While playing various games, participants are consistently reinforcing their mathematical skills and relying on their mathematical abilities to successfully execute each turn. This investigation focuses on a specific card game, *Cardfight! Vanguard*, and the mathematical nature of its design and gameplay.

Throughout this investigation, the following questions were addressed: How is mathematics used in the strategy based card game *Cardfight! Vanguard*? What mathematical concepts arise throughout gameplay? What is the cultural value/significance of these types of card games?

BRIEF BACKGROUND

According to Gilsdorf (2012), "Games can be found in many cultures, and go back a long time in human history" (p. 124). He goes further to organize games into categories that contain the concepts of physical skill, strategy or mental skill, probability, and combinations of those concepts. Games that require primarily physical skill are generally competitions of athletic ability, such as completing a short activity in the least amount of time. As an activity lengthens, more strategy is required of the participant, but the activity can still be classified as a game. An example of a game that requires only strategy is chess, as each player has the same opportunity and the only changes in play of the game are caused by one of the players. In the game of chess, there is no randomness or chance involved. Finally, games that involve chance must have some random aspect that continues throughout. For example, a game in which a fair dice is rolled to determine the winner of each round, there is no strategy required to win the most rounds. The winner of each round is determined only by the roll of the dice, and no strategy can be used. Very commonly seen when examining games are those that contain a combination of strategy and chance. Most games with cards involve drawing cards from a shuffled deck and attempting to use those cards in a beneficial way. In these types of games, "the random aspect comes from the draw of the cards from the shuffled deck, and the strategy comes from deciding how to play the collection of cards that has been drawn" (p. 125). In order to compete in games that involve strategy and randomness, it is important to understand some basic principles of probability. Although the members of a culture may not describe probabilities in the same way they are described in Western mathematics, "cultures that develop games involving probabilities often understand many of these same concepts" (p. 129). Some principles of probability that are important to analyzing card games include: the multiplication rule, permutations and factorials. To calculate the probability of a given event, "the basic principle of probability at use is that *a probability consists of the number of ways a particular outcome can occur divided by the total number of possible outcomes*" (p. 131, emphasis original).

Outside of the different classification of games, there has been a multitude of research on the implementation of games into the mathematics classroom. There are numerous benefits to utilizing games in an academic context. Okpube (2016) claims that "educators and researchers have recommended the use of mathematical games in teaching and learning of mathematics because it is found to sustain and develop students' interest and achievement in mathematics" (p. 94). Games are often considered helpful tools in establishing a higher degree of motivation and interest in the classroom. From review games, to using games to deliver instruction, students tend to be more actively engaged than during traditional lecture classroom settings.

Mathematical games are beneficial to learners for a variety of reasons. Okpube (2016) states that

> They [games] bring joy to the learner, breakdown resistance to learning by reducing tension, clearing boredom, and providing an environment where the student can develop skills and acquire more knowledge. They also stimulate mathematical thinking and generate excitement and spirit of individualism, co-operation and competition. (pp. 94–95)

In his study, Okpube uses a quasi-experimental research design to investigate the difference between the assessment scores of those that learned algebra concepts with the enhancement of games, and a control group taught by conventional methods. His findings found that "the use of mathematical games enhances the learning of algebraic expressions than the non-use of games" (p. 98). The specific games used in this study were a set of card games designed to teach algebra and were successful in raising assessment scores for students.

Although games are often praised as being a hugely beneficial way for students to understand mathematical concepts on a deeper level, there are also drawbacks to this type of activity-based curriculum. Marklund and Taylor (2016) researched the challenges that teachers face when trying to implement a game-based curriculum. The researchers focused on the investigation of two classrooms, a seventh grade class and a fifth grade class, and the implementation of a computer game to supplement curriculum. There were numerous issues that arose throughout the implementation of these games. First, many teachers did not have proficient knowledge of the game in order to be able to successfully facilitate the activity effectively. Additionally, students being at different levels was an obstacle in the classroom. Marklaund and Taylor (2016) concluded that

> The heterogeneity of K–12 students can make classroom sessions difficult to design and monitor as the students who have never played a computer game before needs to be able to collaborate and communicate with students who are very proficient players. (p. 128)

Additionally, the researchers emphasized a struggle that many students face when playing games as instructional tools: staying focused on the content being taught. The authors further explained that "One commonly recurring challenge the teacher tackled during gaming sessions was to bridge the gap between the game content and the details of the subject matter the game is intended to teach" (p.130). These are just a couple examples of the difficulties that teachers often face when attempting to incorporate more game-based activities into the classroom.

METHODOLOGY

Participants

The participants of this study include individuals participating in a *Cardfight! Vanguard* tournament at a local gaming store on a Friday evening. These participants pay a nominal fee to sign up for the tournament and are randomly assigned opponents by the store employees in a Swiss-system tournament style bracket. The specific participants of our study included one player, and the opponents that he matched with throughout the entirety of the tournament.

Methods

This study utilized a case study research design with the incorporation of an activity theory model. The specific application of the activity theory includes a subject: the player, a goal: to win, and his tools: the deck created to support his Vanguard. Vygotsky's activity theory model disputes the more popular Pavlovian theory that action depends on a stimulus. Vygotsky promoted that "human activity is purposeful and carried out by sets of actions through the use of 'tools', which can be physical or psychological" (Hasan, 2014, p. 9). *Cardfight! Vanguard* players set out to specifically use their cards as tools to build a strong enough vanguard to overpower their opponent, fulfilling their goal.

Data collection instruments included artifacts, observations, and unstructured interviews. Artifacts include photographs that were taken of the various cards used throughout the game, specific layout of the gameplay, and games in progress. Interviews were conducted at the gaming store with four different players, and each interview lasted approximately 15 to 20 minutes. These interviews allowed us to collect data about specific playing strategies, preferences of decks, and specifics of the rules of the game.

Procedure

Observations and interviews were conducted at a local gaming store during a Friday night gaming event. One particular participant was followed throughout the tournament, and different interactions were recorded with various pairings each round. Unstructured interviews were conducted throughout the evening with the main player, Sam, and each of his opponents. Data was collected highlighting the rules of the game, and specifically where mathematics was playing a significant role in the decisions made by game players. The constructions of the deck were analyzed, and different strategies were recorded and discussed. Photos were taken of the cards used in different rounds of play, as well as a photo of a description of the game board setup.

DATA ANALYSIS AND SYNTHESIS

Cardfight! Vanguard is a card game based on building decks and implementing strategies to attack opponent's vanguards. The vanguards were the distinguished "main" card that a player establishes at the beginning of gameplay. Players utilize various cards to supplement the power of their vanguard, including attack power and shielding effects. While consistently defending their vanguard and equipping them with sufficient power, they are also initiating attacks against their opponent. In short, if the attack power outweighs their opponent's defense power, a damage point is taken. The game ends when one player accumulates six damage points.

Mathematical concepts are evident in the structure of each player's deck of cards. The main deck consists of 50 cards, sixteen of which are "trigger" cards. Each trigger card enhances a player's hand some capacity. For example, a player might draw a "heal" trigger, allowing scenarios in which health may be restored (by eliminating damage points). There are some limitations to the number of each type of trigger a player may use in their deck (can only use a maximum of four of each name, or maximum of four heals), but inside these limitations, it is player discretion on building each individual deck. Our first main mathematical concept arises in the building of each deck: counting. The players must consistently keep count of how many triggers they build into their deck, and also how many cards of each level they incorporate. Sam (our player) explained that a standard deck consists of one starter "0" card, 16 trigger "0" cards, 14 level 1 cards, 11 level 2 cards, and 8 level 3 cards. With the various breakdown of each level brings our next most prominent mathematical concept that arises in gameplay: probability. With each deck, there are various calculated probabilities of a player drawing each type of card. The combination of using both probabilities, and counting, allows player to consistently alter their gaming strategy to make the most beneficial decisions moving forward.

By combining the mathematical elements of counting and probability, specific strategies of building the deck can be investigated. Although each deck must contain 50 cards, each player deck can be unique. One of these 50 must be a level 0 starter card and 16 must be trigger cards. There are four types of trigger cards that all provide an additional 5,000 units of power along with serving different purposes. The first type is the Draw Trigger, which enables a player to "check" additional cards. The word "check" in the context of *Cardfight! Vanguard* indicates the player draws a card from the deck. The second is the Critical Trigger, which gives a unit additional strength until the end of the turn. The third, the Heal Trigger, allows the player to choose a card from the damage zone and put it in her drop zone. Finally, the Stand Trigger allows the player to choose a rear-guard and change it from resting to standing, allowing the card to be used again if needed. These different roles allow players to customize the types of trigger cards incorporated into their deck to pertain to their specific strategy.

Throughout the playing of the game, cards are drawn from each player's deck in order to assist in a Drive or a Damage Check. When drawing from the player's deck, if the triggers are checked, the player is more likely to be successful in a Drive or Damage Check. For this reason, it is important to keep track of how many cards have been played and how many of those cards are triggers. In our interview, Sam explained that during gameplay, any card that is played is allowed to be "common knowledge" between players. During the observation, Sam mentioned how his moves were directly affected by what his opponent had previously drawn. To look some at the probabilities of pulling a trigger card, we assumed the only card played is the starter card. This would leave 49 cards in the deck, 16 of which are trigger cards. All of these calculations can be repeated throughout the game to assess a player's deck along with the deck of his opponent. Table 4.1 shows probabilities of checking a trigger card for a Drive/Damage Check, Double Drive and Triple Drive.

Table 4.1.
Probability of Drawing a Trigger

# Triggers Checked	Drive/Damage Check (1 Draw)	Double Drive (2 Draws)	Triple Drive (3 Draws)
None	33/49 = .674	(33/49)(32/48) = .449	(33/49)(32/48)(31/47) = .296
All	16/49 = .326	(33/49)(16/48) + (16/49) (33/48) = .449	(16/49)(15/48)(14/47) = .03
Exactly 1	NA	(16/49)(15/48) = .102	(33/49)(32/48)(16/47) + (33/49) (16/48)(32/47) + (16/49)(33/48) (32/47) = .459
Exactly 2	NA	NA	(33/49)(16/48)(15/47) + (16/49) (15/48)(33/47) + (16/49)(33/48) (15/47) = .215
At least 1	**1-.674 = .326**	**1-.449 = .551**	**1-.296 = .704**

While all of the probabilities in Table 4.1 can be calculated if desired, the most important probabilities are listed in bold at the bottom of the table. These can be found most quickly by taking all the possibilities (1) and subtracting the possibilities no trigger cards being drawn. This calculation can be performed quickly during a game, should a player find it necessary. While it would be nice to check all trigger cards, drawing at least one is very helpful. As is also evident in the table, the draws are dependent on each other. When calculating the probabilities of dependent events, the previous events must be considered in the calculations. While it would be quite a bit of work to do these calculations continuously, many players explained that a high frequency of non-trigger cards causes an increase in the probability of checking a trigger card on a subsequent draw. For this reason, players are always keeping track of the number of cards and trigger cards drawn by themselves and their opponent. By using the mathematics of counting, they can make educated decisions for their moves in the game.

Other mathematical strategies arise in the investigation of cards that are *not* considered trigger cards. It is entirely player discretion to make decisions of how to compose the cards in his deck outside of the required seventeen already discussed. Questions arise about how players choose which cards specifically are placed into their decks, which refers back to the research question at hand: how mathematics is used in the strategies of *Cardfight! Vanguard*? Each card level in the regular playing deck, with options 0, 1, 2, 3, determine specific abilities for that card. There are abilities related to triggers and drawing cards, but the most often used characteristics are the power and shield values that each card consists of, and how these values can be added to the vanguard. Table 4.2 summarizes the levels of cards, the symbols they contain, and what that symbol means in terms of gameplay.

Table 4.2.
Card Abilities Based on Level

Card Level	Symbols	Ability Name	Interpretation in Gameplay
0, 1		Boost	Offers assistance to cards played above this card
2		Intercept	Receives assistance from cards played below this card
3		Double Drive	Allows players to perform a "check" on their decks twice in one turn

Card abilities can only be utilized based on where they are placed in the gaming format. Figure 4.1 depicts an example of a single player gaming area. Table 4.3 translates Figure 4.1 into a table with numbers that refer to each yellow and blue circle, showing a specific gameplay position.

All gameplay and card choices are based on supplementing a player's vanguard. Their vanguard begins at a level 0 and is placed in the blue circle depicted in the first row, middle column of Figure 4.1, or position 1 of Table 4.3. Cards can be built on top of the vanguard in sequential order, so a level 1 must be played on a level 0, and so forth. Each ability in Table 4.2 is only beneficial depending on the placement of that card in the rearguard circle (positions 2–6). A "boost" card played in position 4, 5, or 6 depicted in Table 4.3 will boost the vanguard with the amount of power described. However, a "boost" card placed in positions 2 or 3 would provide no additional power. Thus, the mathematical concept of locating is essential to strategies used in *Cardfight! Vanguard*. Similarly, a card with an "intercept" ability is only effective when played in positions 2 or 3. "Double Drive" cards may be placed in any position on the board.

Source: Reprinted from *Final Turn Cardfight!*, by J. Junk (2014). http://finalturncardfight.blogspot.com/2014/10/Cardfight-vanguard-custom-playmat-by.html

Figure 4.1. Example of a gameplay diagram for a single player.

Table 4.3.
Card Gameplay Positions

2	1	3
4	5	6

The limitations placed on each card ability make counting an essential mathematical concept used in the strategy of *Cardfight! Vanguard*. For example, a player deck that consists primarily of intercept abilities would not be strategic, as there are limited positions in which this card is effective. Thus, it is essential for a player to balance their number of each level of cards in order to maximize placement combinations. Once this balance is established, players can adjust to incorporate cards whose specific abilities lend themselves to the player's gameplay strategy.

Mathematical concepts are also prominent when trying to win the game. In order to win, a player must deal his opponent a total of six damage points, which just further enunciates the importance of counting throughout the entirety of *Cardfight! Vanguard*. Each damage point is dealt based on the mathematical concept of inequalities. In order to successfully deal a damage point to the opponent, a player's total attack power must be greater than his opponent's shield value. Figure 4.2 provides two of the artifacts collected in this study and are examples of typical *Cardfight! Vanguard* cards.

The shield value is located on the left margin of the card, and the attack value is located under the main image of the card. Also noted on these two cards are the card levels, 1 and 2, and the appropriate symbols demonstrating the individual card abilities.

Figure 4.2. Artifacts collected in the study as examples of cards used in gameplay.

At its core, this dynamic of winning the game is based on a simple inequality property. Furthermore, these attack and shield values use another basic mathematical computation, arithmetic. Players can add the shield values that are composed of the different cards played on the board, and they can also distribute attack value according to different card abilities, including the boost and intercept characteristics that have been discussed. Additionally, there are often cards that use multiplication, allowing players to distribute multiples of their attack values to the vanguard for its total attack power. The use of these mathematical concepts throughout a highly complex game just further emphasizes the importance of mathematics in a cultural setting.

CLOSING THOUGHTS

Moving forward, it is important to consider how research about card games impacts the way mathematics is taught. To say that table play is just a hobby completely overlooks the large amount of math that is being performed. Although *Cardfight! Vanguard* may not be the best game to introduce to a classroom of students, it contains many similarities to other card games. These card games may also include probability, inequalities and counting. When teaching probability to a group of students, cards are often used in examples of calculating probabilities. Teachers must be sure that students have some background in cards to understand how these examples are applicable. To emphasize these real life applications, teachers should bring more card games into the classroom.

In addition to ensuring that students understand probability examples related to cards, playing card games increases student engagement. Students who actively participate in games containing mathematical ideas are more likely to understand the mathematics behind them. One reason behind this increase in mathematical understanding is that when a student is engaged, he likely cares more about the concept being addressed. Valuing a concept is an extremely important aspect of meaningful learning in the classroom, which affects the level of confidence and motivation to engage in the construction of mathematical knowledge.

In the observations conducted at the game store, the participants were all extremely engaged in the activity at hand, because it held personal value and importance to them. Throughout the game, the speed and accuracy with which they were performing mathematical operations were impressive. These observations just further emphasize that if a person holds intrinsic value for the content being studied, they are more likely to obtain mastery of that subject, which further enunciates one of the most important goals for any teacher. It is not enough to only teach a student content; it is more important to convey the value and significance of the content being taught.

REFERENCES

Community. (2003). In *Merriam-Webster Online Dictionary*. https://www.merriam-webster.com/

Gilsdorf, T. E. (2012). *Introduction to cultural mathematics*. John Wiley and Sons.

Griepp, M. (2016). *Hobby games market nearly $1.2 billion*. ICv2. https://icv2.com/print/article/35150

Hasan, H., & A. Kazlauskas. (2014). Activity theory: Who is doing what, why and how. In *Being practical with theory: A window into business research* (pp. 9–14). THEORI. http://eurekaconnection.files.wordpress.com/2014/02/p-09-14-activity-theory-theori-ebook-2014.pdf

Marklund, B. B., & A. A. Taylor (2016). Educational games in practice: The challenges involved in a game-based curriculum. *The Electronic Journal of e-learning, 14*(2), 122–135. https://eric.ed.gov/?id=EJ1101225

Okpube, N. M. (2016). Card games and algebra tic tacmatics on achievement of junior secondary II students in algebraic expressions. *International Journal of Evaluation and Research in Education*, *5*(2), 93–100. https://eric.ed.gov/?id=EJ1108533

CHAPTER 5

THE MATHEMATICS OF GAUGE STICK IN GAS STATIONS

Anteneh Kibret

ABSTRACT

The intent of this case study is to explore the mathematical idea inherent in the use of gauge stick in gasoline stations. Generally, two methods are used to verify the amount of gasoline pumped in the stations' reservoirs and consequently sold to customers: gauge stick and computer-based operation. The gauge stick method used before and after unloading gasoline into the underground gas tank is carried out by the gas truck drivers and involves measuring volume. The computer-based method, on the other hand, is conducted by customer service representatives (CSR) and is based on computer generated readings. The study sample is comprised of four truck drivers and eight CSRs. All participants were high school graduates, certified as a truck driver or CSR, and came from different cultural backgrounds. Typically, CSRs do not need any special certification if they hold a high-school diploma, however CSRs' hiring requires prior experience and one week of in-store training. Data was collected from different gas stations on site as gasoline truck drivers delivered gasoline and CSRs sold gasoline and groceries.

Evidence-Based Inquiries in Ethno-STEM Research: Investigations in Knowledge Systems Across Disciplines and Transcultural Settings, pp. 89–100

INTRODUCTION

Numerous mathematical ideas emerge from ancient cultural activities that have been widely incorporated using today's computerized tools. Such mathematical ideas were broadly motivated in the day-to-day life practices of people. Within the broad activity of gasoline delivery in the U.S. and other countries, measuring is a crucial mathematical concept that is prevalent in different activities that people undertake as part of the workplace practice. Bishop (1988) describes measuring as encompassing several qualities including comparing, ordering, and quantifying. Typical examples of measurement include measuring distances, volumes, weights, and areas. Interestingly, not all cultures emphasize the accuracy of measuring. In the U.S. gasoline delivery industry, and while the truck drivers have access to pull computerized reports for gasoline levels in the underground gas tank, nonetheless, the truck drivers still must use a gauge stick to read the gasoline level. Gas stations do not let the gasoline truck drivers (GTD) use digital readers to determine the level of gas in the underground gasoline tank. Arguably, the assumed accurate computerized reader helps to determine if the gasoline tank is over-filled or not. The complexity of converting depth unit into a volume and visual estimation to park the truck at the right place near the gasoline tank are some examples of measuring activities carried out by the truck drivers on a daily basis.

It has been clearly established that the act of measuring requires comparing (Burdun & Markov, 1972). Measuring as defined in the English dictionary involves an evaluation or comparison of quantity reproduced by a standard. Historically, ancient Egyptians, having no standard measuring gauge, had to develop a method to measure the Nile's flood level. It is part of human nature that every time we encounter an obstacle, we strive to find the solution. The ancient Egyptians did exactly that. Since the flood had a critical rollover, Egyptian administrators sought a way to estimate crop yields thereby determining the amount of taxes to be paid (David, 2015). Since the amount of harvest depended upon the water level of the Nile, the only way administrators could estimate the future crop product was by keeping record of the flood level of the Nile (Ritter, 2002). The system that the Egyptians used to measure the water level during a flood was the "Nilometer," which resembles the "gauge stick" still in use today. The Nilometer was basically used to record the level of a flood through marks on riverbanks, along stairs leading to the river, on stone pillars or in water wells (see Figure 5.1). The Egyptian administrators wanted to make accurate records on the measurement of depth to collect the fair amount of taxes and the Nilometer provided the tool to successfully achieve this goal.

Source: Retrieved from https://www.aziaminvatat.ro/ro/Nilometrul-Structura-antica-utilizata-in-masurarea-nivelului-raului-Nil-i9637.html

Figure 5.1. Nilometer on Elephantine Island showing marks on riverbanks.

In a similar context, the purpose of measuring the level of gasoline in the underground gas tank is to determine the amount of gas loss if there is a leak due to any of the following conditions: (1) if there is maximum gas skip due to the volatile nature of the gas, (2) if there is a loss of gas due to a broken gas pump, and (3) if there is an untrusted activity like unauthorized pumping of gas by scammers. Therefore, managers from the gasoline delivery company and the store can monitor and regulate delivery activities based on the measured data using the measuring gauge and the computerized reader.

METHODS

Participants of the Study

GTDs and CSRs are the participants who joined the study during the field work. Four gas stations were picked arbitrarily and were approached to be part of the study. The opportunity to observe and interview CSRs was easier than waiting to interview gasoline truck drivers and dealing with the security issue. Eight CSRs, two from each gas station, and four GTDs participated in the interview. For most cases, there were two attendants per gas station. All the GTDs were certified personnel as they already completed knowledge and practical classes. The knowledge part includes rules and regulations on the road. The practical portion includes parking skill in addition to the driving ability. The parking strategy for trucks, in general, is different from the small cars. For example, during backing up in small cars, if steering is conducted counterclockwise, the car moves to the left side. In trucks, the steering moves the truck in the opposite direction. As a result, the truck drivers engage in many measuring activities on the site of gasoline delivery, namely, estimating the width to turn left or right, measuring the parking distance from the gas tank, and knowing the depth of the underground gasoline tank.

The gasoline delivery procedure. Ensuring that the surroundings are safe is a critical priority for GTDs to consider prior to handling the gasoline. Primarily, the GTD needs to be certain that the premises are free from any hazards that could cause fire. After ensuring that the environment is clear, the GTD starts laying down the gas disposing hose making sure that the truck parking position is suitable to reach the designated hole. After laying down one of the hoses, the GTD steps back to measure the distance visually. In most gas stations, there are at least two underground gas tanks: a tank for regular gas, and another for premium. Necessary paperwork is completed first, hence, the unloading process does not start right away. Each manual measurement of the depth of the gas level before unloading is recorded on paper carefully. Per the GTD, each step in measuring is to be conducted with high accuracy. While the CSR usually helps customers in grocery and gas purchase, they are also responsible for monitoring the gas level using the electronic gas level reader.

Methods

The primary method used to collect data is artifact collection including collecting photographs and scraps of papers used by the GTDs. The rationale for using photos is to capture the mathematical ideas inherent in s every step of the gasoline unloading process. Another method used is informal interviewing. Initially, and while the goal of my investigation was to focus on examining the measuring gauge, my observation included encounters that went beyond the measuring gauge. The unit conversion process and the gas vaporization science were some of the eye-catching activities. The gas reading before and after unloading the truck was collected from CSR. Apparently, the CSR data mostly focused on the computer readings. Thus, most data gathered from the CSR are computer generated readings except for some notes, which I collected from informal conversations with the CSR. Of course, the data collecting process at gas stations was not smooth, and many challenges were eminent such as waiting for the arrival of a gas delivery truck and getting the consent of the truck

(a) (b)

Source: Pictures taken by author.

Figure 5.2. (a) Gauge stick, underground tank, disposal hoses, and properly parked truck, (b) Gasoline truck driver using gauge stick.

drivers to be interviewed. During field work, it was clear that almost all the GTDs looked at me suspiciously. To facilitate my interaction at the gas station sites, I spent some time establishing some friendly relationships with CSR that helped me approach the GTDs who were willing to engage in conversations when they saw a familiar face.

After introducing myself, the GTDs started asking their cross-check questions, which was a significant opportunity to continue my informal interview. "Why do you want to study about the gauge?" Even though these questions were initiated by the drivers, it was my interest to provoke such questions. Impressively, having the question in their minds helped get an open-ended explanation. While measuring the level of gas in the underground gas tank, the science of gas handling popped right up from an open-ended discussion. Marco, one of the GTDs who explained the steps of unloading gasoline thoroughly, asked: "Do you know why it is not advisable to pump gas during midday when the temperature rises?" Marco continued, "Gas expands during the hot temperature, and the measuring can show a slight error in the reading. I must put the gauge stick all the way down to the bottom and pull it out very fast. Do you know why?" Of course, my answer was "no", Marco promptly explained: "[because] gas evaporates very fast."

The gauge stick is made up of wood, and it can absorb and release gas quickly. Marco practically demonstrated how the measuring gauge dries out quickly. All his explanations were practical and observable as he highlighted many factors that could impact the accuracy of measuring the gasoline level before and after unloading. Such factors include: (1) the amount of time (how fast the measurement takes place), (2) the chemical property of gas (volatility), and (3) the purchase of gas by customers. The measuring process using the measuring gauge was considered as accurate by most GTD as it does not give a much different reading from the computerized tool. From the CSR side, there was no different action required. After the unloading and the second or final measurement was taken using the gauge, the GTD delivers a handwritten document to the CSR. The computerized measure result prints out automatically after the change in gas level settles. The CSR can look at the computer reading just to confirm the overall difference. The data from the gasoline truck driver and the computer printout values are compared and screened every morning by the store manager or assistance manager (see Figure 5.3).

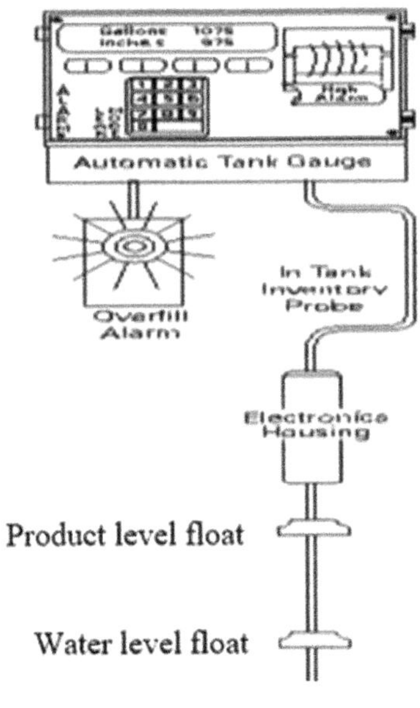

Source: Picture Adopted from the North Dakota UST Operator Training Program Manual.

Figure 5.3. Computerized tank gauge.

Comparing Measurements

Measurement in general refers to the method in which the quantity measured is compared with a quantity reproduced by a standard unit. Comparison measurement is applicable to quantities that can be reproduced by standardized units. The comparison measurement is a preferable method because of its greater precision. The underlying process for measuring levels of water using early Egyptian Nilometer or fuel levels using the gasoline gauge stick involves comparing the linear vertical distance between a reference point (usually the base of a solid base) and the surface of a liquid. With both measuring tools, precise control of the level of liquid is necessary, particularly for inventory management. Using the Nilometer as a standard unit of length, Egyptian farmers were able to communicate with the management about the amount of taxes they need to pay. Taxes were determined by the amount of harvest the farmers collected, and the harvest was determined by the rise or fall in the levels of the Nile. The accuracy of records collected using Nilometer is important and administrators kept the data as accurate as possible. The over flooding of the Nile was disastrous as it washes away crops, and when the river fails to rise, it causes drought and famine. The flood plays critical role in political and administrative policy decisions since taxation depended on the harvest quality.

Following every measurement of depth, the corresponding volume of fuel is obtained from the chart based on the underground gasoline tank size. For instance, for a gauge stick reading of 37 inches, the GTD converted the length measure of 37 inches into volume by referring to the chart appropriate to the tank size. To illustrate, if a gas station has a gasoline tank capacity of 10,000 gallons, reading the related tank size from the chart for 37 inches of heights gives 3517 gallons (see Figure 5.4). I am not interested in showing the relation between the height and the volume at this point. However, if we look at the highlighted ($\Delta(\Delta H)$) part, and the green ($\Delta(\Delta V)$) part, it is obvious that there is almost a negligible difference between the Gauge stick measurement and the computer generated readings (see Table 5.1 and Figure 5.5. Also see Appendices for readings and conversions).

CONCLUSION

In this chapter, I argue that mathematical ideas that emerge in the workplace afford the examination of a broad range of conceptual knowledge and problem-solving strategies. Matang (2002), Davis (1996) and Boaler (1993) have suggested that workplace mathematics is crucial in providing a context-rich problem that enhances societies' problem-solving strategies. Bishop (1988) reminds us that measuring includes comparing, ordering, and quantifying qualities represent solutions for vast social, cultural, and economic problems in ancient times as well as in current times. Too much over flooding of water in ancient Egypt was disastrous as it washed away crops and much of the infrastructure, whereas the river failing to rise caused drought and famine. The flood also plays important political and administrative roles, since the quality of the year's harvest was used to determine the amount of tax to be paid. As a solution, the Egyptians began measuring the Nile's water level to predict the harvest (Ritter, 2002). Social and cultural activities typically present problematic situations that transpire in response to a certain need. Workplace mathematics shows practical solutions to real life problems. For safety consideration and to avoid gasoline

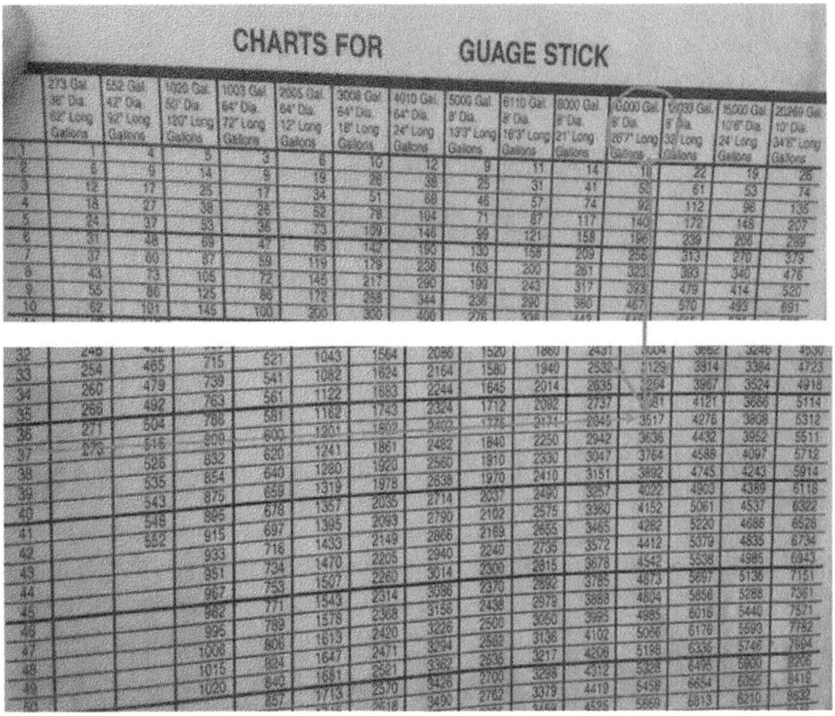

Figure 5.4. Chart depicting gauge stick reading and corresponding capacity level of gasoline tank.

Table 5.1.
Gauge Stick Versus Computer Based Reading

Gasoline	Gauge Stick Reading				Computer Based Reading				Computer Based Reading – Gauge Stick Reading					
	Height		Volume		Height		Volume		ΔH		ΔV		Δ(ΔH)	Δ(ΔV)
	B	A	B	A	B	A	B	A	ΔB	ΔA	ΔB	ΔA	ΔA - ΔB	ΔA - ΔB
R1	37	62	3517	6742	41.9	66.1	4293	7569	4.9	4.1	776	827	-0.8	51
R2	34.5	63	3254	6867	38.1	64.9	3665	7406	3.6	1.9	411	539	-1.7	128
P1	43.5	51.5	4412	5458	48.5	56.7	5214	6346	5.0	5.2	802	888	0.2	86
P2	45	50.5	4542	5328	45.9	48.2	4844	5172	0.9	-2.3	302	-156	-3.2	146

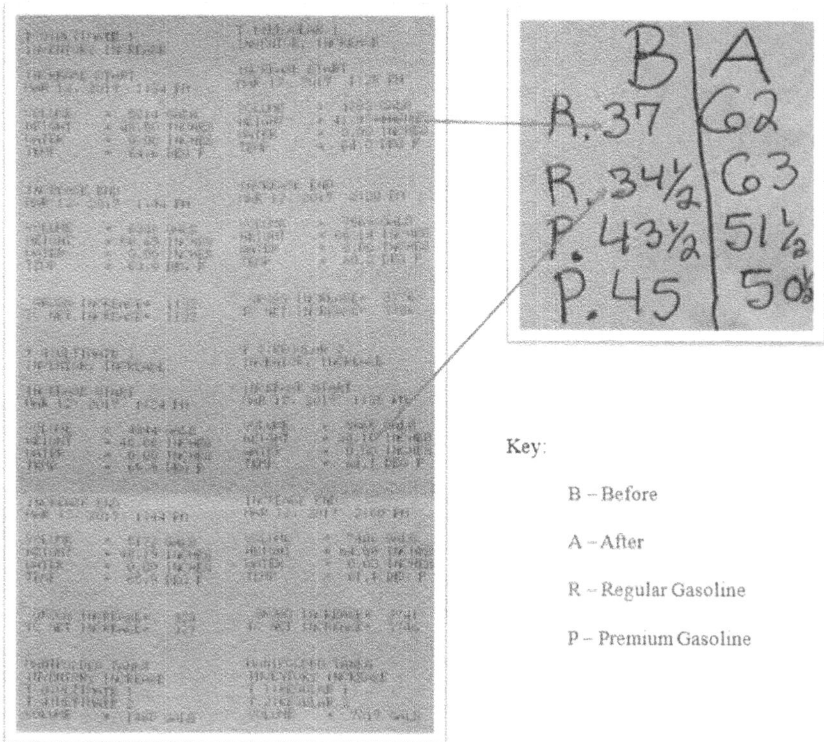

Figure 5.5. Comparing the gauge stick reading (right side) with computer reading printout (left side) at the instance of measuring gauge was in use, it was clear that both methods have very similar measurements.

related problems, like flammability, environmental pollution, and for fidelity of performance in gas stations, a simple measuring tool such as the fuel gauge stick is as effective and accurate as the computer generated records and cannot be impacted or altered by any technical malfunctioning.

REFERENCES

Bishop, A. J. (1988). *Mathematical enculturation. A cultural perspective on mathematics education.* Kluwer Academic Publishers.

Boaler, J. (1993). Encouraging the school mathematics to the "real world" through the integration of process and content, context, and culture. *Educational Studies in Mathematics, 25*, 341–373.

Burdun, G. D., & Markov, B. N. (1972). *Osnovy metrologii* [Fundamentals of metrology]. Izd-vo Standartov.

David, A. R. (2015). *A year in the life of ancient Egypt*. Pen and Sword Archaeology.

Davis, R. B. (1996). Cognition, mathematics, and education. In L. P. Steffe & P. Nesher (Eds.), *Theories of mathematics learning* (pp. 285–299). Erlbaum.

Matang, R. (2002) The role of ethnomathematics in mathematics education in Papau New Guinea: Implications for mathematics curriculum. *Journal of Educational Studies*, *24*(1), 27–37.

Ritter, R. (2002). *The Oxford style manual*. Oxford University Press.

APPENDICES

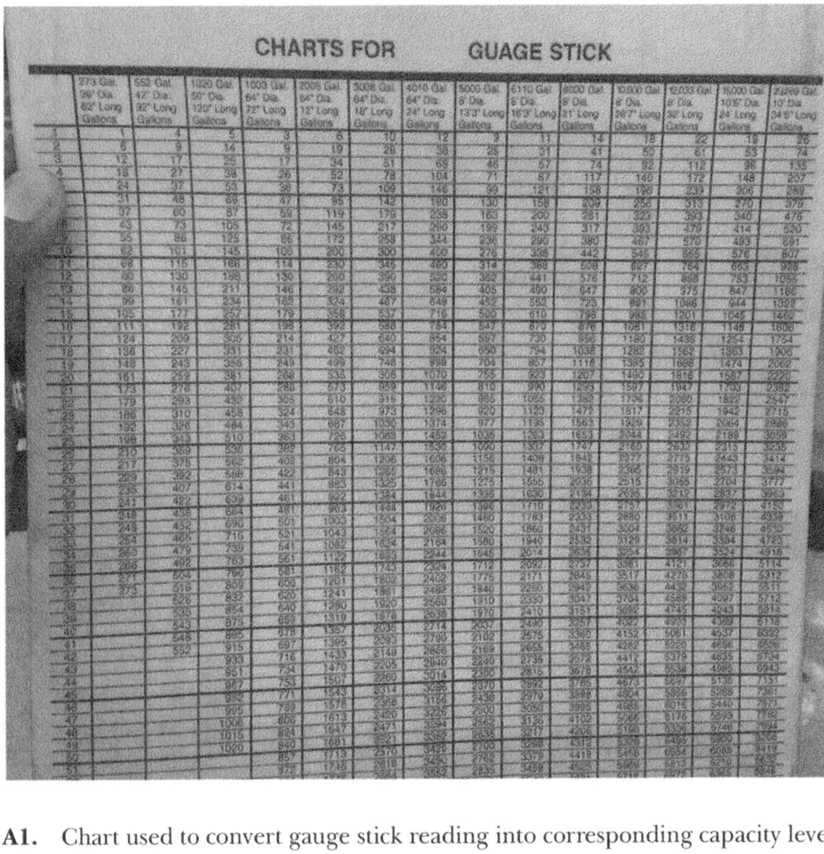

A1. Chart used to convert gauge stick reading into corresponding capacity level of gasoline tank.

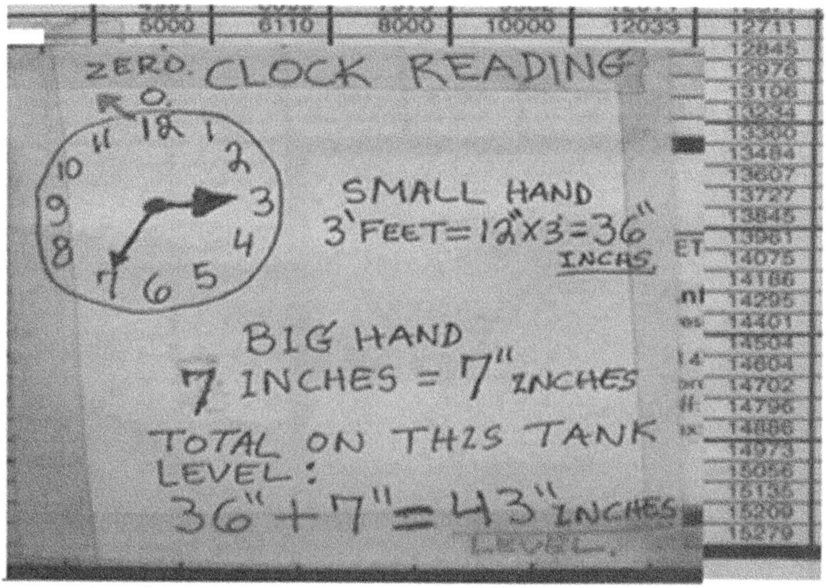

A2. Clock reading way of converting feet into inch unit.

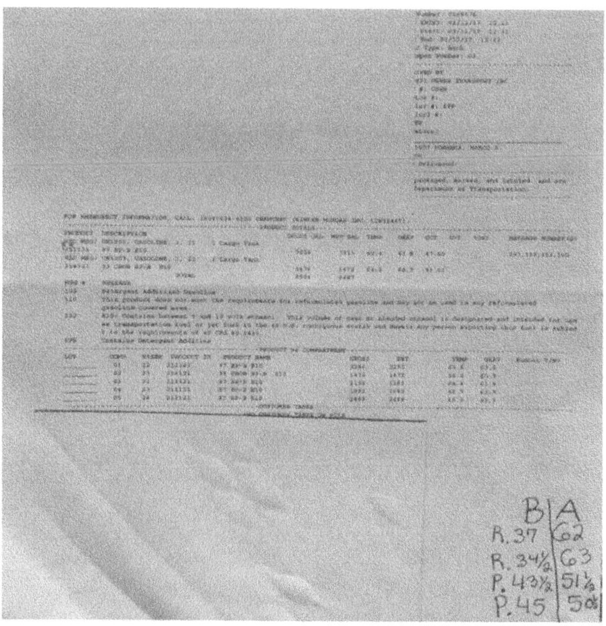

A3. Gauge stick reading,

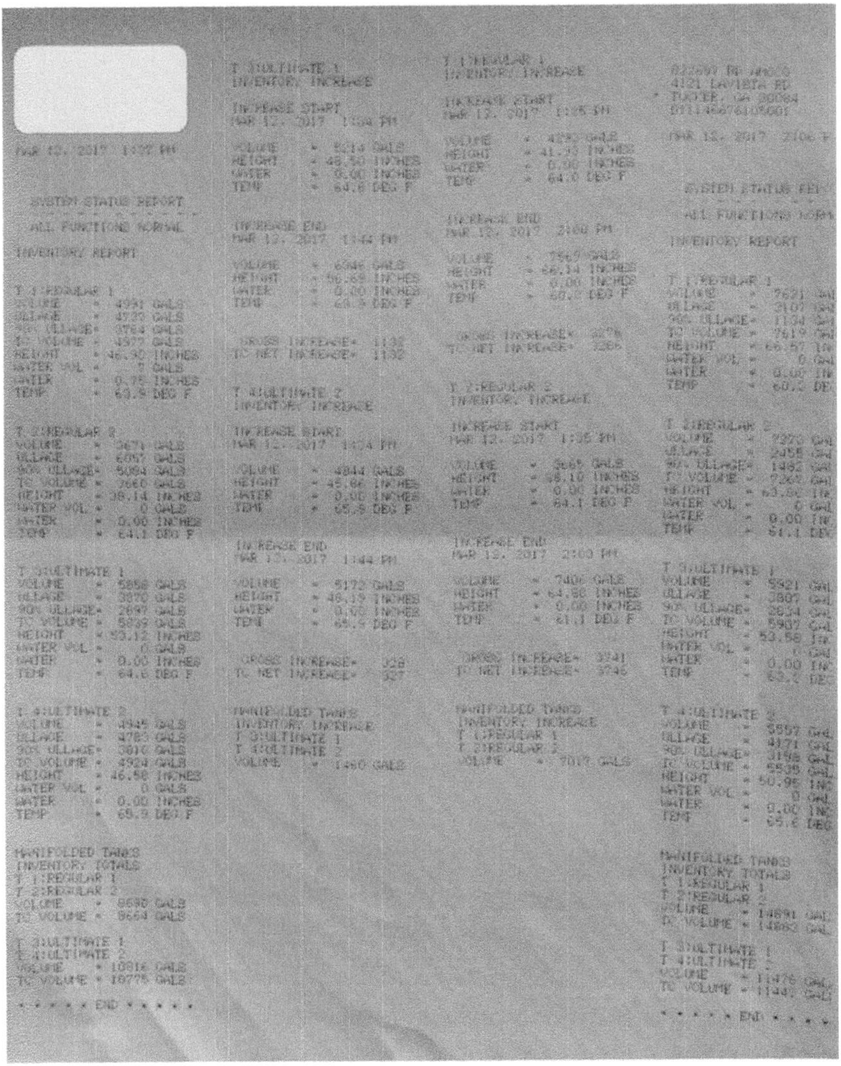

A4. Computer reading printouts.

CHAPTER 6

STUDENT-TEACHERS' VIEWS ON THE AFFORDANCES OF INCORPORATING CULTURAL ARTIFACTS IN MATHEMATICS LESSONS IN SOUTH AFRICA

Erica D. Spangenberg

ABSTRACT

Although the application of mathematics to real life contexts is advocated globally (Geiger et al., 2017), many student-teachers find it difficult to connect mathematics concepts with learners' daily experiences. By incorporating cultural artifacts in mathematics lessons as part of an intervention, student-teachers could make mathematics more relevant to learners' everyday lived experiences. This chapter investigates student-teachers' views on the affordances of incorporating cultural artifacts in mathematics lessons in South Africa. Thus, the research question is: *What are student-teachers' views on the affordances of incorporating cultural artifacts in mathematics lessons in South Africa?* Student-teachers had to execute classroom action research (CAR) in terms of ethnomathematics and cultural artifacts, which involved five

Evidence-Based Inquiries in Ethno-STEM Research: Investigations in Knowledge Systems Across Disciplines and Transcultural Settings, pp. 101–123

processes, namely identifying, planning, acting and collecting data, analyzing and evaluating artifacts. The selected participants of this study were 17 student-teachers enrolled for a Bacholor of Education honors degree in mathematics education. However, only 11 agreed to participate voluntary. They were full-time mathematics teachers enrolled for a postgraduate qualification in mathematics teaching with ages ranging from 23 to 41. Six were male and were female. The student-teachers studied at a specific university in Johannesburg, South Africa. They were from diverse cultural groups, namely Pedi (2), Sotho (2), Venda (3), Xhosa (2) and Zulu (1). The results indicate the following affordances of the incorporation of cultural artifacts in mathematics lessons: (1) Allowing for the teaching of specific mathematics topics; (2) Making mathematics concepts authentic; (3) Leading to better understanding of mathematics; (4) Enhancing enjoyment of mathematics; (5) Stimulating an awareness of mathematics; (6) Encouraging creativity and innovation in problem solving. Research on the use of ethnomathematics to improve teaching practices in South Africa is in its infancy, and this chapter envisages to stimulate discourse in the field.

INTRODUCTION AND BACKGROUND

Mathematics in South Africa is regularly viewed as a subject without any cultural content and values and is often taught as an objective, abstract discipline (Wong, 2005), even though South Africa is a diverse country, rich in various cultures and languages. With its 11 official languages and many different cultural groups, the rich cultural artifacts, customs (and games) and African music provide rich affordances for mathematics education. Nkopodi and Mosimege (2009) argues "mathematics and science are seen as 'western' and are never associated with African culture" (p. 377). However, according to Selin (2000), mathematics forms part of all citizens' everyday lives and is utilized in simple ways, for example, the calculation of petty cash at supermarkets, domestic budgeting, balancing of bank statements and in games. Yet, according to Soanes (2002), mathematics is an abstract science of numbers, quantities and space, which includes abstract concepts (pure mathematics) applied in disciplines such as physics and engineering.

Being a mathematics educator for many years, I have seen how school mathematics in South Africa is taught as an abstract, decontextualized subject in an archaic and uninspiring way with no productive or effective learning taking place. I have also seen some contexts where mathematics teachers have little content knowledge and tend to overemphasize procedures instead of focusing on the understanding of mathematics concepts. Learners are constantly expected to memorize procedures, rules, and formulas without any comprehension, which they forget easily and then start panicking. These instructional routines lead to poor student performance in mathematics.

There is an emerging body of literature that argue that teaching mathematics that is embedded in authentic cultural practices can help learners relate mathematics to their real lives and to enjoy the subject, which might lead to better results (Nkopodi & Mosimege, 2009; Rosa & Orey, 2011; Rowlands & Carson, 2002; Wong, 2005). According to Adam, Alangui, and Barton (2003), ethnomathematics embraces the integration of mathematical concepts and practices from various cultures into formal school mathematics. Ethnomathematics provides learners the opportunity to make connections between mathematics and their various cultures by examining concepts underlying cultural artifacts. Furthermore, according to D'Ambrosio (1990), ethnomathematics explores the ways people in their respective tribes think mathematically. Nonetheless, most South African mathematics teachers reject indigenous knowledge systems (IKS) (Nkopodi & Mosimege, 2009). The same also apply in a life sciences context where a large percentage of teachers do also not have the necessary knowledge or skills to teach IKS, even though they have the social responsibility to enlighten learners of the cultural and practical value of IKS (De Beer & Van Wyk, 2012, p. 4). Besides authentic mathematics being emphasized worldwide (Geiger et al., 2017), many South African student-teachers find it challenging to relate mathematics concepts with learners' cultural practices.

By incorporating cultural artifacts in mathematics lessons as part of an intervention, student-teachers could make mathematics more relevant to learners' everyday lived experiences. Consequently, the research question is: *What are student-teachers' views on the affordances of incorporating cultural artifacts in mathematics lessons in South Africa?*

Student-teachers had to execute CAR in terms of ethnomathematics and cultural artifacts. With reference to the incorporation of cultural artifacts in mathematics lessons, student-teachers employed CAR. In South Africa a scholarship of teaching needs to be promoted, as too many teachers do not engage in either critical reflection or classroom action research. I concur with John Slaughter who said "research is to teaching what sin is to confession. If you do not participate in the former, you have little to say in the latter" (Chmielewski & Stapleton, 2009, p. 53).

LITERATURE OVERVIEW

The Relationship Between Mathematics and Ethnomathematics

According to Wong (2005), mathematicians are born with special talents to reason logically, to manipulate symbols and to enjoy funny ideas and puzzles. Mathematics has been successfully applied in sciences and engineering, which concerns objective reality for hundreds of years. However, its role in modelling social, political, economic and psychological phenomenon involving human interactions and values are underexplored in research. Whereas cultural groups daily use mathematics in an informal way, researchers often undervalue it.

I disagree with scholars like Venkat (2007) who claim that mathematics is a pure academic subject focusing on content and the application of it. Mathematics plays a very important role in many facets of cultural life, for example, in art, games, and music. In my view, the problem with such views of mathematics as an abstract and logic "Western" science is the underestimation of its important role in concrete, real-life authentic situations. However, the national South African school curriculum, the Curriculum and Assessment Policy Statement (CAPS) (Department of Basic Education, 2011, p. 8) connects mathematics to human activity by defining mathematics as:

> a language that makes use of symbols and notations ... a human activity that involves observing, representing and investigating patterns and qualitative relationships in physical and social phenomena ... to understand the world (physical, social and economic) around us.

Wong (2005) divided mathematics into three dimensions, namely content, process and culture. The content includes mathematics subject knowledge and skills. Processes involve reasoning, discovering, drawing conclusions and problem solving in order to makes sense of the world. Culture refers to aspects about mathematics, rather than mathematics itself. For example, the history of mathematics, stories about mathematicians, mathematics in literature and films, philosophy of mathematics, the social affects of mathematics and perceptions about mathematics.

Wong (2005) contended that it is possible to extend the framework of school mathematics by including a cultural perspective, for example, to analyze the mathematics embedded in a design or a game with rich cultural elements. According to Bullivant (1987), culture is a general composition of interdependent, value-driven traditional and current public knowledge encapsulated in behavior and artifacts, which are transferred to new members by signs and symbols from past generations in order to give meaning to future challenges. D'Ambrosio (1989) argued that ethnomathematics includes cultural dynamics and is subordinate to a broader framework of historical events. These events refer to cultural practices, that are never static, but evolves all the time, and highlight the use of mathematics in all spheres of society, for example, artwork, customs, games and music, to name but a few. A cultural artifact or art object is a term used in social sciences, specific anthropology, ethnology and sociology, for anything created by a specific cultural group sharing information about the culture of the creator or users (Bangura, 2012). Thus, a cultural artifact is any object made or formed by a human being representing a specific cultural group (Soanes, 2002). These objects are traditionally found in houses and communities and include tools, artworks or objects, which are of archeological interest. Hence, a cultural artifact is any concrete object or symbol, which is relevant to learners' daily lives and with which they can associate. The use of cultural artifacts assists teachers to apply ethnomathematics in classrooms. According to Leonard (2008), teachers could use cultural artifacts to enhance better understanding of mathematical concepts. For example, teachers can use objects to present real-life lessons, for example, geometric shapes using Ndebele art works (Plummer & Chahine, 2014).

According to D'Ambrosio (1985), research does not focus sufficiently on ethnomathematics due to the widespread belief in the universality of mathematics. However, there is a strong relationship between mathematics concepts and the cultural environment where mathematical ideas emerge. There are communities who employ mathematical ideas, such as counting, sorting and measurement, in ways that differ radically from the mathematics at school. In particular, D'Ambrosio defined ethnomathematics as the mathematics executed by identifiatible cultural groups. The Xhosa and Zulu tribes in South Africa can be examples of such groups. The identities of these cultural groups are mainly determined by the focus of their interests, motivation and specific codes and jargon. Furthermore, ethnomathematics develops and changes continuously due to social changes in communities. One of the problems why the status of ethnomathematics is questioned is due to the Western superiority, which often looks down on rural cultural customs (Nkopodi & Mosimege, 2009). Engagement in indigenous knowledge was seen as witchcraft (especially in fields like ethnobotany and medicinal plant use), and government saw it as their task to alienate the holders of traditional knowledge from this rich oral history (Nkopodi & Mosimege, 2009). D'Ambrosio argued that mathematics is the natural result of evolution of the discipline within a social, economic, and cultural atmosphere.

Ethnomathematics is a study of mathematics developed and practiced by different cultural groups in the world since the ancient times until the modern times. In particular, ethnomathematics is a cultural product, which includes the cultural values of mathematics teaching (Wong, 2005). Rowlands and Carson (2002) argued that ethnomathematics focuses on the executing of mathematics, known as mathematizing. Cultural groups create their own mathematics from their daily lives, rather than from the learning of mathematics in a formal academic discipline. The focus of ethnomathematics has two tiers, namely (1) to investigate mathematical thinking in cultural groups, and (2) to assist mathematics learners to recognize both the universal nature of human creativity and the variety of expression forms in various cultures (Rowlands & Carson, 2002).

On the other hand, ethnomathematics is not the degrading of mathematics as a discipline or abstract knowledge system, but rather attempts to make mathematics relevant to daily life. According to Rosa and Orey (2011), ethnomathematics focuses on the cultural aspects of mathematics and advocates connecting mathematics concepts with learners' daily experiences in order to broaden their understanding of mathematics. As such, ethnomathematics draws on indigenous knowledge systems which is the local knowledge unique to a given culture or society, and which are part of people's lived experiences. Such experiences are nurtured by building connections through physical and bodily interactions embodied cognition"), by direct involvement in real life environment ("situated cognition") or through social interaction with people ("distributed cognition") (Wilson, 2002, p. 625). Despite the fact that IKS are highly valued by the National Research Foundation (NRF) in South Africa who compiled a framework document to support research in the field of IKS and to promote IKS's role in communities (NRF, 2012), most teachers in South Africa were never trained in incorporating ethnomathematics in their classes (Spangenberg, 2015). This highlights the importance of conducting research on the incorporation of cultural artifacts in mathematics lessons.

A Rationale for the Use of Ethnomathematics in the Mathematics Classroom

The teaching of mathematics from a cultural perspective promotes the recognition of cultural heritages and lays the basis to assist learners in their understanding and conceptualizing of mathematics knowledge (Ascher & D'Ambrosio, 1994). In particular, according to Wong (2005), cultural artifacts motivate learners to learn mathematics as it promotes the use of a concrete cultural context for mathematics. Furthermore, the use of cultural artifacts in a classroom creates a motivating learning environment, where learners can learn mathematics without any fear. D'Ambrosio (1989) emphasized the importance of a sociocultural environment in the teaching-and-learning process in order to ensure that the content of mathematics is relevant to learners' daily lives. According to Rosa and Orey (2011), ethnomathematics is a powerful strategy to measure learners' real live experiences and provide learners with the necessary tools to become critical participants in communities.

Furthermore, ethnomathematics assists learners to understand the history of modern mathematics, and how to apply mathematics in their daily lives. The use of cultural artifacts in the mathematics classroom enhances a learner's self-concept as it recognizes the importance of culture in the development of a learner's identity. According to D'Ambrosio (1989), mathematics should be made relevant to learners' needs, cultural values and self-images.

Research Design and—Methodology

An interpretivist paradigm, with an epistemological position that prioritizes student-teachers' subjective ideas (Matthews & Ross, 2010) provided an in-depth understanding into the inquiry. An inductive approach using a qualitative research method was utilized to explore student-teachers' views on the affordances of incorporating cultural artifacts in mathematics lessons in South Africa. Seventeen participants were purposively selected for this study. However, only 10 agreed to participate voluntary. The participants were part-time student teachers enrolled for a Bachelor of Education honors degree in mathematics education at a specific university in Johannesburg, South Africa. They were all full-time practicing mathematics teachers with ages ranging from 23 to 41. Six were male and four were female. They were from diverse cultural groups, namely Pedi (2), Sotho (2), Venda (3), Xhosa (2) and Zulu (1). All of them spoke an indigenous language.

To answer the research question, namely *What are student-teachers' views on the affordances of incorporating cultural artifacts in mathematics lessons in South Africa?*, the student-teachers had first to plan an intervention employing CAR as their major methodological framework to incorporate cultural artifacts in their mathematics lessons. CAR encompasses inquiries done by teachers to improve their own footing in the teaching of their subjects (Walser, 2009). According to Gravett and De Beer (2015, p. 114), CAR "implies an interplay between teaching practice, reflection, collecting of information and specifying learning with a view to improve practice." In particular, *action* describes the strategy or mediation process an educator is following, while *research* shapes an investigation into pedagogy in order to execute the strategy. Thus, CAR refers "to [an] intentional, purposeful and systematic inquiry—finding things out about what you do in a planned and systematic way" (Gravett & De Beer, 2015, p. 344). Hence, the student-teachers' methodology with reference to the incorporation of cultural artifacts in mathematics lessons was guided by the five processes involved in CAR, namely identifying, planning, acting and collecting data, analyzing and evaluating as illustrated by Gravett and De Beer (p. 117) in Figure 6.1.

Student-teachers were requested to first identify mathematics-related curriculum topics in which they could incorporate cultural artifacts. To incorporate cultural artifacts in the teaching of mathematics, student-teachers had to make informed decisions with regard to their teaching

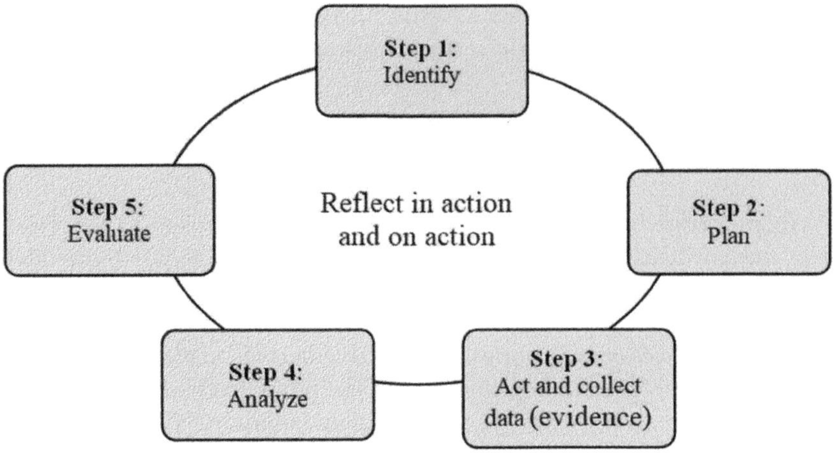

Source: Gravett & De Beer (2015, p. 117).

Figure 6.1. The five processes in CAR.

strategies, especially to make the curriculum relevant, interesting and engaging. As mathematics teachers, who research educational literature for best practices and form discussion groups, CAR provides them opportunities for critical self-reflection (Walser, 2009). Following the steps of McNiff and Whitehead (2002), student-teachers conducting CAR had to start by identifying concerns and reasons for these concerns. Second, they had to plan mathematics lessons in which the specific cultural artifacts would be used. Third, they had to act by teaching the mathematics lessons to learners and collect qualitative data of learners' experiences by observing and interviewing learners during classwork activities. Thereafter, student-teachers had to analyze their collected data. Lastly, they had to evaluate what the affordances of incorporating cultural artifacts in a mathematics lesson were. They had to write their findings up in short reports, which were used as a data-collection instrument for this chapter.

Member checking ensured credibility. Student-teachers were requested to confirm the correct interpretation of the emerging findings. A level of transferability was confirmed by direct quotations from the data and literature evidence. Confirmability was taken into account by adhering to ethical measures.

Results and Discussion

To answer the research question for this chapter, the student-teachers' reports were read thoroughly to obtain complete overview and unrelated data were omitted. An inductive data analysis approach was utilized. Open coding was used to code words and short phrases qualitatively, where after categories were derived. Thereafter, recurring themes were identified. The results of the analysis are summarized in Table 6.1.

Table 6.1.
Themes, Categories, and Subthemes

Theme	Category		Subcategory	Participant
Affordances	**Mathematics Concepts**	**Type of Cultural Artifact**	**Specific Cultural Artifact**	
Making mathematics concepts authentic	Units of measurement Perimeter, area, surface area and volume Number patterns	Tools	Khoi-San tools (medicine and food containers, teacups, and cutlery)	F
	Measurement Surface areas and volumes of cones, cylinders and pyramids Area of circles	Cultural dwellings	Traditional dwellings (*rondavels*)	E I
	Sine rule	Clothing	Zulu female clothing (*Isidwana*)	B
Better understanding Enjoyment Awareness of mathematics	Surface area and volume of three dimensional shapes Transformation	Cultural clothing	Sotho hat (*Mokoro*) Zulu female clothing (*Isidwana*) Xhosa clothing	G B H
	Linear graphs Number patterns Midpoint theory	Indigenous Games	Pedi game (*Diketo*) Xhosa (*Upuca*) Pedi (*Morabara*)	A K D
Creativity and innovation	Geometric patterns	Art work	Venda clay art	C

In this chapter, I describe how the student-teachers viewed the affordances of incorporating cultural artifacts in mathematics lessons. In the following protocols, alphabetical letters are used as names to protect student-teachers' identities and the protocols are from their reports. A limitation to the data was that student-teachers did not indicate how they had used the cultural artifacts in their teaching of mathematics, shape and space and number patterns.

a. Cultural Artifacts Allow for the Teaching of Specific Mathematics Topics

One of the first affordances of the incorporation of cultural artifacts is the allowance of teaching specific mathematics concepts. To teach measurement, Participant F used Khoi-San tools and both participants E and I used cultural dwellings (*rondavels*). To teach shape and shape, Participant F also used Khoi-San, both Participants E and I used cultural dwellings (*rondavels*), Participant G used a Sotho hat (*Mokoro*) and Participant B used Zulu female clothing *(Isidwana)*. To teach the sine rule, Participant B also used Zulu female clothing *(Isidwana)*. To teach transformations, Participant H used Xhosa clothing. To teach linear graphs, Participant H used a Pedi indigenous game (*Diketo*). To teach number patterns, Participant F used Khoi-San tools, Participant K utilized a Xhosa game (Upuca), and Participant C Venda clay art. To teach the midpoint theory in Euclidean geometry, Participant D used a Pedi indigenous game (*Morabara*). The above-mentioned cultural artifacts will be elaborated on in the affordances that follow next.

b. The Use of Cultural Artifacts Makes Mathematics Concepts Authentic

A second affordance of incorporating cultural artifacts in mathematics lessons is that it makes mathematics concepts authentic. Mathematics concepts are linked to real-life contexts. In addition, teachers could create contexts within their classroom, which expect learners to link their social and personal lives with mathematics. A mathematics investigation can include a specific scenario with various tasks, for example, planning of a cultural excursion. One task could be to draw up a budget for the excursion and another task could involve compiling a timetable with various time slots.

Participant F indicated that he used Khoi-San tools, for example, medicine and food containers, teacups, and cutlery as cultural artifacts, to teach units of measurement. He focused in measurement on the choice and utilization of appropriate units, instruments and formulae to quantify the characteristics of shapes and objects. The different shapes were also used to measure perimeter, area and volume (Figure 6.2).

Figure 6.2. Khoi-San medicine containers.

Participant F also stated that patterns on knives (Figure 6.3) were used to explain number patterns. According to him, learners investigated numeric number patterns to establish relationships between variables and to analyze situations in a variety of contexts. Unfortunately, the student-teacher did not indicate how he used the Khoi-San tools to teach the mathematics concepts.

Figure 6.3. Khoi-San knives.

Two student-teachers (Participants E and I) indicated that they utilized Venda traditional dwellings, namely *rondavels* to teach problems in measurement. Participant E used *rondavels* to teach surface areas and volumes of cones, cylinders and pyramids in measurement in 11th grade mathematics (Figure 6.4), while Participant I taught areas of circles in 10th grade mathematics with the same traditional dwelling. In order to calculate the area of a circle, learners should first measure the diameter using the tape measure (measuring from one side of the hut to the other side), and then divide the diameter into two equal lengths, in order to find the radius. According to Plummer and Chahine (2014), a *rondavel* consists of a diameter, height and perimeter. To determine the area of a circle, learners had to determine the radius first by measuring the perimeter by means of a measuring tape. Thereafter they had to divide the perimeter in two equal lengths. Learners had to know that the formula of the area of a circle, namely $A=\pi r^2$, where $\pi \approx 3.14$ and r = radius of the circle. After finding the size of a radius, it was easy to calculate the area of a circle. The *rondavel* is a circular single room building associated in the past, with a thatched roof and used as milk, storage, bed, or living room, which is described, according to Huffman (2009), as a cone on a cylinder or a cone on a drum.

Participant B indicated that she used the *isidwaba* (a female cloth in the Zulu culture) (Figure 6.5) to explain the sine rule in 11th grade mathematics.

Figure 6.4. A Venda traditional house.

Figure 6.5. Clothing of South African Zulu Women (*isidwaba*).

A limitation to the data is that Participant B did not explain how she linked the cloth with the sine rule. However, from her interviews with learners, it was evident that the learners could link the cloth with mathematics. One learner of Participant B mentioned: "I had never thought that traditional dressing could be used in a mathematics classroom.... Now I know even if I am at home, that anything in my house could be linked to mathematics" (boy, 11th grade).

Participant G mentioned he used a South Sotho *mokoro* (a hat with a shape of a cone) (Figure 6.6), which is common in his culture, to teach measurement in 10th grade Mathematics. According to Riep (2005), Sotho weavers use many techniques to create variation on their designs. Mathematical shapes are visible as concentric circles, patterns called *litema* or the plaited grass hats. Riep notes that similar patterns are done on ceramic pots, vessels, beadworks, house painting, and baskets.

Figure 6.6. The South Sotho *mokoro* (a hat with a shape of a cone).

According to Participant 7, measurement included three-dimensional shapes in solving of problems on volumes and surface areas of fixed objects.

Participant 8 used Xhosa cultural clothing (Figure 6.7) to create a context to explain transformations in 8th grade mathematics.

Figure 6.7. Xhosa cultural clothing.

According to Participant 8, transformations include physical manipulation of shapes into new orientations or positions and could assist in distinguishing between translations, reflections and rotations. A second learner of Participant 8 noted: "Now madam, every time I look at cultural clothing, I am always looking for the transformation I can spot. I am no longer just looking at the beauty but now I look for mathematics that I can find" (boy, 8th grade).

Indigenous games were also used in applying mathematics in real-life contexts. Participant A used a Pedi game *diketo* to teach linear graphs in 9th grade. According to her, *diteko* assists learners to understand whether a gradient of a straight line is positive or negative. The game is played with stones. A hole is made in the ground and is filled with a number of stones. While one stone is thrown in the air to be catched, the other stones are removed from the hole. Once the stone is catched, the other stones are being put back in the hole, accept one stone. That stone is thrown again in the air. The process is repeated. The learner who catches the most stones wins the game. As the number of throws increases, the number of caught stones also increases resulting in a positive gradient. A gradient is determined by the ratio of vertical and horizontal variables. The results are indicated in a table where the top row illustrates the x-values and the bottom row the y-values. From the table it can be deduced whether the gradient is positive or negative. Thereafter, the coordinates are indicated in a table. Participant 1 gave the following example.

Two learners competed in a *diteko* game and produced the following result in the table as follows:

Number of throws (x)	1	2	3	4	5	6
Number of caught stones	2	4	6			

a. Complete the table
b. Illustrate the coordinates on a Cartesian plan
c. What type of ratio is indicated in the graph? Provide a reason for your finding.

Participant K used the same game to teach number patterns, but according to her, the game is known as *upuca* in Xhosa.

Participant D used the indigenous game *Morabaraba* (Sepedi) (Figure 6.8) to teach 10th grade geometry, with specific reference to the investigation of line segments that join the midpoints of two sides of a triangle.

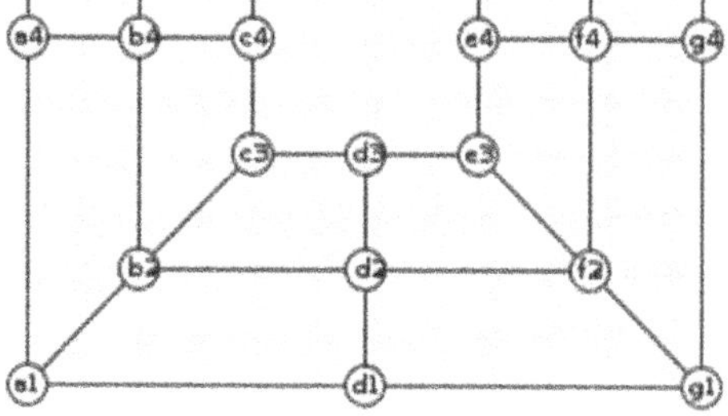

Figure 6.8. *Morabaraba* game board diagram.

According to Participant D, *morabaraba* is played by two learners each with 12 stones or any other objects but are different from that of the opponent in color or shape. Players strive to form a column, diagonals or a row with three stones. A player can form a row, column or a diagonal an opponent's place piece is strategically taken out of the game. This will continue until one player's objects are finished and therefore loosing. Thus, a player is doing two things during the game: block the opponent from forming the row, diagonals or column, while he/she is forming his/hers. Mathematics skills such as counting, symmetry, geometrical shapes, ratio and logical reasoning can be acquired from the game.

According to Participant C, different culture groups have different names for *morabaraba*, namely *umlabalaba* (Nguni), *dizela* (Tsonga) and *izingedo* (Zulu). These games also create a context for the instruction of functions in 10th grade mathematics where learners are exposed to relationships between variables in terms of numeric and graphical representations in the physical world. However, he did not indicate how he used *morobaraba* to teach geometry in 10th grade. Participant J used *umlabalaba* to teach analytical geometry in 10th grade by emphasizing the distance between two points, the gradients, and midpoints, but did not provide any further information. Nkopodi and Mosimege (2009) also noted that the game is played in many Southern African countries such as Lesotho, Botswana, Mozambique, and Zimbabwe.

Lastly, Participant C used Ndebele artwork to teach geometric patterns in 7th grade (Figure 6.9), but again did not indicate how he used it. Artifacts like Zulu beadwork (Chahine & Kinuthia, 2013), or rhythms in African music, hold promise for contextualized mathematics learning.

From the above examples, it is clear that mathematics becomes via the use of cultural artifacts more concrete and contextualized. Abramovich (2002) argued that teachers should start their instruction in a context familiar to learners, before asking questions with regard to mathematical concepts. Mathematics activities from a cultural perspective are more meaningful as such experiences could help learners connect with mathematical ideas expressed in a familiar context. Cultural artifacts can be used to model and solve problems within a particular context.

c. The Incorporation of Cultural Artifacts in Mathematics Lessons Leads to Better Understanding of Mathematics

One of the learners of Participant B acknowledged that the incorporation of the *isidwaba* (female cloth in the Zulu culture) in mathematics lessons is valuable:

Figure 6.9. Ndebele painted walls.

> My view is that if we use cultural artefacts more in mathematics we can understand better, because it's everything we see every day. If you forget at home *uzophinde uyibonefuthi* (you will see it again). So if we use it more in sine rule we are going to pass. (girl, 11th grade)

Cultural artifacts also assisted in distinguishing between mathematics concepts. Learner of Participant H confessed: "Before my teacher used cultural artifacts, I really did not grasp the difference between the term transformation and translation. I totally confused the two terms but now I understand the difference" (boy, 8th grade). Abramovich (2002) explained that the use of cultural artifacts in the classroom activates interactive experimenting with mathematical content and promotes an investigative approach towards mathematics.

d. The Use of Cultural Artifacts Enhances Enjoyment of Mathematics

A second learner of Participant B declared that using cultural artifacts in mathematics increase enjoyment of the subject: "Most of us we really enjoyed that topic, mmm ... mmm I had never thought that traditional dressing could be used in a mathematics classroom. Everyone was talking about it, such things makes mathematics to be live" (boy, 11th grade). A learner of Participant 3 also agreed: "I have enjoyed using clay art to learn geometric patterns" (girl, 7th grade). According to Abramovich (2002), learners experience mathematics in an enjoyable environment where cultural artifacts are used, as something they can do, rather than something they have to memorize, thus, a conversion of knowledge to applications in diverse contexts.

e. Cultural Artifacts Stimulate an Awareness of Mathematics

The incorporation of cultural artifacts in mathematics arouses cognizance of mathematics in everyday living. A learner of Participant H claimed: "Now madam, every time I look at cultural clothing, I am always looking for the transformation I can spot. I am no longer just looking at the beauty but now I look for mathematics that I can find" (boy, 8th grade).

f. The Incorporation of Cultural Artifacts in Mathematics Lessons Encourages Creativity and Innovation in Problem Solving

The incorporation of cultural artifacts in mathematics lessons affords opportunities for creativity and innovation in solving of mathematics problems. One learner of participant C described his experience as follows: "Wow.... It was a wonderful experience having to learn geometric patterns using the clay. I have managed to be more creative and innovative towards designing my own geometric patterns" (boy, 7th grade). Another learner of Participant C noted: "I was actively involved and able to be creative in terms of designing my own geometric patterns that culturally related" (girl, 7th grade). Wong (2005) argued that a good teaching strategy to teach mathematics is to stimulate learners' creativity through the teaching of problem solving. Teachers should set questions proportional to learners' levels of understanding and ask stimulating questions in order to guide learners to the solutions (Wong, 2005). An example of this is Knijnik and Wanderer's (2004) description of how Portuguese tiles, that were popular in the ex-Portuguese colony Brazil, were used in the mathematics classroom to study isometric forms.

g. The Use of Cultural Artifacts in Mathematics Lessons Allows for Cooperative and Active Learning and Develops a Sense of Belonging

The incorporation of cultural artifacts in mathematics lessons stimulates cooperative and active learning as illustrated in Figure 6.10. A learner of Participant C argued: "cultural artefacts could be used to enhance teaching and learning of mathematics.... It encourages learner involvement and active participation among the learners" (girl, 10th grade). Another learner of Participant C noted: "Cultural artefact, I think it ... suggest the kinds of shared beliefs and expectations" (boy, 10th grade), which indicates a sense of belonging.

Figure 6.10. Learners of Participant G actively engaged in investigating *mokoro* (Sotho hat).

Furthermore, according to Siegel (2005), cooperative learning is a teaching strategy that includes participation and group discussions, individual accountability, positive group dependence, social skills teaching and deeply rooted self-reflection. A cooperative teaching strategy is grounded in the view that learning occurs most effectively if learners are active in sharing ideas and working together in completing tasks (Zakaria & Iksan, 2007). According to Cherida (2012), group work develops learners' self-confidence to approach mathematics problems.

Ginsberg and Wlodkowski (2000) claimed that socializing influences cultural experiences which, in return, reflect learners' motivation to execute a learning activity. In particular, the use of cultural artifacts in the mathematics classroom promotes (1) the establishment of inclusivity; (2) the development of a positive attitude; (3) the promotion of meaning; and (4) the generation of competence (Ginsberg & Wlodkowski, 2000). At the beginning of a new topic, for example, sequences and series, a teacher can first introduce an artifact, such as art works, music instruments, cultural games or jewelry, in order to include learners from various cultural groups. Second, a class discussion about how the artifact links with the topic could be used to focus learners' attention to the topic. Third, the discussion could be followed up with a group activity where learners are given the opportunity to experiment with the new concepts. Last, a learner from a group can report back to illustrate how he/she attains the necessary knowledge. Subsequently, all four conditions will be fulfilled.

CONCLUSION

Although the application of mathematics to real life contexts is advocated globally, many student-teachers find it difficult to connect mathematics concepts with learners' daily experiences. I conclude that there are several cultural artifacts, which could be utilized in mathematics lessons, namely cultural tools, dwellings, clothing, indigenous games and artwork. The incorporation of these cultural artifacts in mathematics lessons have several affordances: (1) Allowing for the teaching of specific mathematics topics; (2) Making mathematics concepts authentic; (3) Leading to better understanding of mathematics; (4) Enhancing enjoyment of mathematics; (5) Stimulating an awareness of mathematics; (6) Encouraging creativity and innovation in problem solving; and (7) Allowing for cooperative and active learning and developing a sense of belonging. Ethnomathematics is the mathematics that integrates mathematical concepts and practices from different cultures into formal school mathematics and provides learners the opportunity to make connections between mathematics and their various cultures by examining concepts underlying cultural artifacts. Furthermore, the utilization of cultural artifacts in the classroom creates an engaging atmosphere where learners can practice mathematics without fear and addresses learning needs, cultural appreciation and self-concepts of students. Ethnomathematics also addresses inclusivity, mathematics understanding and mathematical processes and makes school mathematics meaningful.

Further research should focus on how to use specific cultural artifacts in the teaching of mathematics concepts. Although some findings could be supported by evidence obtained from empirical data, the focus of this research did not attempt to try to support or contest other research findings, but to improve teaching practices in South Africa and could therefore not be generalized.

REFERENCES

Abramovich, S. (2000). *Cultural tools and mathematics teacher education*. State University of New York. http://www2.potsdam.edu/EDUC/abromovs

Adam, S., Alangui, W., & Barton, B. (2003). A comment on: Rowlands & Carson "Where would formal, academic mathematics stand in a curriculum informed by ethnomathematics? A critical review." *Educational Studies in Mathematics 52*, 327–335. https://doi.org/10.1023/A:1024308220169

Ascher, M., & D'Ambrosio, U. (1994). Ethnomathematics: A dialogue. *For the Learning of Mathematics, 14*(2), 36–43. http://flm-journal.org//FLMDAmbrosio.pdf

Bangura, A. K. (2012). *African mathematics: From bones to computers.* University Press of America.

Bullivant, B. M. (1987). *The ethnic encounter in the secondary school. Ethnocultural reproduction and resistance; theory and case studies.* The Falmer Press.

Chahine, I. C., & Kinuthia, W. (2013). Juxtaposing form, function, and social symbolism: An ethnomathematical analysis of indigenous technologies in the Zulu culture of South Africa. *Journal of Mathematics and Culture, 7*(1), 1–30.

Cherida, M. (2012, July). *Weaving exploration in the process of the acquisition and development of mathematical knowledge.* Paper presented at the 12th International Conference on Mathematics Education. Seoul, South Korea. http://www.icme12.org/upload/submission/1972_F.pdf

Chmielewski, J. G., & Stapleton, M. G. (2009). The biologists' forum: The undergraduate research experience: It's really not for everyone, students and faculty alike. *BIOS, 80*(2), 53–58.

D'Ambrosio, U. (1985). Ethnomathematics and its place in the history and pedagogy of mathematics. *For the Learning of Mathematics, 5*(1), 44–48. http://www.jstor.org/stable/40247876

D'Ambrosio, U. (1989). A research program and a course in the history of mathematics: Ethnomathematics. *Historia Mathematica, 16*(3), 285–287. https://doi.org/10.1016/0315-0860(89)90026-8

D'Ambrosio, U. (1990). The role of mathematics education in building a democratic and just society. *For the Learning of Mathematics, 10*(3), 20–23. http://www.jstor.org/stable/40247989

De Beer, J. J., & Van Wyk, B.-E. (2012). Inheemse kennis in die lewenswetenskappe-klaskamer: Wetenskap, pseudo-wetenskap, of 'n vermiste skakel? *Suid-Afrikaanse Tydskrif vir Natuurwetenskap en Tegnologie, 31*(1), 1–5. https://doi.org/10.4102/satnt.v31i1.368

Department of Basic Education. (2011). *Curriculum and Assessment Policy Statement, Grades 10–12, Mathematics.* http://www.education.gov.za/LinkClick.aspx?fileticket=QPqC7QbX75w%3d&tabid=420&mid=1216

Geiger, V., Stillman, G., Brown, J., Galbriath, P., & Niss, M. (2017). Using mathematics to solve real world problems: The role of enablers. *Mathematics Education Research Journal*, https://doi.org/10.1007/s13394-017-0217-3

Ginsberg, M. B., & Wlodkowski, R. J. (2000). *Creating highly motivating classrooms for all students: A schoolwide approach to powerful teaching with diverse learners.* Jossey-Bass.

Gravett, S., & De Beer, J. (2015). The teacher as a researcher. In S Gravett, J. De Beer, & E du Plessis (Eds.), In *Becoming a teacher* (2nd ed., pp. 114–128). Pearson Education South Africa (Pty).

Huffman, T. N. (2007). *Mapungubwe: Ancient African civilisation on the Limpopo*. Wits University Press.

Knijnik, G. & Wanderer, F. (2004). The art of tiles in Portugal and Brazil: Ethnomathematics and travelling cultures. In. F Favilli (Ed.), *Proceedings of the tenth International Congress of Mathematics Education, Copenhagen.*

Leonard, J. (2008). *Culturally specific pedagogy in the Mathematics classroom: Strategies for teachers and students*. Routledge.

Matthews, B., & Ross, L. (2010). *Research methods: A practical guide for the social sciences*. Pearson.

McNiff, J., & Whitehead, J. (2002). *Action research: Principles and practice* (2nd ed.). Routledge.

National Research Foundation (NRF). (2012). *Indigenous knowledge systems knowledge fields development (KFD)*. CSIR.

Nkopodi, N., & Mosimege, M. (2009). Incorporating the indigenous game of morabaraba in the learning of mathematics. *South African Journal of Education, 29*, 377–392. http://www.scielo.org.za/scielo.php?pid=S0256-01002009000300006&script=sci_arttext

Plummer, A., & Chahine, I. C. (2014, November). Mathematics lessons from "a place of light". *The Mathematics Teacher: Mathematical Lens, 108*(4), 254–257.

Riep, D. M. M. (2005). *Art on the margins: A reintroduction of the art history of the Sotho of Southern Africa* (Master's dissertation). University of Kentucky, Lexington, Kentucky.

Rosa, M., & Orey, D. C. (2011). Ethnomathematics: The cultural aspects of mathematics. *Revista Latinoamericana de Etnomatemática, 4*(2), 32–54.

Rowlands, S., & Carson, R. (2002). Where would formal, academic mathematics stand in a curriculum informed by ethnomathematics? A critical review of ethnomathematics. *Educational Studies in Mathematics, 50*, 79–102. https://doi.org/10.1023/A:1020532926983

Selin, H. (Ed.). (2000). *Mathematics across cultures. The history of non-Western mathematics*. Kluwer.

Siegel, C. (2005). Implementing a research-based model of cooperative learning. *The Journal of Educational Research, 98*(6), 339–349. https://doi.org/10.3200/JOER.98.6.339-349

Spangenberg, E.D. (2015). Voorbeelde van kulturele artefakte in die voorbereiding van 'n wiskundeles. *Suid-Afrikaanse Tydskrif vir Natuurwetenskap en Tegnologie, 34*(1), 1–9 https://doi.org/10.4102/satnt.v34i1.1310

Soanes, C. (Editor). (2002). *Pocket Oxford English dictionary* (9th ed.). Oxford University Press.

Venkat, H. (2007). Mathematical literacy—mathematics and/or literacy: What is being sought? *Pythagoras, 66*, 76–84. https://doi.org/10.4102/pythagoras.v0i66.82

Walser, T. M. (2009). An action research study of student self-assessment in higher education. *Innovative Higher Education, 34*, 299–306. https://doi.org/10.1007/s10755-009-9116-1

Wilson, M. (2002). Six views of embodied cognition. *Psychonomic Bulletin & Review, 9*(4), 625–636. https://doi.org/10.3758/BF03196322

Wong, K. Y. (2005, July). *Add cultural values to mathematics instruction: A Singapore initiative*. Paper presented at the 4th Asian Mathematical Conference, Nanyang Technological University, Singapore. http://ww1.math.nus.edu.sg/events/AMC/papers-invited/Wong-KY.pdf

Zakaria, E., & Iksan, Z. (2007). Promoting cooperative learning in science and mathematics education: A Malaysian perspective. *Eurasia Journal of Mathematics, Science & Technology Education, 3*(1), 35–39. http://files.eric.ed.gov/fulltext/ED496240.pdf

CHAPTER 7

MATHEMATICAL ANALYSIS OF MUSICAL RHYTHMS USING THE EUCLIDEAN ALGORITHM

Elijah Porter

ABSTRACT

This chapter proposes ways to explore how mathematics can be used to describe and model rhythmic structures inherent in musical practices. It focuses on the use of division, divisibility, evenness, and the Euclidean algorithm in describing musical rhythms.

Envision the following day inside a mathematics class: The teacher gathers all the students, with audio-recording devices in hand, into a line by the door. The teacher escorts the students to the music room or band room. The music teacher or band director is awaiting the mathematics students because the mathematics teacher has already organized this internal field trip. The music teacher/band director takes position at the front of the room; the musicians take the cue and prepare themselves for first sound; the mathematics students follow suit and raise their recording devices into the air. The music teacher/band director begins the count and at the count of three, the worlds of music and mathematics intersect.

Evidence-Based Inquiries in Ethno-STEM Research: Investigations in Knowledge Systems Across Disciplines and Transcultural Settings, pp. 125–132

CONNECTION BETWEEN MATHEMATICS AND MUSIC

The connection between music and mathematics is not a recent awareness. Music is a manifestation of patterns and relationships based upon sound and time. Also, mathematics has been used as a tool to detect, identify, study, and describe patterns of any kind. In this sense, mathematics and music are fantastically intertwined. To illustrate, Mozart composed a waltz using dice to determine critical sequences of sound patterns (Hedges, 1978). Arnold Schoenberg also used mathematics to create a completely new style of music called atonal music. With the advent of technology, mathematics has become a key tool for digital sampling, audio compression, and error correction (Smith, 2007). Each of these elements allow for music to be placed on a portable media player like an iPod. Also, extensive research has focused on studying the relationships between music, and cognitive development (Schlaug, Norton, Overy, & Winner, 2005). Specifically, these studies have found that individuals who excel in music also excel in mathematics. In this chapter, I present several ways to explore how mathematical concepts can be introduced and reinforced through activities inspired by musical contexts.

MATHEMATICAL CONCEPTS: DIVISION AND DIVISIBILITY

In simplest terms, division is the process of grouping the whole of an entity, the dividend, into smaller sets according to selected number, the divisor. In other words, division is an arithmetic operation describing the number of times a quantity can be broken down.

Numerically, this is accomplished by what has been called "long division." For example, if the size of a whole class of students is 17, and it is necessary to group the students in sets of 3, then the dividend for this problem would be 17, and the divisor would be 3. Creating the sets or groups results in five groups of three students, with two students not placed in a group because they are the "excess." In this problem, 5 would be the quotient, and 2 would be the remainder, see Figure 7.1.

The concept of "excess" is important in division because it leads directly to the concept of divisibility. Divisibility is the term used to denote a number's capacity to be divided without producing a remainder—in short, leaving a remainder which is equal to zero. In other words, all the components of the whole are successfully placed into one of the smaller sets. Another name for this is "evenness." Now, the inverse of this condition is that if the remainder does not equal zero, then the condition of divisibility has not been met. A term for this state is "non-evenness."

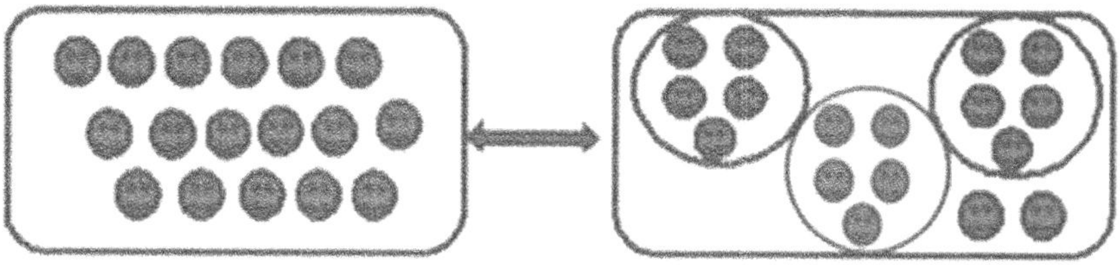

Figure 7.1. Dividing a group of 17 into 3 groups of 5 with a remainder of 2.

APPLYING THE CONCEPT OF DIVISION TO MUSICAL RHYTHMS

An essential tool to analyze musical rhythms involves the concepts of "evenness" and "non-evenness." The pursuit of "evenness" requires analyzing the quantity in each grouping and the remainder, and to manipulate these two components repeatedly until there is no longer a remainder. Without guidance, this iterative process of selecting the correct number can be a daunting task, but Euclid, the renowned intellectual, was able to discover a rule that makes the selection quite efficient and robust. The result of his method highlighted a characteristic quality of the original two numbers. This characteristic quality is the largest number that the original two numbers can be grouped into and "evenness" be achieved. This number is called the greatest common factor (GCF) of the original two numbers.

MATHEMATICAL ANALYSIS OF THE GREATEST COMMON FACTOR: THE EUCLIDEAN ALGORITHM

Euclid created a proficient approach to compute the GCF of any two numbers using the basic mathematical operations of addition and division. Understanding the concept of division, Euclid realized that the remainder in a division problem is always less than the divisor. Euclid noticed that if he performed the division process enough times using increasingly smaller and smaller numbers, he would eventually arrive at two numbers that produced a zero remainder after performing the division. The smaller of these final two numbers would be the GCF of the original two numbers. Euclid determined that if he began with two integers, divided the smaller integer into the larger integer, and recorded the remainder, he could repeat the division process using two smaller numbers, the original divisor and the remainder. This sequence of iterative steps is named, the Euclidean algorithm, in his honor, and is considered one of the oldest numerical algorithms still in common use. The algorithm, which was described in Euclid's seminal book *Elements* (c. 300 B.C.), has many theoretical and practical applications and is used to generate and analyze numerous traditional musical rhythms produced by different cultures throughout the world (Chemillier, 2004; Toussaint, 2005).

There are multiple ways to describe and understand the Euclidean algorithm. Figure 7.2 provides a theoretical representation of the Euclidean algorithm performed on any two numbers, a and b.

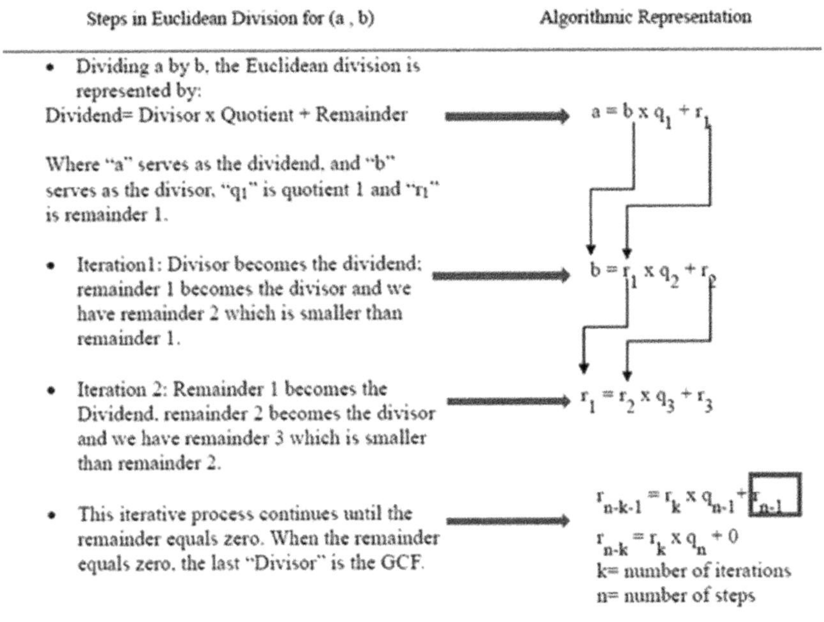

Figure 7.2. Steps for finding GCF of numbers a & b using the Euclidean algorithm.

Figure 7.3 shows a depiction of the Euclidean algorithm performed with the numbers 70 and 42. In Figure 7.3, the last quantity grouped which produced a remainder of zero is the GCF of 42 and 70. Therefore, the GCF for 42 and 70 is 14.

MULTIPLE REPRESENTATIONS OF EUCLIDEAN RHYTHMS

Toussaint (2005) introduced the Euclidean syntax E (a, b), which is interpreted as the Euclidean rhythm of a musical selection that contains "b" total pulses and "a" total beats, which comprises the onsets of a note and where (b-a) represent rests. For example, E (5, 13) represents a Euclidean rhythm with a total of 5 beats, 13 pulses, and 8 rests. Another

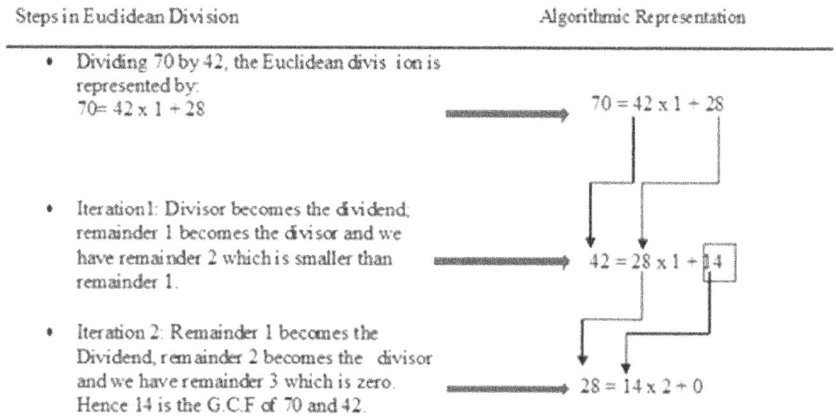

Figure 7.3. Steps for finding GCF of numbers 70 and 42 using the Euclidean algorithm.

representation of Euclidean rhythms involves employing binary codes to describe sequences of beats and pulses (see Figure 7.4).

As illustrated in Figure 7.4, the binary sequence for E (5, 13) could be represented as (1111100000000), where 1s represents beats and 0s represent rests—a silence or an absence of a beat. To convert a binary code into a Euclidean rhythm, we incrementally relocate one zero from the right side of the sequence and place it vertically underneath a one on the left side of the sequence. Upon completion, five of the zeros would be relocated, and three zeros would remain on the right side of the sequence. Next, incrementally relocate one zero from the right side of the sequence and place it vertically underneath a one-zero pair on the left-side of the sequence. We continue this relocation process until all the single-digit numbers have been relocated.

At this point, the orientation should contain three columns of the sequence one-zero-zero, and two columns of the one-zero pair. The Euclidean algorithm only allows one column to be unlike the other columns. At our current place in the algorithm, we have two columns which are unlike the others; therefore, we must continue to relocate columns. From this point, we must incrementally relocate each one-zero pair and place it vertically underneath the one-zero-zero column. When completed, the orientation will contain two columns containing the one-zero-zero-one-zero sequence and one column containing the one-zero-zero sequence. Now, since there is only one column which is unlike the others, we have completed the iterations. The sequence written horizontally is represented

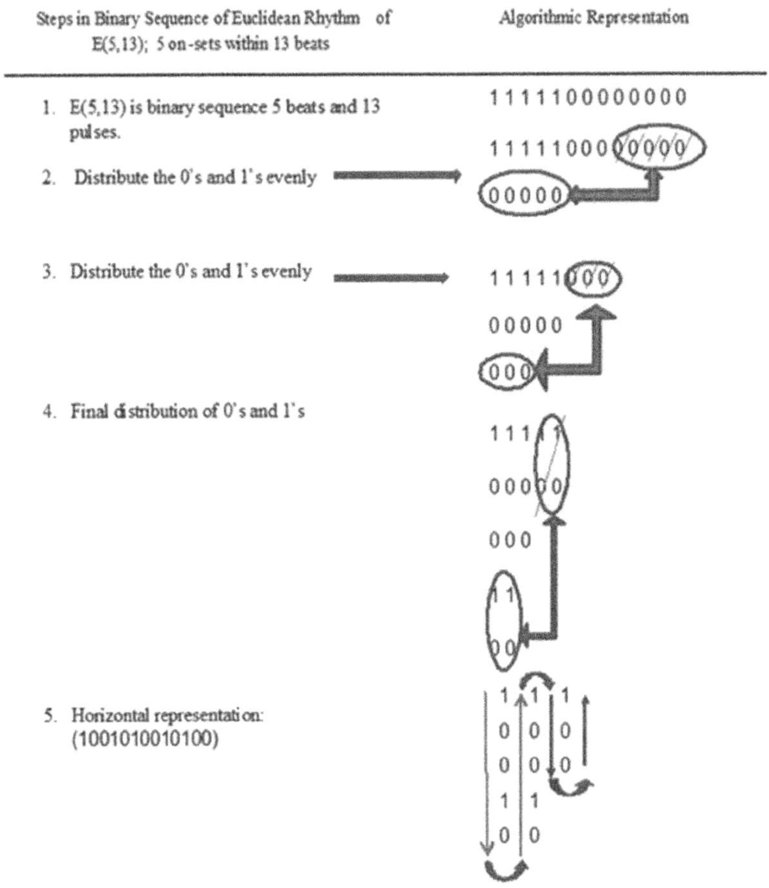

Figure 7.4. Binary sequence for EUCLIDEAN SYNTAX E (5, 13).

as (1001010010100) and would be the Euclidean rhythm. It is important to recall that this syntax represents beat onsets and rests.

Toussaint (2005) argues that a 13-pulse musical selection is not common and explains that pulse counts which are multiples of four are more widespread. Therefore, if we begin instead with E (3, 8) and write the initial binary sequence, we have (11100000). If we correctly follow the iterations of the algorithm, we have the following Euclidean rhythm: (10010010), which contains a "1" in positions zero, three, and six; consequently, beats occur in those position. Due to the fact that its pulse count is a multiple of four, this is a more popular rhythm and is actually the first half of a 16-pulse rhythm played throughout Africa and Brazil.

Another modality for depicting Euclidean rhythms involves using cyclic polygons to visually represent the binary sequence of beats and pulses. In order to convert E (3, 8) with the binary sequence (10010010) into polygonal representation, we need to envision a circular clock which has a maximum of eight positions instead of twelve, because we are evaluating an 8-pulse rhythm. Now, label the top center position as zero and number clockwise, finishing at seven. We then place a dot on each number where an onset exists, specifically where there is a one in the binary sequence (see Figure 7.5).

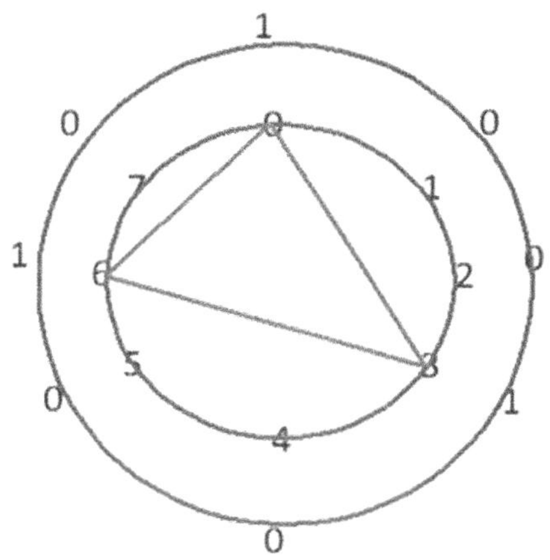

Figure 7.5. Polygonal and binary representation of EUCLIDEAN SYNTAX E (3, 8).

If we now count the number of pulses from one onset to the next and place this number on each side of the polygon, we then would have a visual representation of what is termed by Toussaint (2005) as the inter-onset duration interval or the adjacent-inter-onset-interval vector. In this case the inter-onset duration interval is (332). He defines a rhythm necklace as an inter-onset duration interval which is shared by multiple rhythms but start at different positions in the pattern. In other words, the inter-onset duration interval of one rhythm is a rotation or shifting of the inter-onset duration interval of the other.

It is this concept of the rhythm necklace that was used to determine that there is a commonality among different music styles across cultures (Toussaint, 2002). For example, Toussaint analyzed the inter-onset duration intervals for many of the prominent rhythms of the prominent sub-Sahara African ethnic groups. According to Toussaint (2002), the most popular Euclidean rhythms for sub-Saharan Africa are E (3, 8), E (5, 12), E (7, 12),

E (9, 16). Of these Euclidean rhythms, the absolute most influential is E (7, 12). In fact, according to Toussaint, when E (7, 12) is started on the third onset, it is referred to as the African signature tune.

IMPLICATIONS FOR THE CLASSROOM

There is no doubt that mathematics and music are closely linked. Common elements of music which include beats, rhythms, frequencies, vibrations, meter, and so forth, can be described and understood through mathematics. For the students, exploring the mathematical structures underlying musical rhythms reinforces their understanding that mathematics is much more than a collection of rules and procedures. Additionally, such connections address the traditional question, "When will I use this in real-life?"

REFERENCES

Chemillier, M. (2004). Periodic musical sequences and Lyndon words. *Soft Computing*, *8*(9), 611–616.

Hedges, S. (1978). Dice music in the eighteenth century. *Music & Letters*. *59*(2), 180–187.

Schlaug, G., Norton, A., Overy, K., & Winner, E. (2005). Effects of music training on the child's brain and cognitive development. *Annuals of the New York Academy of Sciences*, *1060*, 219–230.

Smith, J. (2007). *Introduction to digital filter with audio applications*. Book Surge.

Toussaint, G. (2002, July). A mathematical analysis of African, Brazilian, and Cuban clave rhythms. *Proceedings of BRIDGES: Mathematical Connections in Art, Music, and Science*, 27–29.

Toussaint, G. (2005, July 31–August 3). The Euclidean algorithm generates traditional musical rhythms. *Proceedings of BRIDGES: Mathematical Connections in Art, Music, and Science*, 47–56.

CHAPTER 8

THE MATHEMATICS IN GENTLEMEN'S CLUBS

Hannah Oldham

ABSTRACT

This study investigated mathematical ideas that emerge in gentlemen's clubs. Through the lens of ethnomathematics, I investigated the different ways in which mathematics is seen and used by the workers in these establishments. By analyzing the mathematics inherent in workplace practices, I explore the various ways in which mathematics presents itself in this environment and how that can help us better understand informal mathematics in unconventional settings.

INTRODUCTION

Mathematics emerge every day in conventional and unconventional settings. However, many times, the mathematics performed in these unconventional settings is deemed as lesser mathematics or not even mathematics at all. Are we discrediting the mathematics performed in these settings because

Evidence-Based Inquiries in Ethno-STEM Research: Investigations in Knowledge Systems Across Disciplines and Transcultural Settings, pp. 133–142

the setting is unfamiliar or taboo? The study of mathematics is the study of all mathematical ideas, not simply the kind that is taught in schools and using formulas and procedures. Davis and Hersh (1998) offered a succinct definition stating that "Mathematics is the science of making necessary conclusions" (p. 7). While many others have expanded on this definition (which will be discussed later), the core is always the same. If we believe that the heart of mathematics is in the study of problem solving, then of course it lives outside of classrooms and formulas. It is used every day, in every culture, regardless of the formality of its application. In this study, I argue that by studying mathematical investigations in unconventional settings we can broaden our understanding of how mathematics can be used and utilize that knowledge to better understand not only other cultures but our own as well.

CONTEXT OF STUDY

So what exactly is an "unconventional setting"? For the purposes of this study, I describe a *conventional setting* as a space where the majority of people within a certain culture would feel comfortable in and would be viewed as an "appropriate" place for most members of that culture. An *unconventional setting* would be viewed as a place of taboo. It is regarded as a place that bends the status quo and runs counterculture to the norms of the majority. But before we start exploring mathematics in such unconventional settings, we need to define what it is we are even looking for when we explore mathematical ideas in informal contexts outside the classroom.

While Davis and Hersh (1998) provide a succinct definition of mathematics, Bishop (1988) coined six categories that he considered as mathematics and that have a more concrete basis and are easier to observe. These categories are counting, locating, measuring, designing, playing, and explaining. By delineating mathematics into these tangible categories, Bishop argues we can more readily identify how and why mathematics is being used in any setting. Now that we know how to observe mathematics outside of a formal setting, the question arises why do we need to?

Knowledge of mathematics is often viewed as a gatekeeper (Martin, 2007). Being able to practice mathematics opens up opportunities both in education and in many careers. However, this gatekeeper is almost entirely framed in a conventionalist setting. The mathematics that is practiced and valued in unconventional settings is undervalued. It cannot be used as a key for advancement in the same way as formal mathematics. What is the difference between mathematics in these two settings? Is one truly better than the other or is unconventional mathematics undervalued simply because it is different from what we are used to? It is in these questions that we must examine the nature, purpose, and value of mathematics in unconventional settings.

The Most Unconventional Setting

How do we begin to examine the nature of mathematics in unconventional settings? An unconventional setting can be anything that runs counterculture to the dominant culture. What I am most curious about is not only how the mathematics in unconventional settings is used but also how the people who inhabit that setting are viewed. Are their mathematical applications deemed inferior because of the culture they adhere to? Since they inhabit an unconventional setting, is their mathematical knowledge immediately considered to be of lesser value? From these questions, my goal for examining mathematics made me look for a not only an unconventional setting but also a setting where the culture is commonly considered as unfit to do mathematics. One setting that has both of these qualities is gentlemen's clubs. While a gentlemen's club can describe different establishments, for the sake of this research a gentlemen's club will refer to a business wherein women perform striptease for clientele. From a conventional setting, both the clientele and the dancers are commonly viewed as indecent, leading to the notion that the people who inhabit this unconventional setting are unworthy and incapable of practicing mathematics.

PURPOSE OF STUDY

The purpose of this study is to investigate mathematical ideas in gentlemen's clubs. Through the lens of ethnomathematics, I investigated the different ways in which mathematics is seen and used by the workers in these establishments. In this study, I explore the various ways in which mathematics presents itself in this environment to shed light on the nature of informal mathematics in unconventional settings.

The research question that will drive this study is how is mathematics seen and used by workers in the unconventional setting of gentlemen's clubs?

LITERATURE REVIEW

This research is based heavily on understanding the unconventional environment of not only gentlemen's clubs but also the social stigmas attached with the workers and clientele that frequent them. This literature review examines the history of gentlemen's clubs and how they became viewed as taboo establishments. It also examines the culture of the dancers who work at gentlemen's clubs and how they view their role in this unconventional setting.

From Sleazy to Classy

The history of term *gentlemen's clubs* can trace its origins back to 17th century England (Churchill & Klehr, 2000). However, these gentlemen's clubs bear little to no resemblance to the American striptease clubs that this research is about. The English gentlemen's clubs were establishments where wealthy men would go to discuss the happenings of the day from literature and art to politics and gossip. These clubs were isolated to men only and were a common fixture in English aristocracy until the mid-1800s. While similar in name, American gentlemen's clubs did not sprout from their English counterparts. In fact, there was a long period of time (from the mid-1800s to the mid-1900s) that gentlemen's clubs were not common in either county. It was not until the rise of mob bosses in the 1950s and 1960s that gentlemen's clubs even came into public notice in America (Baker, 2010). During this time, these establishments were in no sense viewed as gentlemen's clubs. They were havens of crime and illegal activity. The dancers who worked there were at the end of their ropes and not viewed in a kind light. It is also during this time that the negative connotation associated with strip clubs was formed. Illegal activity combined with crime and supposed sexual deviation painted an image of a completely depraved environment. An environment that is usually pushed outside the realm of cultural acceptance.

It was not until the 1970s that the depraved image of strip clubs began to change in America. The 1970s brought entrepreneurs who saw the business potential in these establishments (Baker, 2010). These entrepreneurs began to remold the image of the seedy strip club into the more refined gentlemen's club. Such new gentlemen's clubs began operating as business ventures and not as meeting places for illegal activities. After all, that would be bad for business. With the removal of criminal activity also came new clientele. These clienteles were looking for an establishment that provided exotic dancing without the stigma associated with the old seedy strip clubs. This is the place where many strip clubs rebranded themselves as gentlemen's clubs in an attempt to remove any connotation with their odious predecessors. Such was the beginning of the gentlemen's clubs we see today. They are business ventures that create a comfortable environment for clientele as well as the dancers who work in them. However, this more sophisticated image of gentlemen's clubs is still rarely projected in the mainstream American culture (Knoll, 2013).

Mainstream American culture places a taboo on public sexual behavior (D'Emilio & Freedman, 1997). Even though gentlemen's clubs are private establishments that enforce an 18 or older age policy, the idea that nudity is being showcased is viewed as indecorous. The continued idea that gentlemen's clubs are havens of crime and indecent behavior continues because of the mainstream fear of investigating the reality of this unconventional environment. For many, the idea that women remove their clothing for money is so unwholesome that the entire establishment is painted in a tasteless light. Such a viewpoint continuously keeps gentlemen's clubs and the dancers who work there out of status quo. Gentlemen's clubs are an uncomfortable topic for many, which leads to the prolonged skewed images of what occurs there, simply because many people are too embarrassed to visit the setting and observe for themselves. These unobserved assumptions extend not only from the establishment itself but also to the dancers who work in them.

The Culture of the Dancers

If gentlemen's clubs are viewed in a disapproving light from the mainstream culture, then the dancers who work there are doubly so. Dancers are painted as harlots or as women who are being taken advantage of by domineering men (Sheila, 2008). Is this image of women in gentlemen's clubs authentic or is it an inaccurate construct created by people outside of the culture? Just like the environment of gentlemen's clubs has been tainted with faulty assumptions, so too has the culture of the dancers in these establishments.

For most gentlemen's clubs, dancers are independent contractors who are allowed to set their own hours and many times their own fees. This perception is in sharp contrast to the idea of an oppressed female submitting to the needs of her male supervisors. One dancer remarked "I control my income, hours, and activities.... I do what I want when I want for how much I want" (Deshotels & Forsyth, 2008, p. 489). Control is a factor in gentlemen's clubs but much of that control is in the hands of the dancers themselves and not the supervisors or the clientele. Dancers use this control to help increase their income. They use cues like wardrobe, age, personality, and eye contact to help them determine where they should focus their attention (Wosick-Correa & Joseph, 2008). In addition to this control, many of these dancers feel a sense of hierarchy over their male counterparts because they choose who they interact with and have absolute control in each of these interactions (Deshotels & Forsythy, 2006). The idea of the exploited dancer seems to be an illusion created by people outside of the culture, perhaps in an attempt to devalue the merits of their line of work.

SIGNIFICANCE OF THE STUDY

Much of the mainstream representation of gentlemen's clubs and the dancers that work there is based on ideas framed on what *might* occur in these places. These perceptions are neither emic nor etic; they are not based on any true observations at all because many people fear going into these unconventional settings and viewing these unconventional contexts. Relevant literature has demonstrated that there is a disparity between how mainstream culture views gentlemen's clubs and what the context is actually like. This is simply due to the fact that there is a shortage of research into what occurs in these establishments. The idea that the clubs and the dancers who work there are distasteful has devalued their work and identity. By observing the culture and environment firsthand, I intend to examine the true culture within gentlemen's clubs. Specifically, through the lens of how mathematics is used and practiced.

As mentioned previously, mathematics is viewed as intellectually challenging and those who formally practice it are heralded for their brilliance. I approached this study by looking for the use of mathematics by the dancers. If I can find mathematical usage and practice from these women then it directly challenges the mainstream notion of inept and unintellectual dancers who work in gentlemen's clubs. Furthermore, I argue that findings of this study can open up discussions for the importance of studying mathematics in unconventional settings that may have been previously ignored.

METHODOLOGY

This research was conducted as a case study at gentlemen's club in Atlanta. I went to the establishment on a Friday night from 9 P.M. to 12:30 A.M. Friday was chosen as it was the busiest night of the week and the time frame spanned from the slowest to busiest times of the evening. This allowed me to observe how the dancers adapted to changing clientele through the night. A case study was chosen because it allowed me to investigate a bounded environment through observations as an etic researcher over a set time period (Creswell, 2013). I purposely chose not to do interviews with the dancers because I wanted to remain an outside observer. I wanted to place my observations in comparison with the current negative perceptions held about these establishments and dancers and that required me to only observe and not to interact. Due to the rules of the club, no videotaping or photography is permitted, so only written notes were taken during the observation. No ethical guidelines were violated as dancers are performing for an audience and assume that they are being observed throughout their shift (U.S. Department of Health and Human Services, 2008). In my observations, I focused on identifying usage and practice of Bishop's (1988) six mathematical ideas.

DATA ANALYSIS

As I entered the gentlemen's club I immediately saw that the layout of the establishment was symmetrical with a clear repeating usage of groups of threes. There were three stages, two rectangular stages framing a circular one at the head. In the center of the three stages was a seating area where all the tables and chairs were aligned in rows where clientele could maximize their viewing of any of the three stages. Each stage had a lighted floor as well as lights directly above whereas the seating area had dimmer lighting which focused all attention on the dancers on the stages (see Figure 8.1).

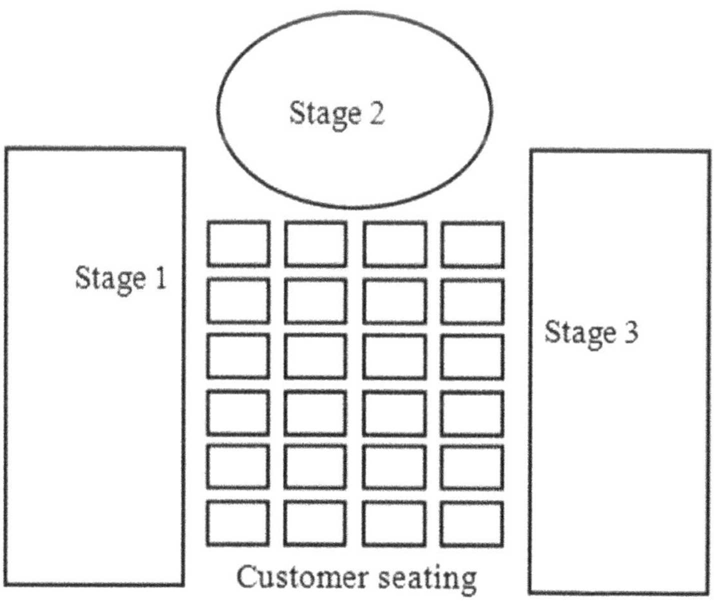

Figure 8.1. A depiction of the layout of the gentlemen's club.

Bishop's (1998) notion of location is clearly evident in the purposeful symmetry of the establishment. The seating is placed in a central location where each client can easily view each stage. The stages form a perimeter around the clientele for maximum exposure while simultaneously placing the dancers on elevated stages under lights making them the focal point of attention. It is also important to note the use of three stages. Throughout the evening, a theme of three was seen in not only the layout of the club but also in dancer rotations and song choices. Why three? What is the importance? Three is a common theme used in many artistic avenues because it can symmetrically frame a central image (Minton, 2007). For example, the circular stage is framed by two congruent stages creating a single focal point while simultaneously creating a symmetrical image.

When it came to the actual dancers, the theme of three also presented itself. Three dancers would get on stage and dance to three different songs before three new dancers would take their place. In between each song the dancers would change places so they each had a place on the left, middle, and right of the stage. Choreographers consistently design routines based on this rule of three because it allows a focal dancer to move in and out of the attention of the audience (Belcastro & Shaffer, 2011; Minton, 2007). Sharing and changing focal attention is extremely important to the dancers who are all vying for the attention of the clientele. At one point in the evening, every dancer on shift came out to the floor. On the Friday evening I was there, 64 dancers were present. Since each dancer has three unique places on three unique stages, there are nine possible locations for each dancer. On a night where there are 64 dancers a permutation of ${}_{64}P_9$ creates 9.99×10^{15} unique dancing combinations on a single evening. Using three locations over three stages combined with a complete rotation after three songs maximizes the diversity of dancers for the clientele. This is important because it allows for a unique show every night.

Later in the evening, starting at around 11:30, the rotations changed. This was due to an increase in the number of customers present. As the club filled up, the number of dancers on each of the stages increased from three to five. So instead of three songs and three dancers there were five songs and five dancers. This creates even more combinations of unique dancing rotations because there are now 15 unique positions yielding ${}_{64}P_{15}$ or 2.09×10^{25} possibilities. Clearly it is impossible to get through every one of these rotations every night but how many rotations can be seen on a given evening? Let us assume that the club is open from 7 P.M. to 2 A.M. which is seven total hours. Based on my observations, the rule of three lasted from opening to around 11:30 or four and half hours. Then from 11:30 until close (two and a half hours), the club followed the rule of five. The average length of each song was about three minutes. Over four and a half hours (270 minutes) that will yield 90 sets of three. Each set has $3^3 = 27$ different locations for each dancer. So, 90 x 27 = 2,430 unique combinations from 7 P.M. until 11:30. The last two and half hours (150 minutes) yields 30 sets of five. Each set now has 5^5 = 125 different locations for each dancer which gives us 30 x 125 = 3,750 unique combinations from 11:30 until 2 A.M. Assuming an average song length of three minutes a customer could watch from 7 P.M. until 2 A.M. and see 6,180 unique dancing rotations.

The above calculations reveal the complex mathematical structures used to maximize the exposure and diversity of the dancers. It also represents a clear usage of Bishop's property of counting, but on a much larger scale. The branch of mathematics dealing with combinations of varying objects is called combinatorics and it is used consistently throughout the establishment (Wilson, 2016). The reason that altering rotations and combinations of dancers is so important is because it increases the enjoyment of the customers. The number of dancers themselves is finite, as dancers cannot constantly be replaced to create a new show every night. By using rotations, dancers can continuously work throughout evenings, weeks, months, and years without repeating a previous combination. A new show is possible every night keeping the entertainment fresh, increasing the enjoyment for the patrons, and increasing the profits for the dancers.

CONCLUSION

When I entered the gentlemen's club, I was expecting to see some dancing perhaps to a routine and some minor observable structure to how the dancers moved around the club. However, what I found was an extremely structured and fine-tuned system maximizing potential based on advanced mathematical properties. The club itself was clean and purposely arranged to emphasize the stages and the dancers. The rotations of the dancers throughout the establishment was based on a complex branch of mathematics called combinatorics that was used to yield the highest number of unique combinations of dancers. Through one night of observations, it became clear that this unconventional setting not only utilizes common mathematics of spacing and structure but also employs complex algorithms daily. These algorithms form the basis of college combinatorics, an advanced mathematical course of study.

Formal mathematics can occur in the most unconventional settings if only we are willing to look for it. This idea is critical to looking past our own assumptions about the role of mathematics inside and outside of the classroom. Just because a culture runs contradictory to our own does not devalue that culture's ability to use and practice mathematics. In many instances, it can exhibit new unique ways of mathematical applications that could strengthen our own mathematical knowledge. Unconventional does not mean deficient and taboo does not mean barbaric. Mathematics seemingly exist in any context no matter how informal or unfamiliar it can be. We must just be willing to see past conventions and be willing to explore areas that may seem beyond our comfort zone. Through this we can evolve and expand our connection to mathematics and the contexts it permeates.

REFERENCES

Baker, A. L. (2010). Gentlemen's clubs and casinos in Las Vegas. *UNLV Gaming Research & Review Journal, 14*(1), 79–91.

Belcastro, S., & Schaffer, K. (2011). Dancing mathematics and the mathematics of dance. *Math Horizons, 3*, 16–20.

Bishop, A. J. (1998). *Mathematical enculturation: A cultural perspective on mathematics education*. Kluwer.

Churchill, W., & Klehr, A. (2000). London's gentlemen's clubs. *British Heritage, 21*(3), 50.

Creswell, J. W. (2013). *Qualitative inquiry & research design: Choosing among five approaches*. SAGE.

D'Emilio, J., & Freedman, E. B. (1997). *Intimate matters: A history of sexuality in America*. The University of Chicago Press.

Davis, P. J., & Hersh, R. (1998). *The mathematical experience*. Houghton Mifflin.

Deshotels, T., & Forsyth, C. J. (2006). Strategic flirting and the emotional tabs of exotic dancing. *Deviant Behavior, 27*(2), 223–241.

Deshotels, T., & Forsyth, C. (2008). Sex rules: The edicts of income in exotic dancing. *Deviant Behavior, 29*(5), 484–500.

Knoll, J. (2013). Girls in the gentlemen's clubs. *Cosmopolitan, 254*(2), 164–166.

Martin, D. B. (2007). Mathematics learning and participation in the African American context: The co-construction of identity in two intersecting realms of experience. In N. S. Nasir & P. Cobb (Eds.), *Improving access to mathematics: Diversity and equity in the classroom* (pp. 146–158). Teachers College Press.

Minton, S. C. (2007). *Choreography: A basic approach using improvisation*. Human Kinetics.

Sheila, J. (2008). Keeping women down and out: The strip club boom and the reinforcement of male dominance. *Signs: Journal of Women in Culture & Society, 34*(1), 151–173.

U.S. Department of Health & Human Services. (2008) *Coded private information of specimens use in research*. https://www.hhs.gov/ohrp/regulations-and-policy/guidance/research-involving-coded-private-information/index.html

Wilson, R. J. (2016). *Combinatorics: A very short introduction*. Oxford University Press.

Wosick-Correa, K. R., & Joseph, L. J. (2008). Sexy ladies sexing ladies: Women as consumers in strip clubs. *Journal of Sex Research, 45*(3), 201–216.

CHAPTER 9

GEOMETRY AT THE KITCHEN TABLE

Jennifer Henderson

ABSTRACT

In this study, I explore how a mathematics teacher can incorporate quilt blocks into classroom instruction to help develop understanding of geometric transformations. I asked my two children to help me develop lessons for use in the geometry classroom. The lessons were developed by watching how my children interact with the materials and engage naturally in in the quilting practice. My ultimate goal in this study was to use quilting practice to create potential lessons that could be implemented in my geometry classroom.

INTRODUCTION

As a quilter and a mathematics teacher, I was interested in exploring ways to incorporate quilting into my instructional practice when I teach transformations in geometry. There are many lesson plans readily available for the classroom using quilting, but none of them were exactly what I wanted

Evidence-Based Inquiries in Ethno-STEM Research: Investigations in Knowledge Systems Across Disciplines and Transcultural Settings, pp. 143–154

to use in my own classroom. For this reason, I wanted to examine how students reacted to the quilting materials and the process of making blocks to see how I could construct a lesson plan for use in my classroom. Cultural aspects of quilting was also a fascination and I constantly search for relevant and meaningful content to integrate in my mathematics lesson plans as I teach my high school students.

In general terms, a quilt is a cover for a bed that is made from two layers of fabrics with batting between the layers that are stitched together. This simple definition does not begin to explore the deep and lengthy history of quilting. Quilts are not only artifacts of art, they also convey historical stories (Paznokas, 2003). In North America today, quilts are such a part of the culture that most households have a quilt, often with complicated geometric designs as part of their pattern (Wares, 2006). Not only are quilts expressions of art and keepers of history, their construction requires a deep understanding of mathematical concepts such as measurement, fractions, geometry transformations, algebra, and mathematical operations (Neumann, 2003, 2005).

CONTEXT OF THE STUDY

In this study, I delineate my journey as I embarked on exploring how a mathematics teacher can incorporate quilt blocks into classroom instruction to help student foster a deeper understanding of geometric transformations. I asked my two children to help me enact lessons for use in the geometry classroom. The lessons were developed by watching how my children interact with the materials and engage naturally in the quilting practice. My ultimate goal in this study is to use quilting practice to create potential lessons that could be implemented to teach geometry concepts. The question that drove my inquiry was: How can I use quilt blocks to help highs school mathematics students have a better understanding of geometric transformations?

A Very Brief History of the Quilt

While the exact origins of quilting are not known, quilting has been in existence since early times in various parts of the world. The first known example of quilting was found on a carved piece of ivory that dated from the Egyptian first dynasty, about 3400 B.C. (Johnson, n.d.). It is assumed that the art of quilting traveled to the New World with settlers from Europe. There are not a lot of records about early quilts, likely because they were such a normal part of family life that they were not mentioned specifically.

The first time quilts are mentioned in the New World was from the end of the seventeenth century; a quilt was listed in the personal household inventory of a sea captain from Salem, MA (Johnson, n.d.).

The early quilts are thought to be purely utilitarian. Their main purpose was warmth and protection. They could be hung over ill sealing doors and windows to keep drafts out as well as laid as traditional bed covering (History of Quilts, n.d; Johnson, n.d.). There are three main categories of quilts: the applique quilt, the whole cloth quilt, and the patchwork quilt (History of Quilts, n.d.). In my readings, I was particularly interested in patchwork quilts and how they have been used by numerous cultural groups in North America. Such background facilitated a sense of how quilt blocks can be used in the classroom to teach mathematics.

The top of a patchwork quilt is composed of smaller pieces of fabric sewn together to create a design. These designs were often made using scraps of fabric that were cut into shapes and sewn together. Often the designs were composed of geometric straight edge shapes such as squares, triangles, and diamonds to make the most of expensive new cloth or even small scraps of cloth. The block-style of patchwork quilts made use of the geometric designs to maximize the number of blocks to make the quilt top. This efficiency was important because, often, many quilts were required to ensure warmth during winter and would need to be quickly made; such requirements were afforded by the square shape, which allowed for quick piecing together of the blocks (History of Quilts, n.d.; Patchwork Quilts, n.d.). The patchwork quilt block patterns became so popular that the patterns were disseminated at large. Some block patterns from this time in North American history are still in use today. These patterns have names inspired by everyday life such as flying geese, the Northern Star, the log cabin, and the windmill (Patchwork Quilts, n.d.). Some of these designs were used by groups in North America to convey meaning beyond artistic freedom in their designs.

Northern Star

It is believed that Native Americans girls in the Northern Plains were taught to quilt by missionaries in the region. Historically, the Native Americans in the region painted their history on buffalo skins. However, buffalo had become extinct so a new way to record history was needed by the Lakota, Dakota, and Nakota cultures in the area (Neumann, 2003). The eight pointed figure known as the Northern Star became a popular image to use in quilting among the Native American cultures (see Figure 9.1). The Northern Star was created by piecing together rhombi, made from reflecting an isosceles triangle, and using geometric rotation to create the eight pointed star. The quilts became a way to represent life events such as births, deaths, and marriages or other aspects of life such as spirituality and community within these Native American cultures. For example, the Lakota tribe thought the colors used in making the Northern Star were important. They used black to represent the rains that come from the west, white to represent the great cleansing wind from the north, red for the morning star that comes from the east, and yellow to represent the summer that promotes crop growth that comes from the south (Neumann, 2003).

Figure 9.1. An 8-pointed Northern Star pattern.

Freedom Quilts

In the 1700–1800s quilts made by enslaved African Americans were used as a secret form of communication to other slaves (Neumann, 2005; Paznokas, 2003). These quilts, called Freedom Quilts, were used to offer coded instructions regarding the Underground Railroad to help slaves escape to the North. The Freedom Quilts with these coded communications were often displayed in windows for the slaves to see and acted as a literal and figurative roadmap to the escaping slaves (Neumann, 2005; Paznokas, 2003). Figure 9.2 (a) shows a pattern depicting the flying geese quilt block pattern, which gave cardinal directions to the fleeing slaves by using different colors to indicate the desired direction. In Figure 9.2 (b), safe shelter was represented by the log cabin quilt block that contained a certain color at its center, and a wagon wheel quilt block indicated the time to pack for the journey to the North is shown in Figure 9.2 (c) (Paznokas, 2003).

Geometric transformations were used to create some of the quilt blocks and Freedom Quilts. The flying geese pattern is created using rotation of

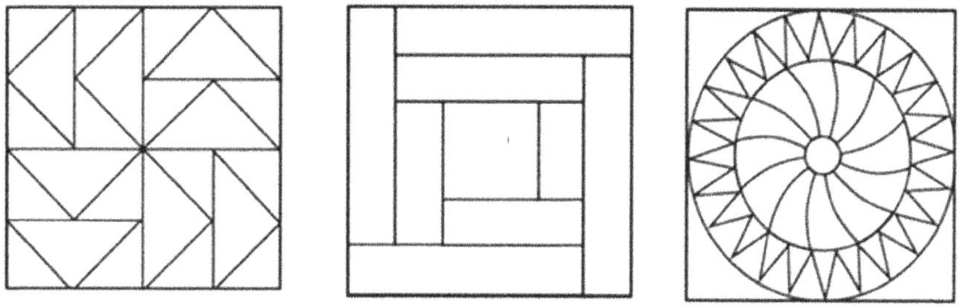

(a) (b) (c)

Figure 9.2. (a) The flying geese quilt block pattern; (b) The log cabin quilt block; (c) The wagon wheel quilt block.

quilt rectangles made up of three triangles to create different patterns. Completed log cabin quilt blocks can create new patterns by using glide reflections, rotations, and translations. In addition to geometric translations, log cabin quilt blocks can be used to explore arithmetic sequences (Packard, 2015).

Half Square Triangle

Half square triangle squares are simple squares created by sewing together two right triangles. These squares are the easiest to make for new quilters. Combining together 16 of these half square triangle squares makes a quilt block. Through a variety of geometric transformations an endless variety of quilt blocks can be created. For example, one designer worked with a programmer who designed a computer program to create quilt blocks. The program was able to generate 72 unique quilt block designs using rotational symmetry alone (see Figure 9.3) (Sewing by moonlight, December 28, 2013). Within these quilt blocks are examples of horizontal symmetry and vertical symmetry.

Quilting in the Classroom

Some of the same mathematical skills required to create a quilt block are also necessary for successful completion of high school geometry. The Common Core Standards for high school geometry state:

Figure 9.3. Half square triangle quilt blocks with rotational symmetry.

> The concepts of congruence, similarity, and symmetry can be understood from the perspective of geometric transformation. Fundamental are the rigid motions: translations, rotations, reflections, and combinations of these, all of which are here assumed to preserve distance and angles (and therefore shapes generally). Reflections and rotations each explain a particular type of symmetry, and the symmetries of an object offer insight into its attributes. (CCSS-M, 2010, p. 74)

The basis of CCSS-M (2010) is for students to apply mathematics in everyday scenarios. A hands-on, real world application of these concepts can be achieved though the making of quilt blocks. Including historical cultural concepts such as the Underground Railroad and the Native Americans to link the quilts to the real world for students will help to enhance the applications of mathematics in everyday contexts.

METHODOLOGY

This study was conducted through self-ethnographic inquiry. Ethnography is a form of qualitative research, which involves studying a topic in a natural setting (Denzin & Lincoln, 2000). Claims from the research are based in constructivism, pragmatism, post-positivism, or advocacy/participatory perspectives (Creswell, 2003). The study was based on my interpretations of how my children interacted with the half square triangles in order to gain insight on how to develop lesson plans to teach geometric transformation in the classroom. Using a constructivist lens, I established an understanding of how students may interact with the lesson materials to help shape how to proceed with developing the lesson plan.

Study Design

The hallmark of ethnographic research is the immersion of the researcher in the community being researched, conducting fieldwork 24 hours a day, 7 days a week, over an extended period of time (Whitehead, 2005). My immersion in the fieldwork environment is fulfilled by the fact that I engaged my children in the practice of quilting over a period of one week. As my children were engaged in the quilting practice, I used observation, interviewing, photographic evidence, and participation as techniques to collect and triangulate data for this study.

Participants: My Children

My children were the participants in this study. A 13-year old White male (given the pseudonym Brandon), and a 15-year old White female (given the pseudonym Laura). Brandon is in middle school in a large suburban county close to a major metropolitan area in the Southeast of the United States. He has been identified as gifted and participates in an accelerated mathematics program in school where he is working one year above grade level, which means he is working at a secondary mathematics level. Laura is a sophomore in high school in the same suburban county as Brandon. She has also been identified as gifted and participates in a high school magnet program focused in mathematics and science. I am a 47-year-old White female who has been a mathematics teacher in the same county as Brandon and Laura for the previous 10 years. I purposefully asked my children to participate in this study because they are enrolled in gifted mathematics courses, which have the same content syllabi as the courses that I teach and one for which the quilt block lesson is being developed.

Data Collection Instruments

I used a variety of data collection instruments, including participatory observation, unstructured interviews, and photographic evidence of the generated designs.

Observation is an integral data collection device in many different kinds of research. Participatory observation has the researcher participating in the activities that are being observed (Whitehead, 2005). By engaging in participatory observation I was able to employ senses beyond just sight to develop meaning from experiences. I observed my children several times during their immersion in the quilting practice. As an observer participant, I also was engaged in the quilting practice myself, for example, creating half square triangle quilt blocks simultaneously as my children are creating their own quilt blocks. I also used unstructured interviews in a conversational form during my observations.

The unstructured conversations occurred before, during, and after Brandon, Laura, and I created quilt block squares out of half square triangles. Our conversations during the quilt block construction process helped direct our thoughts and activities gaining insight and understanding of how to design a lesson for my students. I also recorded our conversations to capture main ideas that emerged during our discussion and to highlight the major thought processes that were elicited within the quilting context. Furthermore, I collected pictures that depict the progression of designs in the created quilt blocks.

Procedure

While sitting at the kitchen table in my home, I gave Brandon and Laura sets of 16 half square triangle blocks made from paper. The only instruction they were given initially was to create quilt blocks that were 4 x 4 half square triangle blocks in size. I watched Brandon and Laura as they designed quilt blocks for the first quilting session. I wanted to see how they would interact with unfamiliar materials. We engaged in unstructured conversation about what they were doing and what their thought process was as they arranged and rearranged the half square triangle blocks.

I participated in quilt block design during the second session. Brandon, Laura, and I sat around the kitchen table making designs. As with the previous observation, our conversation was unstructured. However, I did ask more mathematical questions to Brandon and Laura. I wanted to know more of what their reasoning was as they chose how to design their quilt blocks. I also brought up the idea of geometry, symmetry, rotations, and reflections in an informal way.

Data Analysis

During the first observation I watched as Brandon and Laura created quilt blocks out of half square triangle blocks. Most of the conversation revolved around the process of how to make a quilt block. Even though I have been a quilter, neither Brandon nor Laura remembered me sewing and were not really sure what I meant when I instructed them to create a 4 x 4 quilt block. A common theme from our conversation during their initial engagement with quilting was the need for further clarification as both Brandon and Laura asked about shapes within the block. When I instructed them to make a block, they interpreted what I said to mean only make block shapes within the quilt blocks, the triangle patterns within the quilt blocks were random or not consistent (see Figure 9.4a and 9.4b).

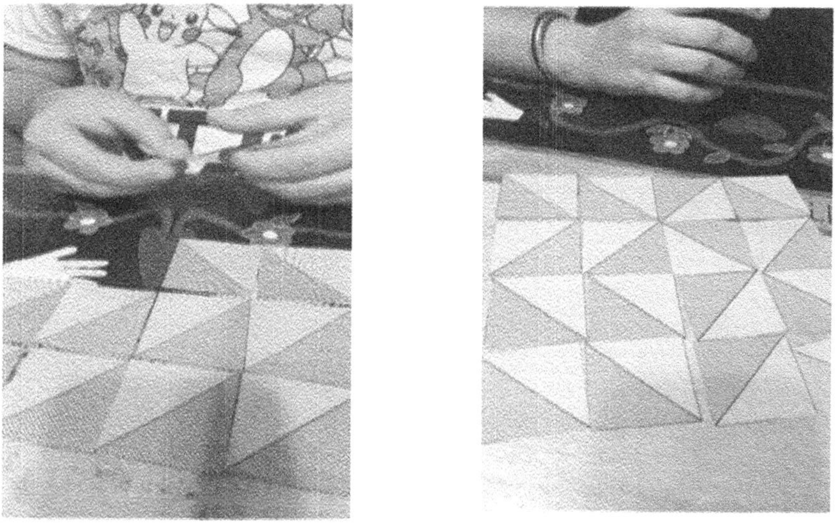

Figure 9.4. (a) Inconsistent pattern use within quilt block; (b) Inconsistent pattern use within quilt block.

Such an exchange of ideas was important because it helped me understand how to build a better concept of what quilts are and a firm understanding of how they are made.

After the first quilt blocks were designed, we stopped creating more blocks and had a discussion regarding quilts, their construction, and their history. I was reminded of the fact that it will be necessary to create a "hook" for the lesson that includes this information before the students receive the material to begin constructing quilt blocks. I underestimated my own children's knowledge of quilts. If Brandon and Laura had trouble understanding what I meant by quilt blocks when they had literally watched me make them I cannot expect students who have never witnessed a quilt being made to understand what a quilt block is or how to make it.

I had chosen not to create a quilt block during the first quilting session because I did not want to influence how my children designed their blocks. Hence, I came to the realization that including some examples of quilt block designs and discussing geometric transformations would be essential components in my lesson plan. As a teacher that views teaching through a constructivist lens, I realized that I needed to activate my students' prior knowledge to help them construct meaningful understanding of mathematics concepts. I did not adequately stimulate my children's prior knowledge of quilting design, which led to much cognitive dissonance with no real ability to accommodate the learning. Under such conditions, my children would more likely disregard learning new ideas had I not scaffolded their learning with hands on explorations of concepts using real life examples.

A common theme that emerged from the second observation was a heavy reliance by both Brandon and Laura on the square or rectangle designs (see Figure 9.3a). Neither Brandon nor Laura initially considered that the triangles within the squares could be used to create a design from color. After discussing the triangles and the difference of the colors of the triangles on the same block, Brandon and Laura used the color difference and the triangles to aid in their designs (see Figures 9.5b, 9.5c, and 9.5d).

FINAL THOUGHTS

Through this study, I was able to begin to develop a lesson on using quilt blocks to teach geometric transformations. Some important ideas about the lesson were brought to my attention as a result of engaging in this inquiry. First, I recognized the importance of introducing quilting in my class. Even

Figure 9.5. (a) A square design; (b) Using triangles and color to drive design; (c) Using color to drive design; (d) Using triangles and colors to drive design.

coming from a home with quilts, my own children had not considered the multiple pieces that went into the construction of a quilt. An introduction into the piecing together of a quilt was a good way to open the lesson. To aid understanding, I plan to model piecing together paper quilt block pieces using different designs so students could try out quilt pieces to build a foundation of quilt blocks.

Second, after modeling with pieces that I created, I will give each student their own set of half square triangle pieces and just let them play with different arrangements and patterns before I explain the lesson any further. This will give them the opportunity to experiment with their pieces for the pure purpose of exploration. After they have had this play opportunity, we can begin to discuss transformations and how such concepts can be exhibited using the half square triangle pieces. Having had the opportunity to interact with their pieces will give the students a structure on which they can begin building their understanding.

Finally, the culminating activity, I planned to use in this lesson is the creation of a class quilt. Once completed, this quilt will hang on the wall in the classroom. To extend the learning, I plan to have the students label the transformations that occur within the quilt to help solidify vocabulary. The construction of the quilt along with the vocabulary identifying the transformations will link together procedural and conceptual fluency, an important component to student learning (National Council of Teachers of Mathematics, 2014).

The purpose of this inquiry was to explore how to use quilt blocks to teach geometric transformations in the secondary mathematics classroom. Through the collection of multiple sources of data using participatory observation, structured, and semistructured interviews, and the collection of photographic evidence I was able to begin to develop a geometric transformation lesson to be used in the secondary mathematics classroom. The intent of this lesson is to aid in the understanding of geometric transformations through the use of quilt blocks.

While lesson development was the intention behind this experience, involving my children in this inquiry made the work more meaningful than initially expected. The conversations started at the kitchen table over those half-square triangle quilt blocks and have extended into mathematical wondering conversations beyond what might have occurred otherwise. My children's view of mathematics has extended beyond what is taught in the classroom. They have pointed out mathematics in the "real" world they might not have noticed previously due to their engagement in the task. They have come to realize that mathematics is beyond formulas and algorithms, it is a personal and subjective lived experience, that transpires in innumerable ways all the time. Math truly is everywhere.

REFERENCES

Creswell, J. W. (2003). *Research design: Qualitative, quantitative and mixed methods approaches.* SAGE.

Denzin, N. K., & Lincoln, Y. S. (Eds.). (2000). *Handbook of qualitative research.* SAGE.

History of quilts: An American folk art. (n.d.). *Quilting in America.* http://www.quilting-in-america.com/History-of-Quilts.html

Johnson, J. (n.d.). *History of quilting.* Emporia State University. https://sites.google.com/g.emporia.edu/center-great-plains-studies/publications-archive/tales-out-of-school/quilting

National Council of Teachers of Mathematics (NCTM). (2014). *Principles to action: Ensuring mathematical success for all.* The National Council of Teachers of Mathematics.

National Governors Association Center for Best Practices, Council of Chief State School Officers (2010). Common Core State Standards Mathematics (CCSS-M). National Governor's Association Center for Best Practices, Council of Chief State School Officers: Washington, DC.

Neumann, M. D. (2003). The mathematics of Native American star quilts. *Mathematics Teaching in the Middle School, 9*(4), 230–236.

Neumann, M. D. (2005). Freedom quilts: Mathematics on the Underground Railroad. *Teaching Children Mathematics, 11*(6), 316–321.

Packard, C. P. (2015). A workshop using the log cabin quilt for teaching math concepts and patterns. *Proceedings of Bridges 2015: Mathematics, Music, Art, Architecture, Culture.* https://archive.bridgesmathart.org/2015/bridges2015-567.pdf

Patchwork Quilts: A timeless art (Patchwork Quilts). (n.d.). *Quilting in America.* https://www.quilting-in-america.com/patchwork-quilts.html

Paznokas, L. S. (2003). Teaching mathematics through cultural quilting. *Teaching Children Mathematics, 9*(5), 250–256.

Wares, A. (2006). Geometry embedded in North American quilt patterns. *International Journal of Mathematical Education in Science and Technology, 37*(5), 608–613. https://doi.org/10.1080/00207390600597518

Whitehead, T. L. (2005). *Basic classical ethnographic research methods: Secondary data analysis, fieldwork, observation/participant observation, and informal and semi-structured interviewing.* Cultural Ecology of Health and Change, Cultural Systems Analysis Group, Department of Anthropology, University of Maryland.

CHAPTER 10

METHOD TO THE MADNESS

The Mathematical Structure of Method Ringing and the Application of Hamiltonian Graphs to Composition

Karen Ratajczak

ABSTRACT

Change ringing or method ringing is the practice of ringing tower bells according to complicated series of mathematical permutations on tuned bells. Each line of numbers is called a change and the guide of what permutation to make in each change is called a method. My study took place at St. Luke's Episcopal Church which has a tower of 10 bells. I analyzed the ringing methods in terms of permutations using cycle notation as a mode of representation. Employing Hamiltonian paths as a means of creating and checking various extent methods, I argue that looking at shorthand notation of methods and more of the language of bell ringing and relating it to formalized mathematics would allow for enrichment opportunities to inform the practices of both bell ringers and mathematicians.

Evidence-Based Inquiries in Ethno-STEM Research: Investigations in Knowledge Systems Across Disciplines and Transcultural Settings, pp. 155–167

INTRODUCTION

Change ringing or method ringing is the practice of ringing tower bells according to complicated series of mathematical permutations on tuned bells. Each line of numbers is called a *change* and the guide of what permutation to make in each change is called a *method*. The ringing of the bells from highest pitch to lowest, treble to tenor, (for example on six bells 1 2 3 4 5 6) is called *rounds*. The rules allowed for the permutations and the probable reasons behind them is discussed in the Context section of this paper. Change ringing is still heard and performed today, however, an understanding of the abstract structure underlying its sound is often unrealized or ignored by those listening who, being used to tuneful music often interpret it as noise.

While change ringing may not have a large knowing audience, it inspires fierce loyalty from its practitioners who have faithfully continued the practice from the seventeenth century to our present day. In order to better understand the practice and its interesting connotation, one should understand some of the basic conventions of bell ringing. By convention, ringers are not allowed any sheet music or memory tools during the ringing practice; they must thus perform all the changes they are ringing by memory. Taking into consideration there are n! [n factorial = 1 x 2 x 3 x 4 .... x(n −1)(n)] possible changes on n bells, this can at times seem an impossible task. If we say that it takes between two and three seconds to ring each change, based on the length of time it takes for the heaviest bell to rotate, it would take about a day to ring all changes on eight bells. Also, by convention, ringers are not allowed to be switched out while ringing. This means that each of those ringers would have to go an entire day without a break (food, water, or bathroom) to complete their sequence. This dedication to the craft is what has inspired so much research into change ringing over the years.

BASIC TERMINOLOGY AND METHODS IN RINGING PRACTICE

Table 10.1 depicts a list of terms that are prevalent in the ringing practice and their descriptions:

In order to conduct the ringing, the ringer must proceed using a set of designated rules as follows:

a. The bells are first rung in rounds and each method ends by ringing rounds.
b. No change may be repeated.

Table 10.1
Terms Used in the Ringing Practice

Terms	Description
Blue Line	A process of learning a method that involves coloring in the path a bell takes in blue to demonstrate the bells movement.
Change	Each row of a method in which each bell is rung once and only once. (On three bells this would be 1 2 3 or any arrangement of these elements).
Extent	The complete ringing of all possible permutations on the set of bells. (The ringing of n! i.e., n factorial changes with no change being repeated).
Leads	The point in a method where a new alteration is introduced. (The point where another permutation is followed to avoid repetition in a method.)
Lead Ends	The point at which a pattern is about to go back to where it was at the beginning of the lead. (The point where another permutation is introduced to avoid repetition.)
Method	Any set of changes that follows the rules of change ringing (see Appendix A for two examples).
Rounds	Bells are rung in order by descending pitch 1 2 3 ...n where n is the tenor.
Treble	The highest pitched and lightest bell marked as 1
Tenor	The lowest pitched and heaviest bell.

c. Each bell may only move by one place when moving from one change to the next.
d. No bell remains in the same place for more than two successive changes.

Two common methods of ringing are referenced in this chapter: the Plain Hunt and the Plain Bob methods (see Figure 10.1).

The Plain Hunt and the Plain Bob methods represent the two earliest methods according to Duckworth and Stedman's *Tintinnalogia* (1671/2006) and are among the first taught to ringers due to the ease with which they can be memorized.

THE MATHEMATICS OF BELL RINGING

Many would probably be surprised to read the bibliography of algebraic group theorists' work and notice the number of articles pertaining to bell

Plain Hunt on Four Bells

```
1 2 3 4
2 1 4 3
2 4 1 3
4 2 3 1
4 3 2 1
3 4 1 2
3 1 4 2
1 3 2 4
1 2 3 4
```

Plain Bob of Four Bells

```
1 2 3 4
2 1 4 3
2 4 1 3
4 2 3 1
4 3 2 1
3 4 1 2
3 1 4 2
1 3 2 4      ← (note that to this point
1 3 4 2^       Plain Bob is the same as
3 1 2 4        Plain Hunt)
3 2 1 4
2 3 1 4
2 4 3 1
4 2 1 3
4 1 2 3
1 4 3 2
1 4 2 3^
4 1 3 2
4 3 1 2
3 4 2 1
3 2 4 1
2 3 1 4
2 1 3 4
1 2 4 3
1 2 3 4
```

^ = the beginning of a lead

Figure 10.1. Plain Hunt on Four bells and Plain Bob of Four Bells methods.

ringing or which credit bell ringing as a source for inspiration. The reason behind this is that bell ringing, or campanology, a practice which uses the practical applications of group theory, predates the formalized study of the subject (White, 1996) Much of the research done on the connection between change ringing and mathematics has focused on group theory and the uses of group theory in bell ringing (Intermont & Murphy, 2011). White's (1996) examination of Duckworth and Stedman's (1671/2006) *Tintinnalogia* and his mathematical explanations for some of the limitations found therein is perhaps the most well-known article on the subject. Its broad appeal lies largely in the accessibility of the language and the use of group theory to explain bell ringing rather than the use of bell ringing to explain group theory. White puts forth a mathematical redefinition of bell ringing rules. He uses rotations of matrices to explain how cosets in Plain Bob methods are formed and how they are associated with symmetric and dihedral groups. White has written extensively on campanology's relation to group theory and is a widely cited authority on the subject.

Other important names in campanology's relation to group theory include R. A. Rankin (1948) who answered a campanological (and mathematical) problem from the 1700s with a theorem, and Richard Swan (1999) who offered a proof for Rankin's theorem. The question, now rather famous in campanological and group theorist circles, was asked in 1741 by a ringer, John Holt, who had tried to create a method that would allow for an extant on 7 bells using only prescribed leads and a method called *Grandsire*. While 7 bells have a total possible 5,040 permutations (7!), he was only able to achieve 4,998. He questioned whether it would be possible to achieve the full 5,040 permutations on bells (Rankin, 1948). While the first answer was put forth in 1886 by Thompson, he solved it by a case-by-case analysis but did not offer a general proof. Rankin based his theorem on the ideas put forth by Thompson and Swan and created a series of proofs that all demonstrated the answer: the highest possible number of changes on a set of 7 bells using only those prescribed methods is 4,998. I have included all three works in my bibliography and leave it to the reader to choose whether to read them.

CONTEXT OF THE STUDY

Before discussing the specific context of my case study, I have decided to include some information about the history of bells relevant to change ringing. According to the North American Guild of Change Ringers (n.d.), there are 34 bells towers in the United States as compared to over 5,000 in Britain (that means there is a bell tower roughly every 18 square miles). This is hardly surprising given the long history of bells in Britain and their import in British culture and history. It also explains why nearly all those towers are located at Episcopal churches. The most recent example of the importance of bells to British culture is the 2012 Olympics in London. Bells were rung to mark the passing of the torch in England and bells were rung for three minutes to announce the beginning of the opening ceremonies. In the ceremonies themselves, a giant bell, the largest harmonically tuned bell in the world, was featured prominently. Much, though not all, of the culturally relevant beliefs surrounding bells have supernatural and religious implications, which is reasonable considering that bells are traditionally housed on religious lands and associated with religious events.

Throughout the medieval and early modern period, there was a belief that ringing bells would keep away demons, the reason they were rung at funerals, and defend against storms. This can be seen most notably on bell inscriptions such as that of a bell found at Erfurt Cathedral which bears the Latin inscription *Fulgur arcens et daemones malignos* [To ward off lightening and malignant demons]. The ritual of bell baptism in Western Europe in the medieval and early modern periods bless the bells in holy water and salt in addition to the preexisting folklore surrounding the blessed sound of bells. According to popular belief in Britain, the bells of Canterbury were said to have rung themselves at Becket's murder in 1170. Bells often were engraved with a dedication to a saint or with religious text to imbue them with spiritual presence. The bells also served a more practical purpose as they were used to alert of invasion or war. This use continued to the modern period. According to the Defense of the Realm Act of 1914, bell ringing was prohibited during wars so that the bells could be used as emergency notification systems in the event of German invasion or bombings.

Bells have long been used in Europe since the fourth century to call parishioners to mass and are still used in Catholic Cathedrals to call for prayers three times a day. The main innovations relevant to change ringers and more particular to Britain began in the reign of Henry VIII and the Protestant Reformation with the advent of industrialization and the reassignment of church materials during the Reformation (descriptions of the reassignment of bells beginning under Henry VIII can be found in Gasquet's *Henry VIII and the English Monasteries: An Attempt to Illustrate the History of Their Suppression).* The tradition of downward facing bells on spindles hung in church towers and rung by deacons who pulled on a rope attached to their mounting were replaced by bells mounted to a quarter wheel with a spindle that served as an axel and a rope attached to the rim of the wheel. As this method of mounting spread, the bells began to be mounted on half wheels. The Protestant Reformation saw the destruction of monasteries and abbeys and their bells being distributed to Protestant churches. The distribution of these bells allowed for churches to have a series of bells rather than just one bell. The addition of multiple bells also drove an innovation in their ringing and many towers moved to a whole wheel system to allow for greater control of the bell while ringing and the addition of a halt allowed the bell to swing in a full circle and the ringer to start and stop the bell at will. This was when the practice of highly structured bell ringing sequences developed, and the true origins of change ringing emerged.

Consequently, ringers began to try variations of sequences for amusement and intellectual challenge. The Ancient Society of College Ringers was founded in the early 1600s and Stedman, a member of said society, wrote the first book on change ringing in the late 1600s. The principles he describes in his book are examples of early group theory. Some even consider Stedman to be the first group theorist despite the fact he did not write his principles mathematically, nor did he show any particular inclination to acknowledge what he was doing as mathematical in nature. Stedman's work became the explanation and exposition of the basic rules of change ringing.

THE STUDY METHODS

My study took place at St. Luke's Episcopal Church in Atlanta, which has a tower of 10 bells. Of the ringers with whom I spoke, all were members of the church at which they rang but I was assured that it was not a necessity at most towers to be a parishioner. It is not uncommon to tour other towers or be invited to ring at other towers. One of the people I interviewed, when asked about the purpose of change ringing, said that it had historically been a means of getting the less religious members of a community to come to church and participate in church life. They also turned it into a drinking game of sorts during the 18th century where ringers who performed well would be bought a round at the pub by those who had performed worst. When asked what the reasons behind the rules of bell ringing were, or what they believed them to be, most pointed to rules 2 and 4 as being more for fun and rule 3 as a physical consideration (there is only so quickly a bell can be rung). Rule 1 was primarily for marking placement or for musicality.

I was able to contact and discuss ringing with six ringers at the bell tower of St. Luke's Episcopal Church in downtown Atlanta, Georgia and five ringers at the bell tower of St. James' Episcopal Church in Marietta, Georgia with various levels of experience. The majority of ringers had been ringing for roughly 10 to 15 years with two ringers having 5 to 10 years and three ringers having 40 or more years. Interestingly, while not all ringers recognized a relation between the mathematics of bell ringing and their outside lives, many had professions or backgrounds that require reasonably high levels of mathematical literacy such as computer science or database administration.

For this study, I chose to do a combination of individual semistructured interviews as well as a questionnaire. Some of the questions asked and were expanded on during interviews include:

1. How long have you been ringing? How do you rate your skill level?
2. What is your profession? Do you transfer any thought processes from bell ringing to your job or vice versa?
3. How would you explain the rules of change-ringing? Do you think there are reasons behind them?
4. How do you keep track of where you are while ringing? How do you think about your position in the change and in the method?
5. How would you explain the terminology of change-ringing in your own words?
6. What is the purpose of change-ringing? (e.g., Its use or meaning)
7. Have you ever tried to ring a pattern backwards? Why or why not?
8. How do you use bobs and singles while ringing? Does it affect how you think about methods and peals?

9. How does your thought process differ from ringing to conducting? (If you conduct)
10. How do you go about composing? (If you compose)

I also made use of published methods, in particular, Plain Hunt on Four Bells and Plain Bob on Four Bells. The use of questionnaires was partly due to convenience and partly because it was better to access a wider variety of people (more than one tower) through an e-mailed questionnaire than attempting to interview numerous individuals located in disparate areas. I accepted the limitations of the questionnaire as opposed to interviewing all individuals and the richness that such discussions would add, to allow for a wider scope of participations. I chose a qualitative approach to focus on the underlying mathematical concepts found in change ringing particularly how to compose methods and describe them using the mathematical language associated with group theory.

Procedure

I conducted individual interviews with the six ringers at St. Luke's and e-mailed questionnaires to the five ringers at St. James's. I used the interviews as a means of examining how mathematic principals were applied by ringers in their way of thinking about ringing. When examining published methods such as Plain Hunt on Six Bells, I explored more the mathematical principals behind the structure of the method. I looked at what permutations took place and how they were written in the language of bell ringers. I also began looking at the practical ability to complete all permutations without repetition given the rules of bell placement change, assuming it was realistically capable of being rung. I focused mostly on Plain Hunt and Plain Bob for this chapter as these are rather simple examples and perhaps more accessible to people unfamiliar with the practice of change ringing.

Because so much research has been done about determining whether an extent of various numbers of bells can be rung, I have focused instead on "mathematizing" the language of bell ringing, that is the writing of methods in purely mathematical terms. Due to time limitations and in the interest of brevity, I examined two of the more basic methods taught to all beginning ringers. These methods are often used in constructing further compositions and can be seen as building blocks to more complex compositions. I begin by discussing Plain Hunt on 4 bells and then briefly extending my analysis to the general case for n bells. I then discuss Plain Bob on 4 bells as an extension of Plain Hunt and as an example of an extent. I have chosen to begin with the 4 bell demonstrations of these methods as they can more easily been written completely (there are only $4! = 1 \times 2 \times 3 \times 4$

= 24 possible changes to write out for 4 bells as opposed to 5! = 1 x 2 x 3 x 4 x 5 = 120 for 5 bells) while being extensive enough to truly demonstrate a pattern. Finally, I demonstrate how a modified Hamiltonian circuit can be used to explain the changes in the methods.

Ringing Methods' Analysis

I analyzed the ringing methods in terms of permutations using cycle notation as a mode of representation. The permutations I refer to are going to act on the positions of the bells, and do not refer to the bells themselves. This means that the permutation (1 2) (3 4) will change the row 3 1 2 4 (which indicates the 3 bell will ring, then the 1 bell, then the 2 bell then the 4 bell) into 1 3 4 2. I also represent permutations such that AB means first doing the permutations of A and then B. S_n and A_n stand for the symmetric and alternating group on n objects respectively.

To write the method Plain Hunt of 4 bells, we first notice that there is a consistent weaving pattern that emerges. The outside pairs move, then the inside pair moves for each change. This indicates a need for products of disjoint transpositions of consecutive numbers, that is, a combination of two permutations: A = (1 2) (3 4) and B = (2 3). By alternating A and B between each line, we generate the method back to rounds. Each row after rounds is acted on by X, XY, XYX, $(XY)^2$, and so forth, until the identity is produced. We know that there can be no more than 8 permutations in this manner as A and B generate a group of order 8. By extending this pattern, we can demonstrate Plain Hunt on any number of bells. For five bells, A = (1 2) (3 4) and B = (2 3) (4 5), for six, A = (1 2) (3 4) (5 6) and B = (2 3) (4 5). Generalization to n bells can be obtained by knowing that for an even number of bells, A = (1 2) ... (n–1 n) and B = (2 3) ... (n–2 n–1), and for an odd number of bells, A = (1 2) ... (n – 2 n – 1) and B = (2 3) ... (n–1 n). These permutations generate a subgroup of S_n, which has order 2n, the dihedral group. In essence, ringers are manipulating cosets of a dihedral subgroup of the symmetric group in an effort to ring an extent. They wish to do so by minimizing the number of permutations used, so that the method is more manageable and easier to remember, and maximizing the length of method, to demonstrate all n! arrangements of elements.

Plain Bob is derived from the permutations in Plain Hunt. Above, I noted that, to a point, Plain Hunt and Plain Bob are the same. The difference at that arrow is made by adding an additional permutation C = (3 4). This addition allows for the remaining sequences to be rung. In bell ringing this would be the beginning of a lead. There are then four leads in Plain Bob, which I have marked with the symbol (^), including the original Plain Hunt pattern, each of order 8 which allow for the complete sequence of 24 permutations. This can be explained mathematically by noting that there would be 3 cosets each of order 8, the union of which provides all of S_4.

When speaking to the ringers, I was told by the church personnel that working bells in a method follow the same pattern but start in a different place, which makes them easy to remember. This got me thinking about what a method would look like graphically and how a method could be composed graphically. Because a Hamiltonian circuit is a graph of a path that visits every vertex exactly once before returning to the vertex from which it started, it is extremely useful for modeling methods of change ringing. Figure 10.2 shows an example of a Hamiltonian circuit moving starting at the lower left corner and then moving to each labelled vertex before returning back to the vertex it began.

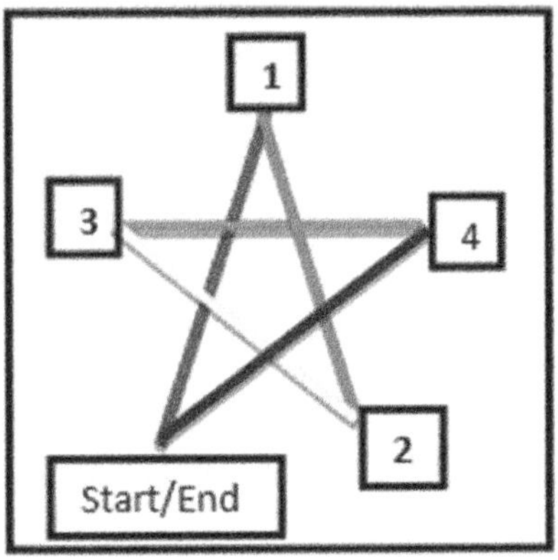

Figure 10.2. An example of a Hamiltonian Circuit.

If we think of each permutation element as a vertex, then each line connecting them is a possible permutation permissible under the rules of bell ringing. We can then use a graphical representation of a Hamiltonian path to compose a method by removing connecting lines until only a circuit remains. Preferably we would use a pattern of permutations that is easy to remember so that ringers would practically be able to remember it.

In Figure 10.3, I depict the Hamiltonian path for a S_4 (a 4 bell extent) demonstrating all possible permutations of four elements regardless of the method. The Figure shows the Plain Bob method on 4 bells in the context of a Hamiltonian circuit. I have included color coordinated lines to demonstrate the various permutations that take place.

Figure 10.4 shows the same Hamiltonian circuit depicted in Figure 10.3 albeit without any red lines because the (1 2) permutation is not found in

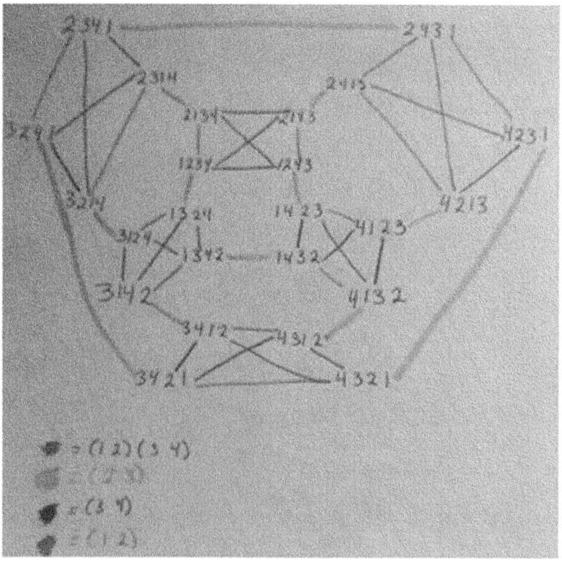

Figure 10.3. The Plain Bob method on 4 bells in the context of a Hamiltonian circuit.

Plain Bob on 4 bells. That is not to say that a circuit could not be created that uses (1 2), other methods exist on 4 bells, but for demonstration purposes I have shown a method I have already discussed. Although there are more aesthetically pleasing ways of arranging the permutations, maintaining as much as possible, the pattern between the two representations to demonstrate their similarities and differences was my primary target.

RECOMMENDATIONS FOR FURTHER APPLICATIONS

Much has been written on the importance and usefulness of relating mathematics concepts to practices that are prevalent in the cultures of students. Kitchen (2001) states that ethnomathematics "places the mathematics of the local culture at the forefront of instruction" (p. 152). It is a field that studies the mathematical ideas inherent in cultural activities that are not formally taught in the classroom. Integrating nuances of cultural practices in the curriculum allows students to establish a practical connection to mathematics as they relate it to their everyday lives. Additionally, it allows for the reinforcement of familial and cultural values that might otherwise be lost in formal education and to allow communities to have a voice that echoes in what their children learn. Harris (1990) describes her approach to ethnomathematics as looking

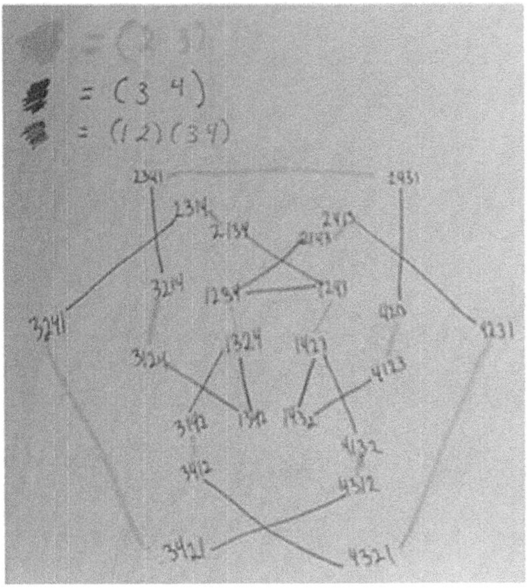

Figure 10.4. The Plain Bob on 4 bells method without the (1 2) permutation.

> at what people do at work, working alongside them when I can, so that I can analyze the mathematical thinking in what they are doing. Many of these people deny that they are doing anything mathematical, either because they do it so routinely that they have forgotten how they worked it out originally or because they have such unhappy memories of school mathematics that they are convinced that they cannot do it. (p. 512)

This reflects a common problem many people have with mathematics, and one that applying concepts to everyday activities could help moderate.

While bell ringing may not be the most commonly known practice, it is one that can be held as testament of mathematical thought predating formalized mathematics theorization. Bell ringers were doing group theory before Lagrange was even born. Perhaps applying some of the advancements made to group theory since then could help solve compositional problems in bell ringing and would bring about ideas that could further our understanding of group theory. While I looked at Hamiltonian paths as a means of creating and checking various extent methods, there are various ways of looking at permutations in group theory that could lend themselves to compositions using only prescribed methods. I think looking at shorthand notation of methods and more of the language of bell ringing and relating it to formalized mathematics would also be worth further studying and would allow for enrichment opportunities to inform the practices of both bell ringers and mathematicians. Another potential of looking at bell ringing is to expand on the idea of the blue line technique that many use to remember methods and examine the patterns made by the movement of the bells as if it were a woven pattern specifically zooming in on the movement of each different colored "thread." Such observations would expand and enrich existing literature on the subject of weaving and examining each thread in weaving as a means of creating complete permutation sets.

REFERENCES

Duckworth, R., & Stedman, F. (2006). *Tintinnalogia, or, the art of ringing wherein is laid down plain and easy rules for ringing all sorts of plain changes*. Project Gutenberg. (Original work published 1671)

Harris, M. (1990). Symmetry and dissymmetry in mathematics education: One view from England. *Leonardo, 23*(2), 215–223.

Intermont, M., & Murphy, A. (2011). An application of group theory to change ringing. *The College Mathematics Journal, 42*(3), 223–228.

Kitchen, R. (2001). The sociopolitical context of mathematics education in Guatemala through the words and practices of two teachers. In B. Atweh, H. Forgasz, & B. Nebres (Eds.), *Sociocultural research on mathematics education: An international perspective* (pp. 151–162). Lawrence Erlbaum Associates.

North American Guild of Change Ringers. (n.d.). *Towers and handbell groups*. http://www.nagcr.org/afftower.html

Rankin, R. A. (1948). A campanological problem in group theory. *Proceedings of Cambridge Philosophical Society, 44,* 17–25.

Swan, R. G. (1999). A simple proof of Rankin's campanalogical theorem. *American Mathematical Monthly, 106*(2), 159–161.

Thompson, W. H. (1886). *A note on Grandsire Triples.* https://www.ringing.info/thompson_1886.pdf

White, A. T. (1996). Fabian Stedman: The first group theorist? *American Mathematical Monthly, 103*(9), 771–778.

CHAPTER 11

AN EXAMINATION OF THE ECONOMIC, MATHEMATICAL, AND SOCIAL-PSYCHOLOGICAL STRUCTURES OF A BARGAINING TRANSACTION

An Exploratory Case Study in Morocco

Kori Lloyd Hugh Maxwell and Patrice Parker Waller

ABSTRACT

The purpose of this study is to explore, using structural analysis, potential economic, mathematical, and social-psychological processes that emerge in a bargaining transaction. This topic came about in a cultural immersion experience as part of a course in ethnomathematics which took place in Fez, Morocco. Bargaining is a key method of doing day-to-day transactions in Morocco, which differs from most places in the United States. We employed the use of an exploratory case study to determine how participants conducted

Evidence-Based Inquiries in Ethno-STEM Research: Investigations in Knowledge Systems Across Disciplines and Transcultural Settings, pp. 169–191

the negotiation process and what meanings they gave to their actions as they engaged in the bargaining practice. To aid our efforts, we triangulated data collection using three methods: Participant observation aided by video-taped interactions and negotiations, researcher's introspection with collection of artifacts including pictures, and written reflections. As a result of our analysis, major themes emerged comprising behavior patterns, attitudes, and interactions in the market that highlighted the mathematical aspect of the bargaining transaction. Findings of this study focus on the critical role that mathematics plays in empowering people to make informed decisions in various situations and to negotiate effectively to earn a living.

INTRODUCTION

Pervasive in many facets of our daily lives are negotiations, and more specifically, bargaining. *Bargaining* and *negotiations* are terms that are used interchangeably at times. In general usage, the term *bargaining* often seems to refer to the interaction between individuals over some sale or purchase (Kramer & Messick, 1995). However, Rubin and Brown (1975) argue that *negotiations* tend to indicate interactions involving broader, more "complex social units" such as unions, organizations or nations, and so forth. According to Lebow (1996), "[b]argaining is a search for advantage through accommodation. When people, organizations, or states bargain, they try to glean information about others' preferences, influence their estimates of their own preferences, and exchange proposals in search of mutually acceptable agreements" (p. 9). Throughout this chapter, we use the terms bargaining and negotiations synonymously and we adopt Rubin and Brown's definition of bargaining as "the process whereby two or more parties attempt to settle what each shall give and take, or perform and receive, in a transaction between them" (p. 2).

In this chapter, we hypothesize that underlying the process of effective bargaining is an array of economic, mathematical, and social psychological considerations inherent in bargaining relationships. These considerations are depicted in Figure 11.1.

Lebow (1996) highlights three necessary but insufficient conditions the involved parties must adhere to in order to ensure a productive bargaining process: (1) they must believe that their interests would be best served by an agreement; (2) they must have conflicting preferences regarding the terms of their agreement; and (3) they must have incomplete information about each other's preferences.

With these three conditions in mind, it would be clear to understand how parties come together for a mutually beneficial purpose. If one party concludes that they would not be better off from bargaining, then they would have no incentive, and therefore no bargaining would occur. If both parties

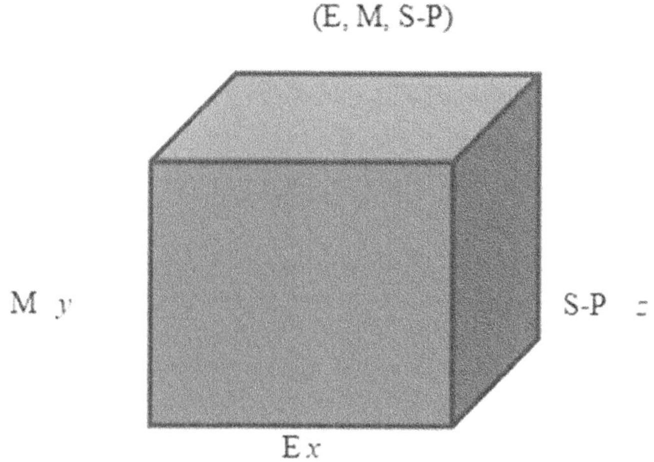

Figure 11.1. Authors' depiction of macroscopic 3-dimensional model of bargaining.

have the same preferences or are trying to achieve the same outcome, there is no need for bargaining. Similarly, if either party knows of the other's preferences and related outcomes, bargaining is again unnecessary.

Rubin and Brown (1975) cite a number of structural determinants to effective bargaining, namely

> the presence and absence of audiences, the availability and role of third parties; the number of parties involved; coalitions; the location of the bargaining site; physical arrangement of the site itself; variations in the types of threat and communication options possible; time limitations and the duration of the bargaining session; the number of issues at stake and the prominence of alternative solutions. (p. 37)

Although bargaining relations have received a great deal of attention in social psychology, we have scarce empirical evidence on the major mathematical considerations that underpin a bargaining transaction. Goldreich and Pomorski (2011) argue that in a transaction, little is known about what determines when negotiations start, which agents are more likely to propose the initiation of bargaining, whether the identity of the party initiating bargaining influences reaching an agreement, or whether the initiator's pay-off is systematically affected.

The purpose of this study is to explore, using structural analysis, potential mathematical processes that emerge in a bargaining act. The study further highlights economic and social psychological aspects that underlie bargaining and examines systematic and successive patterns of interactions between parties that converge towards resolving a conflict related to a purchase transaction.

BRIEF REVIEW

Bargaining is a central issue in many areas. The early practice of price negotiation has, to a great extent, been replaced in modern societies by the posting of prices. Such practice is less time consuming and is a tremendous advantage in retail markets especially when thinking of having to negotiate individual prices for every item to be purchased at the supermarket (no doubt this would be a pain-staking task!). Cason et al. (2003) continue to assert, nevertheless, that bargaining (or haggling as it is popularly called) has not entirely disappeared and that, beyond time costs, it is not clear whether bargaining or a fixed/posted price is a more efficient method. The bargaining process imposes fewer barriers than the "take-it-or-leave-it" offer allowed in the posted prices institution, which may make haggling more efficient. Herrmann (2003) argues that although bargaining is used in the purchase of high-priced items, Americans are ambivalent about the practice of bargaining. In the United States, garage sales are one of the few venues where Americans bargain for low to moderately priced goods, however this type of negotiation is not unique only to garage sales. Many small, local shops and businesses would also negotiate prices based on customer loyalty.

Historically, both the bargaining and posted price market institutions are of special interest as haggling spans the gap between premarket gift exchange and early organized marketplaces. While bargaining has tended to be displaced by more complicated or structured forms of exchange like the posted offer (North, 1991), it still holds a major share in the modern world (for example, car dealerships and wholesale transactions) as well as in traditional economies like Morocco (Geertz, 1979). Furthermore, Cason et al. (2003) highlight the fact that time costs are higher for bargaining than for posted prices and that the expansion of retail from sole proprietor "Mom and Pop" shops to chain stores leads to agency problems that favor posted prices. These factors account for the shift into posted offer in modern economies, especially for standard, small ticket items. With the advent of new and improved technological methods including the internet, these factors may be conquered in modern economies. Arguably, bargaining is reemerging in e-commerce with internet sites because sellers may believe it will afford them a greater share of the surplus (Cason et al., 2003).

A question that arises, then, when thinking of negotiations is whether or not losses are as a result of insufficient time to conduct negotiations. But as data suggests (Roth et al., 1988), the losses are not reduced neither by longer negotiation periods nor by subjects' experience. On the other hand, the losses could be due to "deadline effects" in bargaining. Roth (1988) also alluded to previously conducted bargaining experiments where parties frequently reached an agreement within the last few seconds of the trading period, despite the length and experience of those involved. Perhaps, sometimes, traders hold out for their counterparty to concede in the last few seconds and that concession does not materialize.

Herrmann (2003) explains: "As with much economic behavior, bargaining can be seen as more or less socially engaged" (p. 241). The author further highlighted positive bargaining qualities that are valued in the United States. These qualities include decisiveness, knowledge of consumer goods and prices, and skills in negotiating and getting a deal. For Westerners "the price and quality of goods are related [as] sellers may become insulted when shoppers bargain for their possessions because they see it as literally devaluing their goods" (p. 242). Such characteristics hold as particularly true in a country such as Morocco especially with handcrafted items that are considered to be of exceptional quality.

With social engagement, another important factor to consider in bargaining are the inherent psychological influences. Malhotra and Bazerman (2008) explain: "the effort to positively influence another party's attitude towards a given idea or proposition without changing the incentives or objective information set of the other party" (p. 5) is a crucial part of the negotiation process. Both parties have to socially engage with and convince the other that their price is the best and that it is a valid and worth-while investment. Moreover, Malhotra and Bazerman argue that this type of influence involves "leveraging an understanding of psychological biases and heuristics" to facilitate the framing of, in this case, initial prices and counter offers, in such a way that increases their appeal to the target audience; craftsmen and/or shoppers.

At the onset, it is highly recommended that parties involved in the negotiations be comfortable and feel like they achieved a fair agreement given their expectations. According to Hahn Training (2008), building rapport and establishing a trustful relationship with parties we negotiate with is a key element to any successful negotiation. Moreover, the author noted that it would be beneficial to have an understanding of the negotiating partner's style of behavior in order to adjust or revise how to approach that individual.

Salient in every negotiation is the establishment of value proposition. Hahn Training (2008) highlighted that more often than not, parties fail to determine the real value for their goods and so proceed to sell on price, get stuck in providing concessions, and then are forced to reduce their price as a result. The assumption here is that the counterparty comprehends the value of items because a feature may be obvious to the seller (which does not mean it will be obvious to the buyer or that the buyer will have the same appreciation for it is worth).

When considering negotiations, the issue of power arises from both the buyer and seller's perspectives. Power negotiating, as defined by Ilich (1980) is "[t]he ability of the negotiator to motivate an opponent in a manner that is favorable to the negotiator's negotiating objectives" (p. 2). Ilich also pointed to the literal meaning of the word power, which is the ability to accomplish objectives. So if this literal definition of power is to accomplish objectives, then it is important to be mindful of the fact that, for goals in negotiations to be accomplished, one must motivate others. In every instance of negotiation, there are four conceptions of power that are vital (Ilich, 1980). Two pertain directly to the buyer; namely, the buyer's conception of the extent of their own negotiating power and that of their opponent (the seller). Similarly, the other two pertain directly to the seller; namely, his or her conception of the extent of their negotiating power and that of their opponent (the buyer). Ziegler (2013) indicates that the early definitions of power, as the ability of a party to move another party toward an intended direction, has gained prominence over the years. Despite this prominence, the definition has been improved by focusing on the determinants of the outcome rather than the outcome itself. So power should be thought of in terms of the way in which one individual uses his or her resources with another individual so as to bring about a change that leads to a desired outcome (Ziegler, 2013).

Negotiations have one element in common; conflict will be present to some extent (Hahn Training, 2008). Some parties will exhibit aggressive (at times overbearing/overpowering) behavior, becoming actively involved in conflict situations and others will exhibit "people-oriented" behaviors, seeking a peaceful, mutually beneficial and accommodating solution which values the relationship over the actual negotiations. We have witnessed first-hand evidence of both aggressive and people oriented behavior in our immersion experiences while conducting the study.

THE STUDY

The research idea came about because of a course in ethnomathematics, which examined the application of mathematics from another culture's perspective. As a part of this course, we were to participate in a cultural immersion trip to Fez, Morocco. Founded in the 9th century and home to the oldest university in the world, Fez replaced Marrakesh as the capital of the kingdom in the 13th–14th centuries. The urban fabric and the principal monuments in the medina—palaces, residences, mosques, and fountains date from this period. Despite the transfer of the political capital of Morocco to Rabat in 1912, Fez has retained its status as the country's cultural and spiritual center (The United Nations Educational, Scientific and Cultural Organization [UNESCO], 2009). Considered to be one of the most extensive and best conserved historic towns of the Arab-Muslim world, the Medina of Fez has unpaved urban spaces, which conserve the majority of its original functions and attributes. Not only does the Medina represent an outstanding architectural, archaeological, and urban heritage, but it also transmits skills, culture, and a lifestyle that persists despite the effects of the evolution of modern societies (UNESCO, 2009).

On this trip, we were not only excited to experience a new perspective on how mathematics and culture are interconnected, but we were also excited to completely immerse ourselves in the language, food, and customs of Fez. Mathematics was evident in so many aspects of the Moroccan culture and we were informed that bargaining is a key method of doing business in Morocco, which differs significantly from most places in the United States. As we strolled around in the markets, we were involved in many bargaining acts which allowed us to participate in the bargaining transactions and observe how our colleagues interacted in this space.

Our theme was "everything is mathematics" and so we were poised to find the mathematics in all situations we encountered. We came upon a plethora of hand-made, intricately crafted items such as woven silk fabrics, embroidery on napkins and tablecloths, carpets, leather products, and jewelry. As we engaged in price negotiations in order to procure items from craftsmen in the medina, we closely focused on the potential mathematical ideas that emerged during our bargaining interactions with the sellers.

In this study, we aim to uncover how craftspeople and/or sellers utilized *their* mathematics to negotiate their business practice and how we, as outsiders/foreigners used *our* mathematics to engage as one of the major parties in this practice. Our objective was to explore the mathematics that is evident in the bargaining process and that results in a series of computational processes in which instantaneous decisions need to be enacted with regards to purchase of items. The major research question that guided the study is: How does the negotiation process emerge as craftspeople determine their prices and how do we, as shoppers, arrive at a counteroffer? While our visits in the old city of Fez included multiple sites such as brassware shops, carpenters' workshops and tiling factories, in this chapter, we will focus on transactions that occurred during a visit to a traditional textile factory that weaves and sells various fabrics and scarves.

Context: The Textile Shop

The bargaining episodes took place in a textile shop in the medina located in one of the old buildings along the pathway, and you would pass it if you did not know exactly where it was because it seemed tucked away in a hidden section of the area. The shop comprised a large open space that contained all the fabrics, with a worker towards the back operating his heavy machinery to construct the different items. This shop had two floors displaying and selling different items on each floor. On one side of the main floor, there were bags of different sizes and colors and on another side, there were scarves in every color of the rainbow and of different types of material. There were also tables located all around the shop, on which additional scarves and materials were exhibited. Most interestingly, facing the shop's entrance, there was a traditional wooden vertical-shaft textile loom that holds the warp threads under tension to facilitate the interweaving of fabrics with a variety of colors and designs.

In the upstairs portion of the shop, there were articles of traditional Moroccan clothing of all sizes and a fitting room set up for customers to try on the clothes. Clothing shops flourish in the medina as most of the fabrics worn by the local people are produced locally. The weavers in these textile shops handle many materials including wools of different qualities, cotton, silk, and flax. Interestingly, most of the shops still use rather rudimentary looms to weave fabrics for curtains, cloaks, dresses, and scarves.

METHODOLOGY

The Study Design

We employed the use of an exploratory case study design; a shortened approach to examine situations with considerable uncertainty (Morra & Friedlander, 2009). The rationale for using this method was to explore how participants engaged in the negotiation process and what meanings they gave to their actions with the goal of getting a better understanding of what actually occurs.

Sample/Participants

Our sample participants included six Moroccan artisans ranging in age from mid-30s to mid-40s. They were all males with a range of workplace skills and experience. Upon entry into the store, the master craftsman, who spoke the best English, introduced himself and showed us around the store, the items that were being sold, and the entire process of hand weaving the fabrics. The master craftsman who is seemingly the most experienced among the artisans spoke to us with authority and confidence when explaining the products. Another craftsman who was operating the wooden textile loom, demonstrated how the materials were used. A third craftsman took us on a tour of the second floor where there were traditional clothes displayed and there was yet another craftsman who dealt solely with the purchases. The other two craftsmen were on hand to assist with selecting items as they were stationed next to these products. The welcoming encounters we experienced with the craftsmen were pleasant. Despite the language barrier, the craftsmen were very friendly and willing to help as best as they could. They also had jovial personalities, often communicating with each other in their native Arabic.

Data Collection Techniques

Data was triangulated by using three methods: Participant observation aided by video-taped interactions and negotiations, researcher's introspection with collection of artifacts including pictures, and written reflections (see Figure 11.2).

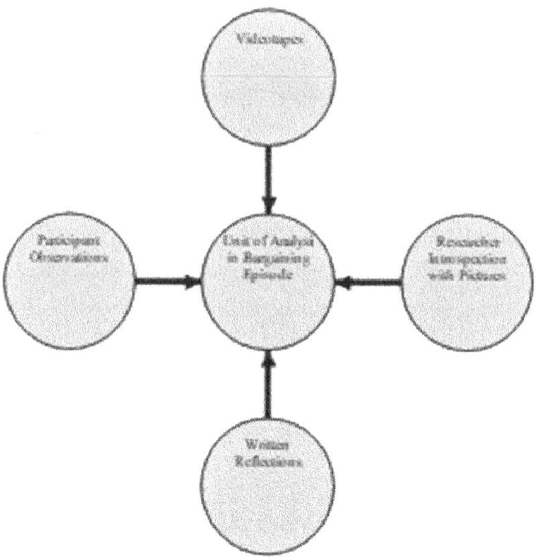

Figure 11.2. Sources of data triangulation in the bargaining episode.

Procedure

We thought it to be crucial to record our negotiation process when we were about to purchase an item and document our experience of interacting with the various sellers of the items. We noticed that some sellers would try to convince us of their price and would interact with other sellers around them throughout the negotiation process. We also observed how other shoppers interacted and negotiated with the same group of sellers and we compared our experiences. Since the items we were looking at were hand-made, we thought it was important to investigate just how these items were created and to explore the hand and mind tools that are dubbed necessary to create the intricate designs.

As part of our negotiations, we considered the quality and labor time that was needed to create the items since they were all authentic merchandise. We provide below a description of a bargaining episode that we encountered in the medina market.

The Bargaining Episode

A climate for bargaining was prepared by the seller. Approximately three craftsmen approached and greeted our group and invited us inside their store. They offered us the traditional Moroccan tea and prepared us for a demonstration of their crafts by one of the craftsmen who was very knowledgeable about the process. This demonstration included a tour of the store and details as to how their crafts were created and by whom, how durable or of high quality the items were, and how much time and effort went into the creation of these items. Our next step involved the beginning stages of the decision to purchase. Here, we browsed through the store to identify anything that caught our eyes. Finally, we put the items we were interested in together in a pile and negotiated a price for all of them instead of individually asking prices for each item. Since we went individually to bargain for the items we wanted, our strategy was to allow those of us in our group who had the least amount of items to be the first to bargain with the sellers. At any point in the bargaining transaction, it was obvious that there were three interrelated aspects that influenced the direction of the negotiation process: economic aspect, mathematical aspect, and social psychological aspect.

DATA ANALYSIS AND RESULTS

The unit of analysis in this study is the bargaining transaction, which we define as the coming together of two or more parties in an attempt to settle what each shall exchange between them. In this transaction, our recorded videotapes, pictures, participant observation accounts and our personal introspections and reflections were examined in order to investigate our bargaining experience. Upon our return to the United States, data was transcribed and coded for patterns and themes that stood out. Video footage was transcribed and reoccurring themes were interpreted. While reflecting on this process, we came across the idea that sales and clearance prices are a lot similar to the negotiations that were made in Morocco. For example, in Morocco, we visited a shop that was known for its hand-made threads including very high-quality silks. The price quoted for a scarf in this shop was about twice what one would pay for it in the United States. So as the negotiators, we immediately thought of countering with an offer that was half price. Instead of actually going through this process in stores in the United States, the seller simply places items on sale or on clearance. This allows the consumer to purchase items for prices lower than the original without the hassle of bargaining.

We employ three multilayered, multistage, multidimensional analyses of a bargaining transaction: (1) economic analysis, (2) mathematical analysis; and (3) social psychological analysis (see Figure 11.3). Stage 1 was the preparation stage where, posing as customers, we prepared for the negotiation with the seller. Stage 2 involved the actual bargaining episode and Stage 3 culminated in the bargaining agreement. The characteristics of bargaining relationships have shown to be prominent in our experience. The bargaining act seemingly consists of a particular structure and involves a multiplicity of mathematical applications and social psychological actions, including behavior patterns and attitudes. During the bargaining experience, we noticed at least two participants with conflicting interests who voluntarily enter into a special kind of implicit agreement to achieve a particular goal (Rubin & Brown, 1975). The activity in this special agreement concerns the exchange of one or more specific resources and the resolution of issues among the concerned parties. Rubin and Brown (1975) noted that a presentation of proposals or demands is made by one party and evaluated by the other party, which then leads to concessions and counterproposals.

We carried out the analysis along the three dimensions, namely, economic, mathematical, and social-psychological. As a result of analysis, major themes emerged comprising behavior patterns, interactions in the market and the mathematical aspect of the bargaining act.

Economic Analysis

Economic considerations of bargaining include the level of knowledge of the quality of the products (information asymmetry), product substitutability, number of buyers and sellers in the market, relative opportunity

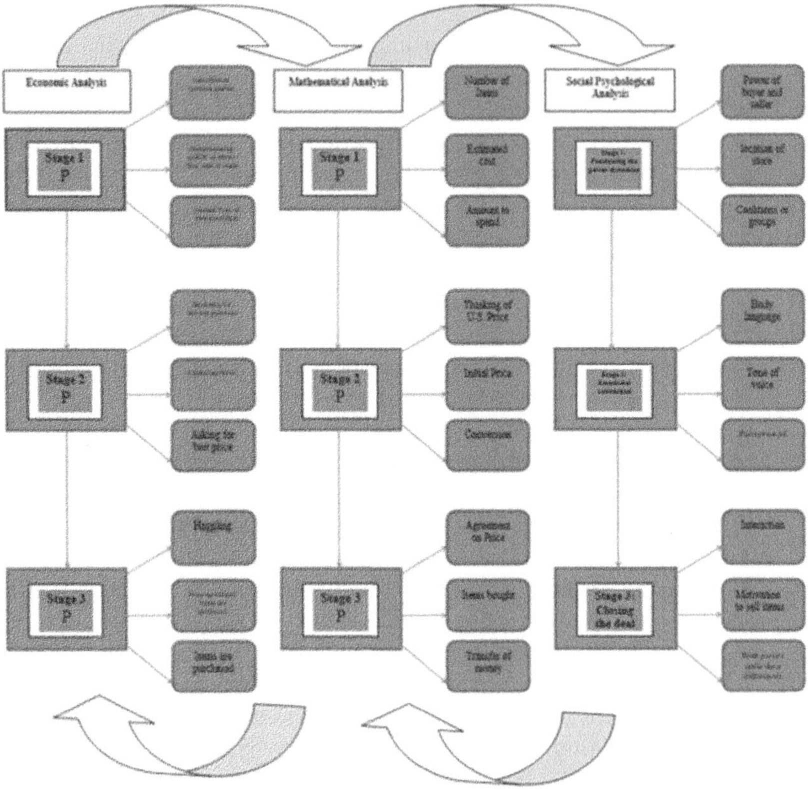

Figure 11.3. Multilayer stages in analyzing a bargaining act: Structural, mathematical and social.

cost of time of parties involved, and nature of the transaction (i.e., whether it is a potential one-time transaction involving a tourist buyer or a potential repeat customer). Used car market may present a classic example of information asymmetry where the seller of a used car arguably knows better than a prospective buyer whether it is a reliable used car or a "lemon."[1]

As the information gap expands, the rationale for bargaining will be more evident. When a product has many close substitutes (e.g., breakfast cereal), buyers will display higher price elasticity of demand reflecting the fact that there exists many options (substitutes) to choose from. On the other hand, when a product is unique or does not have many close substitutes (e.g., insulin), consumers will display low price elasticity of demand reflecting the fact that there are not many alternatives to the product. Therefore, everything else equal, bargaining for breakfast cereal will be more competitive than that for insulin. As the number of sellers increase, the bargaining potential (prospect of "shopping around") also increases for buyers, and as the number of sellers decrease, the bargaining potential also decreases for buyers. Opportunity cost of time of prospective buyers is negatively correlated with bargaining (or haggling). According to Desai and Purohit (2004), buyers with lower opportunity cost of time are hagglers and non-haggling buyers are the ones with higher opportunity cost of their time. When prospective buyers are tourists, sellers may behave more opportunistically with them than with customers with the potential to become regular customers in the future. This common knowledge implies that one-time customers (e.g., tourists) may face higher average asking price than customers with prospect of repeated future transactions. Therefore, bargaining can be more important for a one-time buyer than for a repeat buyer.

As suggested in Akerlof (1970), commodity of bad quality will eventually drive commodity of good quality out of the market. The intuition is that the seller has more authentic information about the quality of the merchandise than its prospective buyer does. This information gap prompts a risk aversive buyer to suspect that actual quality of the merchandise on sale is most likely worse than what is touted by its seller and therefore bid a price commensurate with the suspected quality of the merchandise. Facing this low bid price, the seller feels discouraged and consequently withdraws high quality products from the market and offers only below average quality products for sale. Thanassoulis (2004) articulates a model and contends that a profit-maximizing seller is better off channeling haggling energy not to focus on one good, but to keep bargaining over a range of substitute goods. The Fez city sellers' preference for encouraging prospective buyers to bid over a bundle of goods rather than over each individual merchandize item may also be as much strategic as customary.

Mathematical Analysis

A closer examination of the bargaining scenario reveals a significant mathematics concept that underlies the purchase transaction, namely proportional reasoning. From a mathematical modeling approach, we represent a direct transaction situation using proportionality as follows: if 1 scarf costs 200 dirham, 3 scarves would or should cost 600 dirhams (3 X 200) (aee Table 11.1).

However, this is not the case in the bargaining scenario where there is uncertainty in the final prices that ensures a successful purchase. It is evident how different the initial offer and counteroffer were and how many bargains it took for both the buyer and the seller to agree. It also took a considerably long time (4 minutes and 35 seconds) to reach an agreement (see Figure 11.4).

Table 11.1.
Mathematical Proportional Reasoning Scenario

	Scarves	Money (*Dirham*)
Mathematical Proportional Reasoning	1	200
	3	600

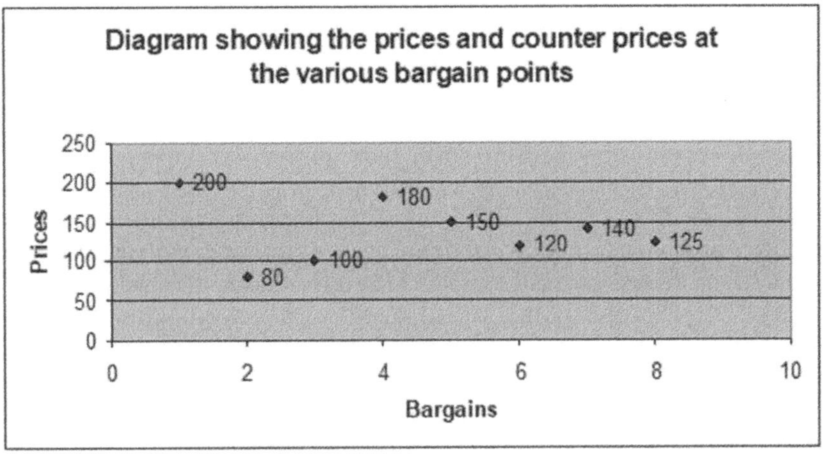

Figure 11.4. Different prices offered at each bargaining stage in the second author's experience.

With the interactions in the market, we were exposed to several craftspeople creating their items. After reflecting on visiting the women who made the various items and what was learned from this visit with them, the first author in this scenario considered the quality of the material, the people who made it and the time it took to complete it and felt that nothing would be lost by paying the final amount that was offered (150 dirhams) as it was the midpoint between both the seller's initial offer and the buyer's initial counter-offer (see Table 11.2).

Table 11.2.
The Bargaining Scenario

	Scarves	Money (*Dirham*)
Bargaining	1	200
	3	150

Getting the best deal or the lowest price was not top priority in this case. The first author felt confident in the fact that he was getting quality and authenticity so there was no need to negotiate further. The buyer was also mindful of the workers and the time and effort it took to create these pieces and did not want to devalue their work in anyway, especially after developing a feeling of empathy towards them and how hard they have to work to make a living. According to Hahn Training, (2008), this is an example of a more "people-oriented" approach, focusing more on the value of the relationship rather than the actual negotiations. In this experience, the number of bargains was half that of the other experience documented previously and an agreement was reached more quickly (in about 1 minute and 23 seconds) (see Figure 11.5).

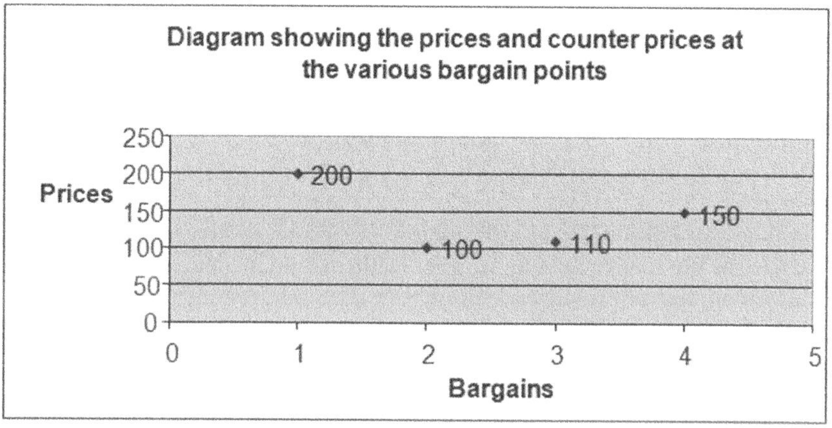

Figure 11.5. Different prices offered at each bargaining stage in the first author's experience.

The mathematical aspect theme emerged in both bargaining experiences as we considered and calculated how much we were willing and could afford to spend and also considered how much these items would be worth in the United States when making our decision.

Social-Psychological Analysis

This analysis involved an examination of participants and researchers and the interaction between them during the bargaining transactions. Behavior patterns were noticed in the second author's bargaining episode as aggressive stances towards the negotiations were frequently taken, favoring the price over the relationship (Hahn Training, 2008). In this experience, the second author and seller settled on a price that was closer to the second author's offer. A few questions arose. What made the seller finally agree to sell it at that price? Could he have been tired of negotiating? Did he realize that the second author was not going to budge? We argue that in this scenario, a person will concede when they acquire an understanding of their counterpart's personality and behavior and if they feel like their counterpart is not going to waver, they will sell it at that price instead of risking the loss of a sale. The second author wanted to get her money's worth in this negotiation because the scarves were not made of pure silk and were much simpler than the other scarves obtained by our colleagues. Power was exhibited in many ways in these situations. From the perspective of the craftsmen and/or seller, power peaked through sometimes in their unwavering stance on prices for particular items. From our perspective as the buyers, we exhibited power by negotiating prices until we were presented with those with which we were comfortable. This power also showed itself in the fact that both parties were confident enough in their price that they were willing and ready to walk away if a favorable compromise wasn't attained. Mathematical power, as quoted by Bishop (2010, p. 157) from the *Standards*,

> denotes an individual's ability to explore, conjecture and reason logically, as well as the ability to use a variety of mathematical methods effectively to solve non-routine problems.... In addition, for each individual, mathematical power involves the development of personal self-confidence. (p. 5)

Power is evident in both scenarios. We had two different approaches but we both came to an agreement we were satisfied with, which, it seems, is related to the level of interest in the item. Power is not necessarily awarded to the one who received the best price in the encounter. Power was exhibited in a willingness to concede after being made aware of the quality of the items, which made the first author confident in his purchase. Power was also exhibited in a refusal to concede as the second author was focused on getting a good deal. We were therefore confident enough in our mathematical abilities to agree on and be satisfied with the final price.

Figure 11.6 further highlights the amount of negotiation that occurred for the first author and second author. In bargaining episode1, the first author had only two instances of negotiations; the initial offer and a counteroffer, before a mutually satisfying agreement was reached. On the other hand, the second author negotiated for her item through eight instances of offers and counteroffers before an agreement was settled on. It is also important to note that the scarves being negotiated in both bargaining episodes were of different quality and materials and that point was highlighted in the negotiation process.

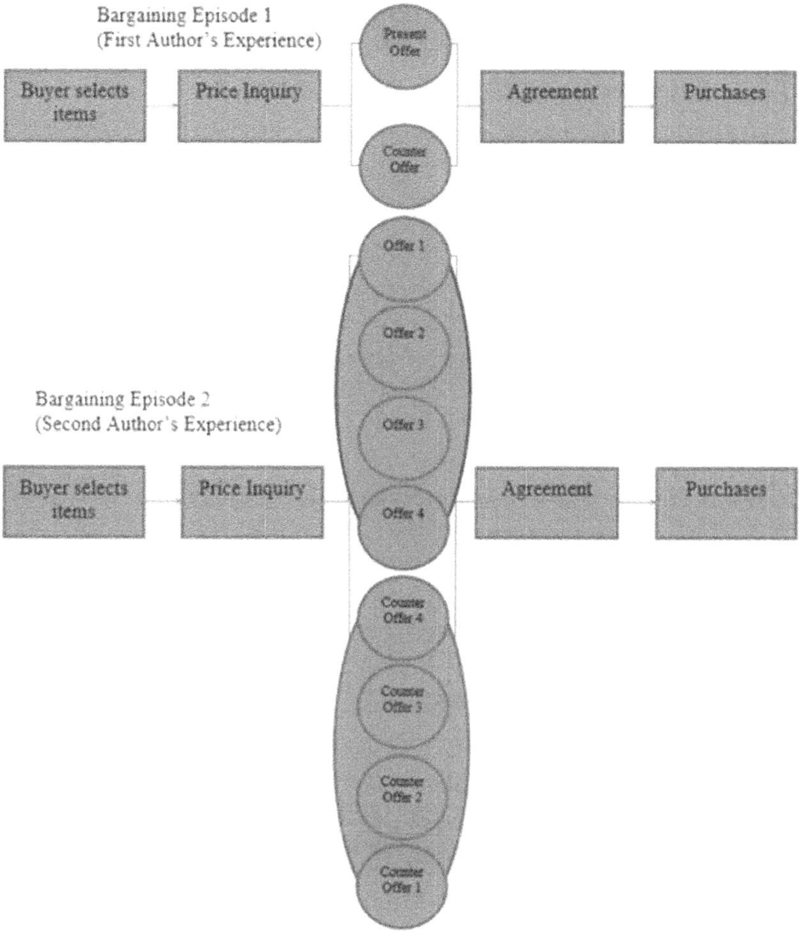

Figure 11.6. Dissecting a bargaining transaction with all parties involved.

Below are two sample transcriptions (vignettes) of our videos, both of which were taken in the scarf store in the Old City of Fez. The first transcription involves the buyer (first author), the seller (Hasad, a pseudonym) and the translator (third author):

> *{After gathering all the items to be purchased, they were presented to the seller to determine how much he was willing to sell them for}*
>
> ***Hasad:*** *I'll give you for 200 dirhams (referring to the scarf)*
>
> ***Third author (addressing*** *second author): How much do you want to pay for this?*

first author: Umm <Third author: I know you want it free> [laughs]

first author: 100

[Third author and Hasad discuss in Arabic (Hasad shakes his head)]

***Third author**: How much can you go up?.... 10 dollars? <first author: Yes, 10 dollars> ... you mean 110 dollars? < first author: dirhams> 110 dirhams.*

***Hasad**: You're price is like your hand on my neck (gestures as if someone is choking him) [laughs]*

***Person off camera**: It's a scarf that we could get probably for about 7 dollars in the states. First author just offered to pay 15 dollars for it and they said no. So, 7 dollars in the states, 15 dollars here ... they want ... they want ... umm 8, 16, 24, 32 ... they want almost ... they want almost 35–40 dollars for the scarf.*

(Towards the side, third author and first author discussed the cost of the scarf and settled for 150 dirhams).

Our second transcription includes our buyer, the second author; the seller, Hasad (pseudonym) and our translator, the third author:

***Second author**: How much for these? (showing the scarves she wanted to purchase)*

***Hasad**: 200 dirhams (looking at the different scarves)*

***Second author**: 200 dirhams for this scarf right here ... and it's not even silk ... I mean all the way. It's like cotton ...*

***Hasad**: Yea ... it's by hand.... <Second author: No..no...no...no...no> No? You don't like it?*

[Pause]

***Second author**: I'll give you ... (calculating in her head) 80 ... <Hasad: for what?> for these two (showing him the ones she wanted).*

***Hasad**: You'll take me with you? [laughs]*

***Second author**: Sure you can come ... you got a passport? [laughs] Come on! ... can you make it through customs? [more laughs]*

***Hasad**: I'll give you a good price.*

***Second author**: You just gave me 200 for this.... <Hasad: You bargain more than her (referring to the third author)> ... You said 200 dirhams, I said 80 dirhams ... <Hasad: You said 80 dirhams each> No, I said 80 ... 8 ... 0 ... (drawing the number in the air).*

Hasad: *for both?*

Second author: *For both*

Hasad: *No (shaking his head)*

Second author: *You said 100.... <Hasad: I said 200 dirhams (holding up 2 fingers)> for 1....<Hasad: for both> ... for both.*

Second author: *Ok, well the highest I'll go ... the highest I'll go will be 100 dirhams for both.*

Hasad: *No ... for this one (shaking his head and referring to only 1 of the scarves).*

Second author: *So you're giving me the original price?*

Hasad: *180 for both <***Second author***: No> 180 for both <Second Author: No>*

Hasad: *You're last price is how much? You're last price ...*

Second author: *My last price is 2 for 1.*

Hasad: *2 for 1? You want to take the price for 1 and get 2?*

Second author: *Yes*

Hasad: *Very clever!*

Second author: *That's how they do it in the sates ... right? (asking other U.S. colleagues off camera)*

Hasad: *In the states you don't bargain, they fix the price there.*

Second author: *And I'm glad you mentioned ... in the states we don't bargain, we have regular price, we have sale price and we have clearance price. Clearance ... 2 for 1!*

Hasad: *You want me to give you my last price?*

Second author: *What's your last price?*

Hasad: *150 dirhams.... <***Second author*** (shaking her head): No, we can't ... we don't have a common ...> How much you want to ... give me a price for both (scarves) and we'll see.*

Second author: *I'll pay in American dollars for it....<Hasad: Yes! How much you want to....> Hold on...wait....*

Hasad: *I think 900 is 120 (converting dirhams to US dollars)*

Second author: *I was thinking more along the lines of about ... 25*

Hasad: *25 what?*

Second author: *American dollars*

***Hasad**: Come on … these prices are not to buy … they are for looking and spending time (to him, these prices indicated that she wasn't interested in really buying anything). You know, she bargains more than you … (talking to the third author) [laughs]*

***Third author**: She's my student!*

[Hasad and the second author both try to convince the third author of their prices]

***Second author**: I said 2 for 1*

***Hasad**: I said 2 for 150*

***Third author**: you're saying 2 for 200?*

***Second author**: He's saying 2 for 200, I said 2 for 100*

***Third author**: 2 for 100 … (thinking)*

***Hasad**: I say 150, my last price for this 1, it was 150 dollars*

***Third author**: I say 2 for 150.… <Hasad: I said 150 … I agree with her>*

[Now they compare the other scarves that were purchased against the ones the second author wanted to buy. It was noted that some of them are made with different materials.]

***Second author**: 120*

***Hasad**: What? <Second author: For 2> 120 dirhams? Never! (shaking his head) … come on … why are you bargaining … give me 140!*

***Second author**: 125*

***Hasad** (after mumbling in Arabic): Ok (and hands it over to be processed).*

From these two scenarios, it can be seen that the first and second authors both had two different styles and ways of interacting with the same seller. The scarves we intended to purchase were of different quality materials and that was brought up in the negotiation process. The results section will have further analysis of these transcriptions.

DISCUSSION

According to Hermann (2003), most bargaining models start with the maximum price the seller is willing to sell his items and the maximum price the buyer is willing to pay for said items. A negotiation will then occur as to how both parties will reach an agreement on price. In this study, the craftsmen pointed out that their items are all hand-made and so they are priced based on the labor time, materials, and quality. We, the buyers, also countered with a price we were willing to pay based on what we thought it would cost us in the United States.

During our negotiations in the market, it was also interesting to see the vital role that language played. Valero (2008) identified

> language and the way in which people phrase the world is not an innocent act but an act through which the construction of the world is affected. Language formulates ideas and meanings that regulate individual and collective action and that, at the same time, make possible that action. (p. 45)

The vital role of language was glaring to us because we did not speak the native language and so, these craftsmen would talk to their counterparts in their native tongue and then come back to us with a price. It seemed that they were discussing what they considered to be profitable and then presented us with that amount. By the same token, D'Ambrosio (2004) indicates that a culture's language is an important activity as "culture manifests itself through jargons, codes, myths, symbols, utopias, and ways of reasoning and interfering. Associated with these we have practices such as ciphering and counting, measuring, classifying, ordering, inferring, modeling, and so on, which constitute ethnomathematics" (p. 197).

Valero (2008) again writes that

> the mathematics curriculum and the teaching of mathematics are not devices and processes in charge of the transmission of a highly valued knowledge. They are social practices that, through the transformation of knowledge from one field of practice to another field of practice, helps regulating the action of students, their thinking frames and their possibilities of participation and exclusion from participation in the social world. (p. 53)

The practical application of mathematics in the Moroccan culture was important for us as educators to observe and participate in because we too often hear that students do not feel that learning mathematics is useful to them in their lives. This experience allowed us to have readily applicable evidence of how mathematics is not only useful in the Moroccan culture but also pertinent to the United States as well.

There was also an instance of bartering, where one of our colleagues exchanged items in her possession for purchasing the product. Another fact to consider was that we were clearly identifiable as tourists/foreigners, which begs the question; would these craftsmen charge the same price to local residents? It is doubtful. The residents would probably be more aware of what these items were worth and so would be able to identify any overpricing that they might try to sneak by us foreigners. In addition, local residents may not be seen to have as much money as foreigners, so the sellers would offer a price based on what they think their potential buyers can and will be able to afford. Nonetheless, we did what we could to make the best deals for ourselves. We also observed an encounter our colleague had at the carpet store. We were all there negotiating prices for carpets but were in different areas. Our colleague was by herself negotiating prices and our other colleagues were in a group. The prices that she was able to negotiate for her carpets were markedly higher than the price they were giving the group. So another question arises; did she get a higher price because she was negotiating by herself, and female? And our group had three people and a male? It is indeed possible. It is also possible that our colleague empathized with the craftspeople and did not feel the need to negotiate the price further.

Most importantly throughout our experience in Fez markets, we realized how vital it is to have knowledge of products we may encounter in any market before beginning the negotiation process. In the United States, a buyer does not usually go to a car dealership and make an offer prior to obtaining some information about the value of the vehicle. Although the goods we were negotiating for in Morocco were not as grand as buying a car, the same principle still applies. Arguably, the buyer must be knowledgeable about the item and the market from which they are buying in order to minimize overpayment for goods and services.

FINAL THOUGHTS

We learned a myriad of things from this cultural immersion experience. Not only the fact that mathematics is evident in various cultural contexts, but that we have more in common with this *other* culture than we previously thought. We were in an environment where negotiation was the way of the people and we had no other choice but to adapt. Being in this experience, however, highlighted the fact that the U.S. also has instances where negotiation is the dominant method. For example, in the case of garage sales or even small, local businesses which are willing to negotiate prices for their loyal customers. Mathematics empowers people; it allows them to make decisions and negotiations for their living. Mathematics really is everywhere and in everything as it "is a cultural phenomenon. It reflects the culture it serves and, in turn, is shaped by that culture. Cultural trends in mathematics are constantly evolving" (Swetz, 2009, p. 13).

NOTE

1. In today's used car market particularly in the developed countries, some institutions such as Carfax.com, autobytel.com, and so forth. have emerged to address this information asymmetry problem. But information asymmetry in this market in developing and emerging countries is still pervasive.

REFERENCES

Akerlof, A. G. (1970). The market for "Lemons": Quality uncertainty and the market mechanism. *The Quarterly Journal of Economics, 84*(3), 488–500. https://doi.org/10.2307/1879431

Bishop, A. J. (2010). *Mathematical power to the people*. Harvard Educational Review.

Cason. T. N., Friedman, D., & Milam, G. H. (2003). Bargaining versus posted price competition in customer markets. *International Journal of Industrial Organization, 21*, 223–251.

D'Ambrosio, U. (2004). Ethnomathematics and its place in the history and pedagogy of mathematics. *Classics in Mathematics Education Research*, 194–199.

Desai, P.S., & Purohit, D. (2004). *"Let me talk to my manager": Haggling in a competitive environment.* https://doi.org/10.1287/mksc.1040.0045

Geertz, C. (1979). *The bazaar economy in sefrou: Meaning and order in Moroccan society*. Cambridge Press.

Goldreich, D., & Pomorski, L. (2011). Initiating bargaining. *Review of Economic Studies, 78*, 1299–1328.

Hahn Training. (2008). *The psychology of negotiation.* http://www.champtrainer.com/documents/the_psychology_of _negotiation.pdf

Herrmann, G. M. (2003). Negotiating culture: Conflict and consensus in U.S. garage-sale bargaining. *Ethnology, 42*(3), 237–252.

Ilich, J. (1980). *Power negotiating: Strategies for winning in life and business.* Addison-Wesley.

Kramer, R. M., & Messick, D. M. (1995). *Negotiation as a social process.* SAGE.

Lebow, R. N. (1996). *The art of bargaining.* The Johns Hopkins University Press.

Malhotra, D., & Bazerman, M. (2008). Psychological influence in negotiation: An introduction long overdue. *Journal of Management, 34*(3), 509–531.

Morra, L. G., & Friedlander, A. C. (2009). *Case study evaluations.* World Bank Operations Evaluation Department.

North, D. (1991). Institutions. *Journal of Economic Perspectives 5*, 97–112.

Roth, A., Murnighan, K., & Schoumaker, F., (1988). The deadline effect in bargaining: Some experimental evidence. *American Economic Review, 78*, 806–823.

Rubin, J. Z., & Brown, B. R. (1975). *The social psychology of bargaining and negotiation.* Academic Press.

Swetz, F. J. (2009). Culture and the development of mathematics. An historical perspective. In B. Greer, S. Mukhopadhyay, A. Powell, & S. Nelson-Barber (Eds.), *Culturally responsive mathematics education* (pp. 11–43). Routledge.

Thanassoulis, J. (2004). Haggling over substitutes. *Journal of Economic Theory, 117*, 217–245.

United Nations Educational, Scientific and Cultural Organization. (2009). *Medina of Fez*. http://whc.unesco.org/en/list/170/

Valero, P. (2008). *Discourse of power in mathematics education research: Concepts and possibilities for action. PNA, 2*(2), 43–60. http://funes.uniandes.edu.co/568/1/ValeroP08-2851.PDF

Ziegler, O. (2013). *EU regulatory decision making and the role of the United States*. https://doi.org/10.1007/978-3-658-00054-7

CHAPTER 12

THE "ART" OF SEEING MATHEMATICS

Kimbeni Mansion

ABSTRACT

How can we change what we cannot see? How can we change the perceptions of what a mathematician looks like and what qualifies as mathematics if all we see is a Western perspective of mathematics? Until my introduction to ethnomathematics, I had only been exposed to the history and implementation of mathematical ideas from a Western perspective. I had never examined mathematical concepts from a cultural viewpoint. A few months ago, I met an amazing artist named Steve R. Allen. Mr. Allen is recognized internationally for his bold use of color and ability to create in a variety of styles, including portraits, sports art, and Afrocentric masterpieces. Like myself, Mr. Allen has a unique origin story which lead him to be the talented artist that he is today. In this chapter I highlight aspect of Mr. Allen's work and connect his artwork to mathematics.

Evidence-Based Inquiries in Ethno-STEM Research: Investigations in Knowledge Systems Across Disciplines and Transcultural Settings, pp. 193–206

CAN YOU SEE ME?

I am a Black, female mathematician. As a very young child, I loved mathematics. I could add, subtract, multiply, and divide before I entered kindergarten. My kindergarten teacher realized early that my mathematical skills were beyond those of my peers and arranged for me to attend first grade mathematics classes. In my first-grade mathematics class, I was placed in the highest ability group and excelled. After first grade, I was placed in the gifted program, where I continued to excel academically. I was a born mathematician.

In middle school, I ended up being the only female in a gifted mathematics class. Not only the sole female student in the class, I was only one of three Black students in the class. My gifted mathematics teacher made her views of me being in the class very apparent. She always reminded me that I do not belong in the class and often belittled me in front of the other students. After a while, I began to see myself and my mathematical abilities through her eyes. Following her advice, I left the gifted class and enrolled in an accelerated mathematics course. Needless to say, with the proper encouragement from my accelerated mathematics teacher, I became one of the top achieving students in the course.

In a few years later, I attended a magnet high school for gifted students. High school was like middle school all over again. Being one of only a few Black students in each of my classes, my teachers did not consider me as "smart." Was I really that bright? My teachers hardly saw "me" at all.

Fast forward 17 years later and after receiving a Bachelor of Science degree in mathematics and teaching middle and high school mathematics, I returned to graduate school where I completed a master's degree in mathematics education and pursued a PhD in teaching and learning with a concentration in mathematics. I once again, constantly encounter those who question my status as a mathematician. Why is it difficult for others to see me, a forty something Black female, as a mathematician? I now know the answer to that question. It is because many do not know that "I" exist. My calling as an educator, as a mathematician, and as a Black woman is to show the world that not only do I exist, but many like me exist. My charge in this chapter is to shift the perception of what a mathematician can be and what counts as mathematical ideas.

HOW CAN I MAKE YOU SEE ME?

How can we change what we cannot see? How can we change the perceptions of what a mathematician looks like and what qualifies as mathematics if all we see is a limited perspective of mathematics? Until my introduction to ethnomathematics, I had only been exposed to the history of mathematics from a Western perspective. I had never examined mathematical concepts from a cultural viewpoint. Ethnomathematics focuses on the human creation of mathematics, investigates the reasons why humans create mathematics, and leads to the creative exploration of the implementation of mathematical ideas in diverse cultures.

But what is ethnomathematics? D'Ambrosio (1985) identifies ethnomathematics as "the mathematics which is practiced among identifiable cultural groups, such as national-tribal societies, labor groups, children of a certain age bracket, professional classes, and so on" (p. 45). He also describes it as a study between the history of mathematics and cultural anthropology. As described by Mukhopadhyay, Powell, and Frankenstein (2009), mathematics is a discipline rooted in cultural practices. Cultural mathematical practices acknowledge "realization that many cultures and peoples contributed to the development of mathematics highlights its universal appeal and importance" (Swetz, 2009, p. 39). Ethnomathematics is a study that respects contributions of "others" and the mathematical achievements of all. By observing and appreciating cultural practices of others and oneself, we can self-identify with the study at hand and position mathematics as relevant and meaningful.

Since "mathematics is a human endeavor" (Swetz, 2009, p. 39), I argue that students must be provided with opportunities to explore the human aspect of mathematics, which involve the creation of mathematics from different perspectives around the world. As noted by Swetz (2009), today's educators must mesh a variety of cultural inputs, such as those used to identify individuals, families, communities, and schools. As an interdisciplinary endeavor, ethnomathematics acknowledges the connection between mathematics and culture and, as a result, humanizes mathematics by highlighting the many applications of mathematics.

WHAT COUNTS AS A MATHEMATICAL IDEA?

Cultural activities use mathematical ideas and thinking that can be described by Bishop's (1988) six categories: counting, locating, measuring, designing, playing, and explaining. *Counting* can be described as a relationship that establishes a one to one correspondence of a set of objects with a set of numbers. *Locating* describes the way that cultures interpret and represent special structures. Some cultures use directional language to describe location while others use geographical features and landmarks to describe location. *Measuring* includes the processes of comparing, ordering, and quantifying qualities. The emphasis on the accuracy of measurements vary from culture to culture. *Designing* encompasses how humans create objects from the physical, social, or intellectual world as well as the creation of objects that are not physical. Cultural designs are as varied as cultures themselves. *Playing* does not only involve games, it involves strategy, probability, and may also include other cultural activities that have important religious and social significance. *Explaining* describes how people create abstractions and formalisms. Explaining also includes the comparing and contrasting of ideas and objects.

How Can We See Mathematical Ideas?

Art is defined as the expression or application of human creative skill and imagination, typically in a visual form such as painting or sculpture, producing works to be appreciated primarily for their beauty or emotional power. We can see artistic creations. With 21st century success in the United States relying heavily on interest and performance in STEM education (President's Council of Advisors on Science and Technology, 2010), it is imperative that today's students acquire the skills required to embrace their creativity, collaborate with others, and generate innovative solutions in STEM fields.

By combining the visual and creative aspects of art, STEM becomes STEAM. The addition of art to science, technology, engineering, and mathematics, allows students to develop pathways for self-identification, self-expression, and self-motivation through the creative expression of mathematics interwoven in the arts (Land, 2013). Placing "A" (art) into STEM provides additional opportunities for students to explore the artistic aspects of STEM fields that embrace innovation and creativity. As Land (2013) describes, "adding the arts into the STEM equation can re-invigorate the platform, providing not only an interesting approach, but also opportunities for the self-expression and personal connect new generations crave" (p. 548).

We Can See Art!

A while ago, I met an amazing artist named Steve R. Allen. Mr. Allen is recognized internationally for his bold use of color and ability to create in a variety of styles, including portraits, sports art, and Afrocentric masterpieces. Like myself, Mr. Allen has a unique origin story, which led him to be the talented artist that he is today. He feels that it is important to share his life story with youth, especially those who have a background similar to his own upbringing:

I feel it is very important to share my life story with the younger generations. My life in particular can serve as an example of beginning with very humble circumstances at birth and achieving notoriety internationally and historically. The importance of my story being shared with the youth stems from what seems to be little or no connection to our history and its role in helping us maintain our identity as a "people." It is also important to see actual living people that the young people can interact with on a personal level to share how things actually were "back in the day." To share actual life experiences.

Figure 12.1. Mr. Steve Allen standing in front of his birth home.

Mr. Allen entered the world on July 10, 1954 in Lillington Crossroads, North Carolina. He was born in a modest shack that was originally used as an army barracks that housed soldiers during World War II (see Figure 12.1). He lived there with his mother, father, and two older brothers until he moved as an infant to Fuquay-Varina, North Carolina. Eventually the family, now including a younger sister and brother, moved to Raleigh, North Carolina. At the age of 17, Mr. Allen married, received a GED, and began raising a family of his own.

Mr. Allen preformed several jobs including a stint in the military, working for the federal government, and working as a technician for Canon

Copiers. In response to the racial riots in Los Angeles centered around the Rodney King decision, Mr. Allen decided to move back to the South. He relocated to Atlanta where he resumed his career as a copier technician. As the only Black technician in the Atlanta area working on high-end equipment, Mr. Allen faced such overt racism that he decided to quit his job. It was then that he decided to give painting a chance after a fortuitous visit to the National Black Arts Festival. After viewing many of the paintings at the festival, Mr. Allen said, "Well hell, I can do that!" Thus, began his career as an artist. His career as an artist bloomed rapidly. In less than four years from the time he quit his job as a technician, he would become the official artist for the host city of the Atlanta 1996 Centennial Olympic Summer Games. His work hangs in many prestigious art collections and as the official artist for the host city of the Atlanta 1996 Centennial Olympic Summer Games, he was commissioned to create *Uniting Colors of the World*, an eighty feet by twenty-eight feet mural, as the City of Atlanta's official commemoration of its hosting the Centennial Olympic Summer Games.

In an effort to give back to the Black community in the United States, Mr. Allen became involved with one of the largest historically Black college and universities football classics, the 40th and 41st Anniversary Bayou Classics where he created the "Look of the Game" images that continue to elevate the marketing efforts of the Bayou Classic to heights never before reached. Nonetheless, Mr. Steve Allen's accomplishments reach far beyond sporting arenas as he was appointed as the Cultural Ambassador to the United States, representing the Democratic Republic of Congo. In 2013, The Smithsonian Institution *National Museum of African American History and Culture* acquired four of Mr. Allen's paintings as part of its Founding and Permanent Collection. Although his sports related pieces propelled him to international acclaim, Mr. Allen is now focusing on the type of art he first began to find joy in creating, paintings which reconnect him to the Motherland of Africa. During a visit to Mr. Allen's studio, I realized how "mathematical" his art actually is.

Mr. Allen: The Ethnogeometer

As a mathematician, I see mathematics in many places and my engagement with ethnomathematics research expanded my view of what qualifies as mathematics and who qualifies as a mathematician. Although Mr. Allen admits to being "good at math" when he was in school, he does not think of himself as a mathematician.

Mr. Allen recalled the mathematical talent of his grandmother and other members of his family. He told me the story of how his grandmother was such a skilled seamstress and accurate measurer that she would sew perfectly fitting garments using only a few carefully taken measurements and a paper bag pattern. It is obvious that Mr. Allen inherited his grandmother's acute measuring skills:

> I think I received my mathematical ability from both sides of my family. The women on the maternal side of my family were actually quite good with numbers. In fact, two of my mother's sisters rose to prominent positions in the world of banking. The ability of my paternal grandmother, Iola McLeod Allen, to work with mathematical things was well known. This ability was actually quite prevalent among many members of my family. I speak about her ability to make suits, men and women, of clothing by hand without the use of 'store bought' patterns. She had the ability to see an item of clothing and then lay it out on newspaper, or something similar, and reproduce what she saw. There were many examples of this type of thing by a number of people in my family.

On a visit to the art studio of Mr. Steve R. Allen, he began to explain his design process in the creation of his latest masterpiece, *Alkebu Lan*. As we discussed the creation of his latest masterpiece, he first explained the idea behind the piece. "The idea of *Alkebu Lan* is to depict Black women as the 'goddesses' they most certainly are. The name *Alkebu Lan* itself means 'Mother of Humanity' " (see Figure 12.2).

Figure 12.2. *Alkebu Lan* poster.

To begin his creative process, Mr. Allen finds the solitary point that he actually measures with a ruler, the center of the blank canvas. Like the ancient Greeks who use only a straight edge and a compass in geometric constructions, Mr. Allen uses his ruler only (he does not use it for measuring after finding the center of the canvas) and a "compass" to create his pattern. His "compass" is one of an artist, it is comprised of a rubber tipped paint brush with a pencil attached to it by an adjustable piece of wire (see Figure 12.3).

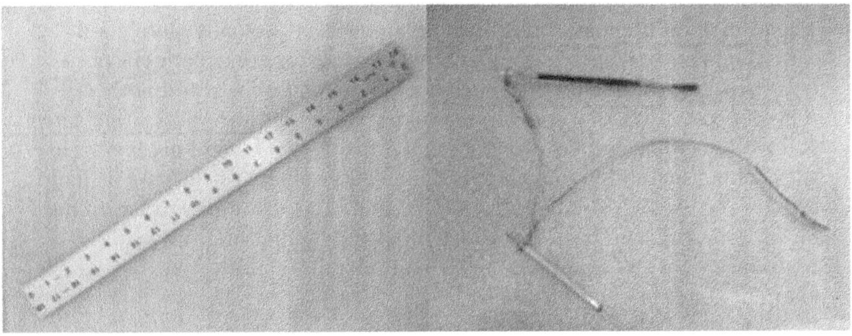

Figure 12.3. Tools employed to complete artwork.

He goes on to explain the beginning of his process, "This painting started with a simple form and eventually ending with a very complex form created totally from layering forms upon one another." Mr. Allen's" simple form consists of three almost perfectly symmetric figures which he sketches free hand (see Figure 12.4).

Figure 12.4. *Alkebu Lan*—First layer of color.

> After applying the color, I covered the entire canvas with masking tape and then I drew initial designs which were something I came up with from another painting. I drew large and small repeating patterns. I didn't have any

> direction that I was going into particularly. There was no particular reason I chose that design. I thought to repeat the design and actually made a grid system on the tape starting at the center of the painting.

Mr. Allen relies heavily on his grid system to guide his constructions. The grid system changes with each layer of masking tape he applies. He uses only his keen special sense to apply individual row of masking tape.

Figure 12.5. (a) *Alkebu Lan*—First "grid" and strip pattern, (b) First cutting, (c) *Alkebu Lan* after first layer of tape is removed.

"After I drew the designs I cut them out with an x-acto knife." With the precision of a surgeon, Mr. Allen cuts away pieces of his design drawn on the masking tape to reveal a magnificent strip pattern design (see Figure 12.5 (a)). He then applies paint to the uncovered design and removes the remaining masking tape (see Figure 12.5 (b)). He further explains, "The next step was to paint the designs and remove the tape for the next step in creating the image—the painted designs with the masking tape removed" (see Figure 12.5 (c)).

Mr. Allen's grids take many fascinating forms and shapes during his creative process. Some are created with masking tape and some are hand drawn. "I added concentric circles and triangles to the design to complete the pattern. The taped design here is actually something from another painting of mine. I repeated a merged flip design from the corner of one my other paintings." After the masking tape is removed to achieve the desired design, a layer of paint is applied to the uncovered sections of the artwork (see Figure 12.6).

Mr. Allen creates yet more grids using varying widths of masking tape and geometric shapes to achieve his desired effects (see Figure 12.7). "I then created a diamond pattern using various widths of masking tape. I did this by dividing the canvas from the center. By doing this I just have to measure various distances from the center point and tape from that measurement to the corresponding measurement."

Figure 12.6. (a) *Alkebu Lan*—Second "grid", (b) *Alkebu Lan*—Paint applied to second "grid."

Figure 12.7. *Alkebu Lan*—Another "grid" and paint applied.

Figure 12.8 depicts Mr. Allens' design strategy,

> Why is this tape ¼ inch wide? I used ¼ inch masking here to create a vertical rectangular pattern. I did this because if I covered the whole canvas with masking tape I would actually have to cut these shapes out individually. By using this tape, I made these little squares and cut them. From the squares, I made rectangle like bricks by cutting and removing alternating pieces of tape.

Many of the symbols and shapes incorporated in Mr. Allen's artwork have a deeper meaning that just being astatically pleasing. In this piece, he used designs and ideas based on African culture. "I used a combination of intersecting circles, squares and triangles to creat a 'halo' effect to help depict the divine nature of the main subject, women as goddesses."

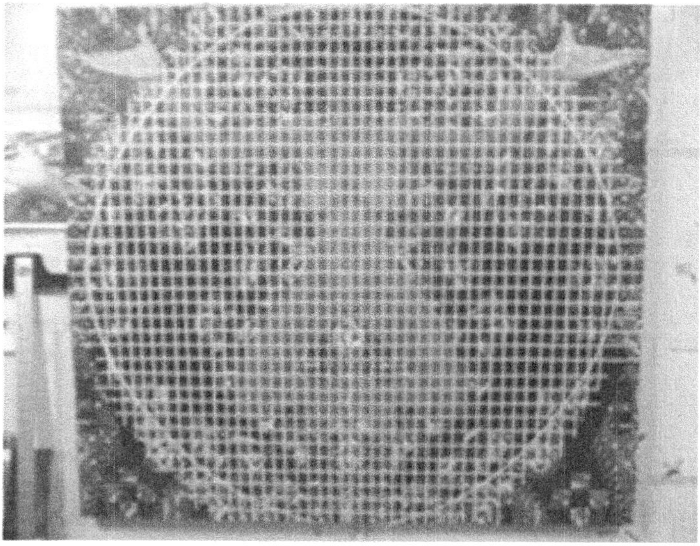

Figure 12.8. Depicts Mr. Allens' design strategy.

He further describes the underlying symbolism of his work as shown in Figure 12.9.

> The symbol depicted here is the Adinkra (Ghanian) symbol NYAME YE OHENE which means the "majesty and supremacy of God." It is to symbolize a connection with God to state that God is the foundation for all. I used the grid pattern to begin creating perspective because I wanted to add depth to the painting visually and metaphorically. I also started this because the image appeared flat, too two dimensional.

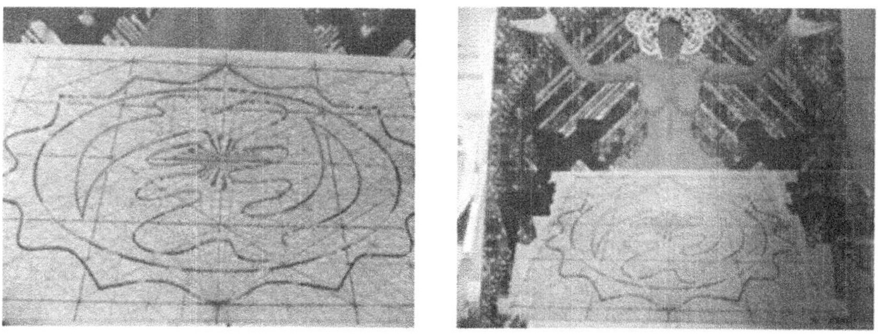

Figure 12.9. (a) NYAME YE OHENE symbol, (b) NYAME YE OHENE symbol created to add depth to the painting.

Alkebu Lan is a complex network of geometry, design, color, and creativity. I asked Mr. Allen where his inspiration comes from, he responded, "It just comes." He believes that the inspiration for his pieces come from a higher power and is very modest about his creative genius. Layer after layer, Mr. Allen creates depth and designs which culminate in astonishing form (see Figure 12.10).

> At this stage of the painting the idea came to create a pattern that truly used every basic shape (square, rectangle, triangle, circle and oval) in the design. It also has a feeling of symmetry and asymmetry at the same time.

Figure 12.10. *Alkebu Lan*—More strip patterns.

Mr. Allen is very aware of the use of symmetry and asymmetry in his work (see Figure 12.11). He argues,

> The symmetrical aspect of the painting presents itself in an image to the left and to the right, but the asymmetrical aspect of the piece begins to manifest in the details of the images. They are generally balanced with regards to the figures, but the details in the individual figures are different when observed up close. There are also meanings and symbols in the details inside each figure. These similarities and differences become more apparent when the paint is applied."

The metaphorical and visual depth Mr. Allen creates in his piece are as astonishing (see Figure 12.12). His understanding and ability to add depth and perspective to his creations confirms his heightened special sense.

Figure 12.11. (a) *Alkebu Lan*—Designs drawn on figures, (b) Tape removed from painted figures.

"This aspect of the masking process was solely done to create another level of separation, depth if you will, and for each figure to begin to stand separate and apart but also remain a part of the whole."

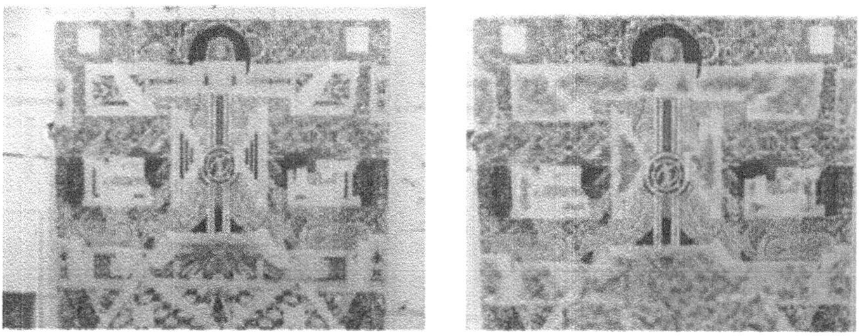

Figure 12.12. *Alkebu Lan*: (a) Creation of more depth, (b) Paint applied to depth creating design.

NOW HE SEES!

By the end of our time together, Mr. Allen experienced a transformation that I yearned to unfold; he gained a new appreciation of his artwork from a mathematical prospective. He not only saw himself as a mathematician, he saw the mathematics of his art in a different light. From the tools of his trade, to his use of geometrical concepts, to his unique spatial sense, Mr. Steve Allen is a ethnogeometer. His final words of our meeting and looking at his *ALkebu Lan* poster (Figure 12.13), he noted, "I guess I am good at math."

Figure 12.13. *Alkebu Lan* poster.

REFERENCES

Bishop, A. J. (1988). *Mathematical enculturation: A cultural perspective on mathematics education*. Kluwer Academic.

D'Ambrosio, U. (1985). Ethnomathematics and its place in the history and pedagogy of mathematics. *For the Learning of Mathematics*, *5*(1), 44–47.

Land, M. H. (2013). Full STEAM ahead: The benefits to integrating the arts into STEM. *Procedia Computer Science*, *20*, 547–552.

Mukhopadhyay, S., Powell, A. B., & Frankenstein, M. (2009). An ethnomathematical perspective on culturally responsive mathematics education. In B. Geer, S. Mukhopadhyay, A. B. Powell, & S. Nelson-Barber (Eds.), *Culturally responsive mathematics education* (pp. 65–84). Taylor and Francis.

President's Council of Advisors on Science and Technology (2010). *Prepare and* inspire: K-12 education in science, technology, engineering, and math (STEM) for *America's future*. http://www.whitehouse.gov/sites/default/files/microsites/ostp/pcast-stemed-report.pdf

Swetz, F. J. (2009). Culture and the development of mathematics: An historical perspective. In B. Geer, S. Mukhopadhyay, A. B. Powell, & S. Nelson-Barber (Eds.), *Culturally responsive mathematics education* (pp. 11–41). Taylor and Francis. https://www.georgiastandards.org/Georgia-Standards/Frameworks/8th-Math-Unit-1.pdf

CHAPTER 13

THE ANCIENT CHINESE WAY OF SOLVING HIGH DEGREE POLYNOMIAL EQUATIONS

Iwan Elstak

ABSTRACT

The main objective of this chapter is to highlight the special place and role of indigenous knowledge in Ancient China. The history of the ancient Chinese way of solving high degree polynomial equations is an extraordinary example of more than 1,500 years of indigenous Chinese development and creativity. The history of the mathematical sciences in China covers more than 2,000 years during which there was no significant influence from outside. We know that there was Indian—Bhuddist (Plofker, 2009) and Islamic influence (Li & Du, 1987) and starting around the 17th century Western Europeans attempted to introduce new methods of mathematics into China. The mathematics developed in China is a unique body of texts, tools, concepts, symbols, and systems that represent an indigenous Chinese achievement that has few parallels. Despite this impressive development of mathematical knowledge, the mathematical culture of China does not get the attention it deserves. Merzbach and Boyer (2011) in their *A History of Mathematics*, first published

Evidence-Based Inquiries in Ethno-STEM Research: Investigations in Knowledge Systems Across Disciplines and Transcultural Settings, pp. 207–247

in 1968 and still in print today, devoted no more than 12 pages out of 668 pages to the history of mathematics in China while 14 pages were dedicated to just the work of Newton alone. Recent literature on the history of mathematics, such as *The Crest of the Peacock* by George Gheverghese Joseph (2011) takes a more balanced approach to the contributions of diverse cultures to mathematical thought. Out of 560 pages Joseph devotes 122 to Chinese mathematics. This chapter is dedicated to teachers who want to deepen their understanding of Chinese indigenous mathematical knowledge systems. The aim is to provide evidence from historical documents regarding the practical implications of the Chinese body of work and to describe the numerous methods used, in order to provide a rather global perspective on mathematics. It is also dedicated to students who are interested in studying the history of mathematics beyond Western contributions

INTRODUCTION

Many eminent Chinese scholars have contributed to the impressive body of Chinese mathematics. While some of the work of important contributors has been lost, nonetheless, the number of extant classical works produced in China is large. Since ancient times, much of the accumulated mathematics knowledge was collected, documented and reported over a long period of time. For example, Figure 13.1 sketches the chain of books and mathematicians who contributed to the creation of the tools, concepts, techniques, insights, and solutions to solving polynomial problems of any degree with any type of coefficients.

EXTRACTING SQUARE ROOTS, THE ANCIENT CHINESE WAY

This chapter is dedicated to teachers who want to deepen their understanding of Chinese indigenous mathematical knowledge systems. The aim is to provide evidence from historical documents regarding the practical implications of the Chinese body of work and to describe the numerous methods used, in order to provide a rather global perspective on mathematics. It is also dedicated to students who are interested in studying the history of mathematics beyond Western contributions. The focus is on the techniques of the ancients: how did the ancients attain such excellence without the tools that we have today? How could they be certain that what they did was correct? Why did their methods work? In general: how did they do their mathematics? It is not possible to answer these questions relying solely on the ancient Chinese techniques. Contemporary algebra is therefore used to clarify various steps of the ancient mathematics.

Mathematician	*Book*	*Period*
Author Unknown	Jiu-zhang suan shu (Nine Chapters on the Mathematical Arts - NCMA)	100 BC – 100 AD
Sun Zi	Sun Zi suaniing - The Mathematical Classic of Sun Zi	220 – 420 AD
Liu Hui	Commentary on NCMA	260 – 270 AD
Liu Yi	Discussions on the Old Sources	1050 AD
Jia Xian	Mathematical Manual	1150 AD
Qin Yiushao	Mathematical Treatise in Nine Sections	1247 AD
Li Ye	Sea Mirror of Circle Measurement	1248 AD
	New Steps in Computation	1259 AD
Yang Hui	A detailed Analysis of the Mathematical Methods in the NCMA	1261 AD

Source: The books by Liu Yi and Jia Xian are lost but part of their work was discussed in Yang Hui's works (Li & Du, 1987).

Figure 13.1. Timeline of development of the methods for solving polynomial equations

We will illustrate nuances from the Chinese indigenous mathematics by showing how to solve high degree polynomial equations with the tools that were developed in China over a period of 1,500 years. The evolution of the methods starting with extracting square and cube roots to solving high degree polynomial equations is sketched. The chain of development starts very early and we pick it up at the systematic extracting of square roots.

In his book, *The Transformation Of Mathematics In The Early Mediterranean World: From Problems To Equations*, Reviel Netz (2004) explains why we cannot simply transform ancient texts into contemporary mathematical language: we may lose sight of its importance for the ancient audience. He emphasizes that to understand ancient mathematics we need to become proficient in the mathematical language of the ancients (p. 3). We will therefore try to explain the Chinese system by focusing on their tools, diagrams, symbols and lines of reasoning to gain insight into the underlying ways of how they conceptualize their mathematical ideas. To keep some connection with the language that we use today, we will use conventional numbers, and intersect the explanations with some algebra to maintain a link with what we know.

The Chinese Numeral System

By 200 B.C. the Chinese culture had produced a complete system for all arithmetic operations such as addition, subtraction, multiplication and division. Rods and counting boards—a special mathematical tool—were used for calculation since 500 B.C. Their numeration system was based on place value with base 10. With this system of numerals one was able to write any number of any size (Lam Lay Yong & Ang Tian Se, 2004). The basic rules that we use today are most probably descendants of this method.

Counting boards and rods. Figure 13.2 shows a counting board with columns to indicate place values. The counting board needs not be an actual board.

Any flat surface could function as a counting board. The columns drawn are for explanatory purposes and were NOT drawn on the actual counting board. The horizontal lines are usually visible on the board, but the dotted

10000	1000	100	10	1.	
		2			*Shang*
4	3	6	7	4	*Shi*
2	4				*Fang fa*
1		1			*Xia fa*

Figure 13.2. A counting board with columns to indicate place values.

vertical lines have been added to clarify the place value of digits. The ancient counting boards did not have these vertical lines. The names of the different mathematical tools used in the process are shown on the right. *The numbers are written as contemporary numerals.*

A representation of the ancient Chinese numerals first as single digits (Lam Lay Yong & Ang Tian Se, 2004) and then as actual numbers will look as shown in Figure 13.3.

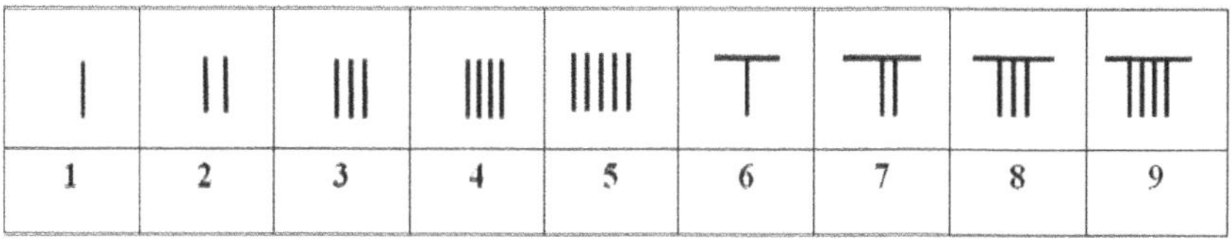

Figure 13.3. Representation of ancient Chinese numerals and actual numbers.

In the system used by the Chinese mathematicians to represent numbers on the counting board the place value was also indicated by writing digits on the board alternatively in horizontal and vertical positions. For example: the number 656 would be placed on the board as (see Figure 13.4).

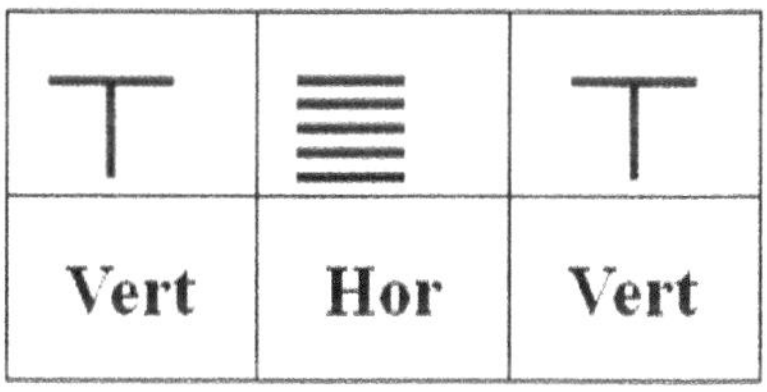

Figure 13.4. Numerals with vertical and horizontal positions.

The digits are rotated to distinguish each one from its neighbor. Otherwise a number like 42 would be impossible to read properly on the board because there would be six straight rods on the board which could be interpreted as 33. The digit on the right started in a vertical position. The placeholders, our zeros, were simply empty placeholders, that is, open spaces in the ancient system. This open space was named "kong" which means empty. Eventually it became a dot (Yong & Se, 2004).

The written language in China had special characters to indicate 10s, 100s, 1,000s, and so on. For example, the number 55 would be written (as a word, or character) not as a number symbol, by writing the character for five, followed by the character for 19 and then followed by the character for five. Furthermore, the number 531 would be written as a word by first writing the character for 5, then followed by the character for hundred, then the character for three followed by the character for 10, and finally the character one. In that sense, the rod numerals were different from the written numbers: the rod numerals on the board did not require special signs to indicate place value because of the rods, the rotation of the numerals when moving to higher place values, and the presence of empty spaces or dots to indicate the absence of a particular power of 10.

Another example is the configuration of 9604, note the open space for the absence of the 10s digit. Also, the rotation of digit "9" has the vertical rod on top instead of on the right or left of the four horizontal rods.

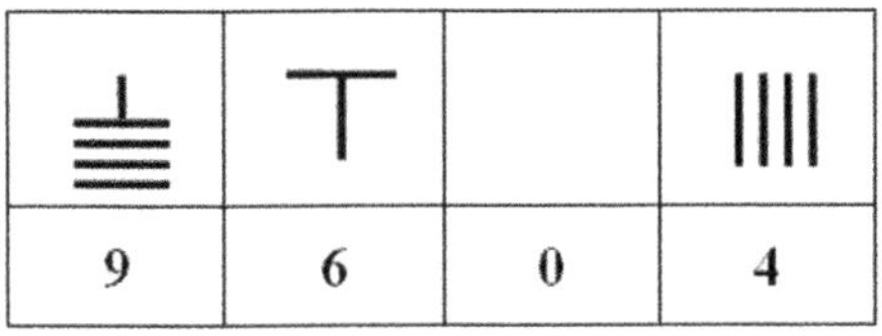

Figure 13.5. Rod numeral for 9604.

The rods were made of bamboo, while some rods were made of ivory or even jade (Lam Lay Young, 2004). They were square in shape. When the number "5" is written as a single numeral it has five rods. In the numeral for six, or eight, it is represented by a single rod, perpendicular to the other rods. For example: 𝍰 This symbol represents eight, with one rod for 5 and three vertical rods for the three.

The Extraction of Roots of Degree Two and Higher

Very early on the Chinese developed notations for fractions, including proper and improper fractions, and positive and negative numbers, the nine chapters on the mathematical arts which date from around 100 A.D. give a detailed account of fractions and their rules. The mathematical classic of Master Sun Zi from approximately 280–473 A.D. (Yong & Se, 2004, p. 28) describes fractions in detail. After creating the tools for executing addition, subtractions, multiplications and divisions the Chinese mathematicians expanded their techniques to include the extraction of square roots using the same counting boards and the same rods to represent numerals in a place value system rooted in the powers of 10 (Yong & Se, 2004, pp. 93–112). The motivation behind the extraction of roots comes from three areas:

1. Surveying, a science highly developed in China from the earliest times. The Gou-Gu theorem, known in the west as the Pythagorean theorem was a major technique employed in this field.

2. Astronomy and calendar construction that relates to observing the stars and their movements (The classic Zhoubi Suanjing or *The Arithmetical Classic of the Gnomon and the Circular Paths of Heaven* contain examples of the "gnomon" (= a T-square or carpenter's square) and its use for calculations of heights, and depths and shadow lengths of the vertical segment.
3. Calculating the length of a side or the radius of a sphere given the volume or the area of a certain object (Li Yan & Du Shiran, 1987).

Steps to Extract Square Roots

To illustrate the methods used by the Chinese to extract square roots we will use as an example the square root of 83674. This number is not a perfect square, so its root will be an irrational number. We will show that the method used by the Chinese works even in those cases. The ancients are thereby followed in their actions as closely as possible. Only by retracing on our own the steps they took, can we appreciate the techniques of the ancients properly. The student who goes through these steps should be aware that the ancients did not have our symbolic machinery. This awareness may enable him or her to act and think like the ancients.

Step 1. To find out what $\sqrt{83674}$ is we start by placing a rod numeral on an actual board or some surface that serves as a board that holds the rods (Figure 13.3). The rods on the board represent numerals with place value. Each of the digits has a separate symbol in the ancient Chinese system. One must understand that in Chinese the number words and the number symbols are identical (Young & Se, 2004). A number word like five in English is a word, the numeral 5 is a separate symbol. In ancient China, the word for five and the symbol for 5 were identical. However, on the counting board the numeral 5 would be constructed with rods on the surface. The separate written numeral 5 *for calculations* did not exist as such in China for a very long time. For a fuller description of the traditional, ancient numeration system we refer to Yong and Se (2004) and G. G. Joseph (2011).

In Figure 13.6, the number from which the root will be extracted is on the board, together with the technical names of the various rows. The names of the rows may shift during the calculations. The Shang is the row that will contain the results of the calculation of the square root. The word Shang means "dividend" reminding one that the method and the technical names for the rows evolved from earlier methods for finding quotients. The Shi is the row that contains the number you begin with and the numbers that you work with in the process after each step. The Fang fa is the "divisor of the square." We will try to explain where these technical terms came from

10000	1000	100	10	1.	
					Shang
8	3	6	7	4	Shi
					Fang fa
					Xia fa

Figure 13.6. A counting board with the starting number and names of rows.

and how they fit into the whole story of the extraction of the root. Xia fa is the "carrying rod" the rod that indicates what the place value is of the various numbers that are being calculated. In this example the carrying rod starts in the column under the 8. As mentioned before, on the actual board the columns are implied but not visible. They are shown here to make the explanation easier to understand.

Yong and Se (2004) quotes Yang Hui, the Chinese mathematician who in 1261 wrote, *A Detailed Analysis of the Methods of Computations*, as follows:

> When the xia fa (the carrying rod) is below the units of the "shi" number, this determines the units of the root; if it is below the hundreds of the "shi", this determines the tens of the root, when it is below the ten thousands this determines the hundreds of the root, when it is below the thousand thousands, this determines the thousands [of the root]. (p. 95)

The quote can be explained as follows: if a number is less than 100 but positive, then its root has an integer part with one digit. If a number is less than 10,000 but more than or equal to 100, its root has an integer part with two digits. If a number is less than 1,000,000, but more than or equal to 10,000 its root has an integer part with three digits.

In our case the number has 5 digits and therefore the 6 represents the hundreds, the 8 represents the 10,000s. Because of this, the 8 is the first digit to be considered. The 8 in the 10,000 place means that the first digit of the square root must be a 2 for the hundreds of the root, because 3 for the hundreds would lead to a square of at least 90,000! The determination of the first digit, the "2," is based on knowledge of the squares of integers. We then place the digit 2 in the hundreds place, in the "Shang" row, which is the results row. Another digit 2 is placed under the 8 from the Shi row (see Figure 13.7).

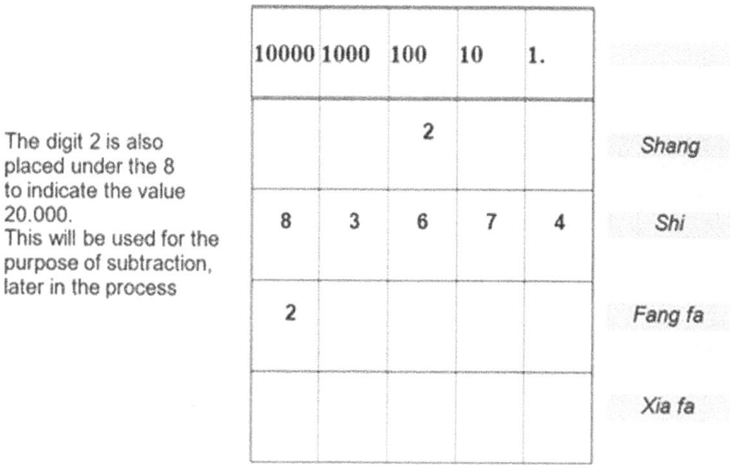

Figure 13.7. The board after determining the first digit: First digit found from $2^2 = 4$ and $3^2 = 9 > 8$.

Step 2. We multiply the digit 2 from the "Shang" by the 2 under the 8 to get 4. Subtract 4 in the column from the 8, leaving 4. This is in fact the subtraction 83674 - 20,000 x 2 = 43674. In the same step, the 2 from under the 8, is doubled and moved one step to the right. The "Xia fa" resides in this step under the 2 of the hundreds. Figure 13.8 shows how step 2 is executed.

Step 3: The calculation of the 10s digit in the ancient system. This step will include some algebraic calculations that are connected to the steps from the ancient Chinese method. Since our symbolism does not reflect the ancient Chinese way of doing mathematics we need to include modern mathematical language to bridge the gap.

Through systematic probes with digits from 0 to 9 we want to find the largest digit for the 10s place that leaves a positive difference for

$$43674 - \{4000\ (\text{digit}) + 100 \times \text{digit} \times \text{digit}\}.$$

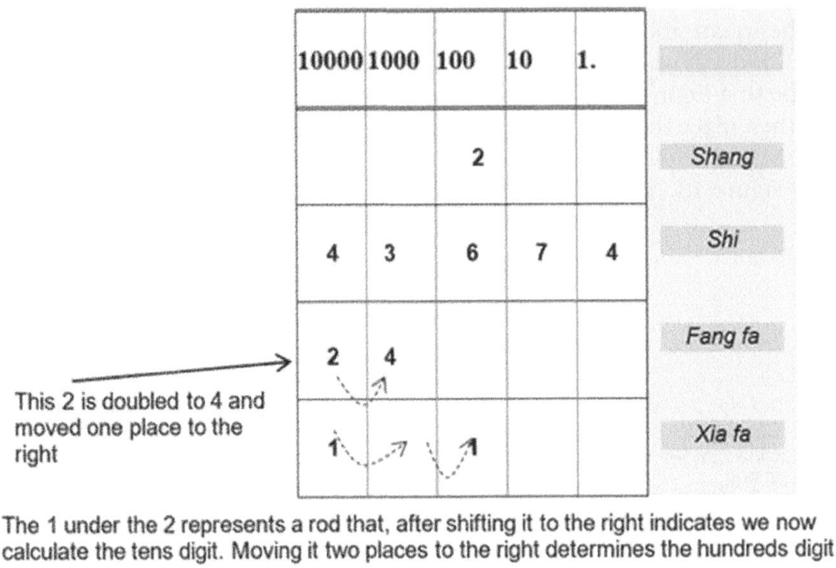

Figure 13.8. Step 2: Doubling the 2 and moving the 1 rod by 2 steps.

To calculate the 10s digit we take the 4 from the second column from the left (it represents 4,000) and multiply the digit 4 by a digit. We will start with the 9 digit, so the multiplication 36, with an actual value of 36,000. Next we take the chosen digit 9 again and multiply it by the 1 from the "Xia fa" which represents 100. The result is 900 and we multiply this, again by digit 9. The two numbers are then added: 36,000 + 8,100 = 44,100. Subtracting this 44,100 from the 43,667 in the Shi row results in a negative number: – 426. Because the outcome is negative we will discard the choice for 9 as the 10s digit. The next digit in our probe is 8:

$$43674 - \{4000 \times 8 + 100 \times 8 \times 8\} = 5274.$$

The digit 7 is our next choice:

$$43674 - \{4000 \times 7 + 100 \times 7 \times 7\} = 10774.$$

We conclude that the digit 8 is the best choice: higher choices for the digit, like 9, leave a negative number and lower choices like 7 give a number that is much larger than the value for "8." So we select 8 as the largest digit for the 10s place. By selecting the largest digit for the 10s place in the numeral we make sure that the next digit is of a lower place in the numeral (see Figure 13.9).

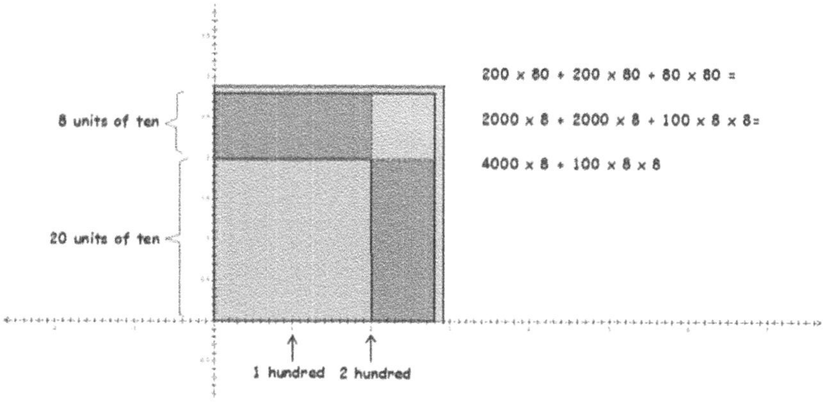

Figure 13.9. A geometric illustration of the digit by digit calculation.

The square in Figure 13.9 shows the area of 83,674 units, reduced to fit the page. The 1 mark in the diagram represents 100 units. It therefore belongs in the column for hundreds on the board. The width of the brown rectangles represents 8 units of 10. The T-square – shaped remainder of the initial large square represents the geometric image of the calculation:

$$43674 - 4000 \times 8 + 100 \times 8 \times 8.$$

The slender rectangular shaped figure that is not included in this difference is the part that will be used to find the next unit's digit. When the Chinese mathematicians did their calculations, they knew that one had to double the 2000 from the multiplication 200 x 80 to extract the full share connected with the 10s digit. This principle allows the method to be used even when we need fractions. In modern terms: after finding the first hundred's digit 2, the Base-10 numeral for the root has the form (like the Chinese numerals of ancient times):

$$\sqrt{83674} = 200 + n \times 10 + m_1 \times 1 + m_2 \times \frac{1}{10} + m_3 \times \frac{1}{100} + \cdots.$$

(n and m_i are all digits between 0 and 9). An example to clarify this: suppose we have $\sqrt{42532} \approx 206.2328 \ldots$ This is the same as

$$\sqrt{42532} = 200 + 0 \times 10 + 6 \times 1 + 2 \times \frac{1}{10} + 3 \times \frac{1}{100} + 2 \times \frac{1}{1000} + 8 \times \frac{1}{10.000} + \cdots$$

So $n = 0$; $m_1 = 6$; $m_2 = 2$; $m_3 = 3$; $m_4 = 8$, ... Finding the digits was the goal of the ancient method. If we let "$\underline{a}$" stand for the part of the root above 200 we have:

$(200 + a)^2 = 83674$ .... (1)

Expand:

$40000 + 400\boldsymbol{a} + \boldsymbol{a}^2 = 83674$ ....(2)

$400\boldsymbol{a} + \boldsymbol{a}^2 = 83674 - 40000 = 43674$ (3)

This leads to:

$400\boldsymbol{a} + \boldsymbol{a}^2 = 43674$.............. (4).

To determine the digit for the 10s the $\boldsymbol{a}$ is written as = $10n + \boldsymbol{b}$, where $0 \leq n \leq 9$. The n must be an integer and $0 \leq \boldsymbol{b} < 10$ ($\boldsymbol{b}$ is in fact the part that contains units and the irrational fraction of the square root of 83674). The use of place value and the role of the digits to get to the root are noteworthy. In the search for the digit it makes no difference if we keep the lower parts that contain the units and the fractions, or not. We just need to be aware that the equality is replaced by an inequality.

When you substitute the form $10n$ + b in equation (3) for $\underline{a}$, this follows:

$400 \text{ x } (10n + \boldsymbol{b}) + (10n + \boldsymbol{b})^2 = 43674$..................... (4).

Expanding the square, collecting like terms and separating terms with n lead to:

$4000 \text{ x n} + 100 \text{ x } n^2 + 400\boldsymbol{b} + 20\text{N}\boldsymbol{b} + \boldsymbol{b}^2 = 43674.$ .... (5).

The genius of the method is that the parts with N (integer digits) are isolated from the rest of the root. We can state:

$$400\boldsymbol{b} + 20n\boldsymbol{b} + \boldsymbol{b}^2 = 43674 - (4{,}000\text{N} + 100n \text{ x } n).$$

Probes. Because the right-hand side must be positive, and N must be one of the digits 0–9, the highest digit that makes the right-hand side positive can be found by probing digits 9–0 and determining when the smallest positive difference is obtained. We have to probe less than 9 digits!

Probe 1: Try $n = 9$; 43,674 – (4,000 x 9 – 100 x 9 x 9) = 43674 – 44100 = – 426. $n = 9$ is too high.

Probe 2: Try $n = 8$; 43674 – (4000 x 8 + 100 x 8 x 8) = 43674 – 38400 = 5274

From here the selection for the second (10) digit is complete: it must be 8. The 38400 is subtracted from the 43674 to yield 5274. This number is placed as the new "Shi." A row is inserted for the preparation for the next (units) digit calculation. After calculating the digit 8 for the 10s, it is recorded in the Shang row for results. Of course, the 8 represents 80 and therefore it goes in the second column in the Shang row.

Calculation of the units. For the calculation of the units a temporary row is added (called the Lian fa (= side divisor, later to be explained for its geometric origin) (see Figure 13.10).

10000	1000	100	10	1.	
		2	8		*Shang*
8	3	6	7	4	*Shi*
		4			*Fang fa*
			8		*Lian Fa*
		1			*Xia fa*

Figure 13.10. Inserting the Lian fa ("side divisor") for the steps of finding the unit's digit.

In the Lian fa, the same digit from the 10s (= 8 in our case) is placed in the 10s column. The Xia fa still resides in the hundreds column as a "1." The Fang fa and the Lian fa are combined as follows: 40 + (2 x 8) = 56 (Figure 13.10). The Lian fa is doubled and added to the Fang fa: the 4 and the 8 are in different place values by one space. So, the adding of 4 and 16 should be done as indicated: 40 + 16 = 56. This number is placed in the columns of the 100s and the 10s.

If we compare the procedure with equation (5) and study the $400\boldsymbol{b} + 160\boldsymbol{b}$ that will emerge after the $n = 8$ has been determined we will see that this leads to the number $400 + 160 = 560$ as the new coefficient of the $\boldsymbol{b}$ in equation (5). The $4000n + 100n^2$ has been subtracted from the equation resulting in the right-hand number of 5274. In the ancient system of calculations, this was done in two steps: first they subtract the 400 and then 160 is subtracted, making the total subtraction 560. Here we collapsed both steps into one subtraction (see Figure 13.11).

Double the 8 to 16. Then add to 40 to get 56

10000	1000	100	10	1.	
		2	8		Shang
	5	2	7	4	Shi
		=5 4	6		Fang fa
		+ 16	8		Lian Fa
		1			Xia fa

Figure 13.11. The units' calculation.

This correlates as follows to present algebraic practices with equations:

$4000 \text{ x } n + 100 \text{ x } n^2 + 400\boldsymbol{b} + 20nb + b^2$ 43674................(5). $(n = 8)$

$400b + 160b + b^2 = 5274$. Or $560b + b^2 = 5274$.

This is reflected on the counting board in Figure 13.12 that shows the unit's spot of the root still open. We see that the Shi is at 5274, the Fang fa at 56, standing for 560 and the Lian fa is at 8, standing for 80. In Figure 13.13 the Xia fa has been moved to the unit's column to indicate that the units are being determined. To select the units, the following calculations are carried out: take the 56 from the Fang fa representing 560 and add one-digit N from 9–0 to the number. For example: if we choose N = 9 we get 569. Then we multiply the number 569 by the same digit 9: 569 x 9 = 5121. We select the largest digit for which the product as described is less than or equal to 5,274. In our case the choice is $n = 9$. After selecting $n = 9$ for the last digit we can proceed to the decimals of the root (see Figure 13.14).

10000	1000	100	10	1.	
		2	8		Shang
	5	2	7	4	Shi
		5	6		Fang fa
			8		Lian Fa
		1			Xia fa

Figure 13.12. The starting point for the units' step.

10000	1000	100	10	1.	
		2	8		Shang
	5	2	7	4	Shi
		5	6		Fang fa
			8		Lian Fa
		1		1	Xia fa

The Xia fa is moved two steps to the right
to indicate the units' calculation

Figure 13.13. Board ready for the units' calculations.

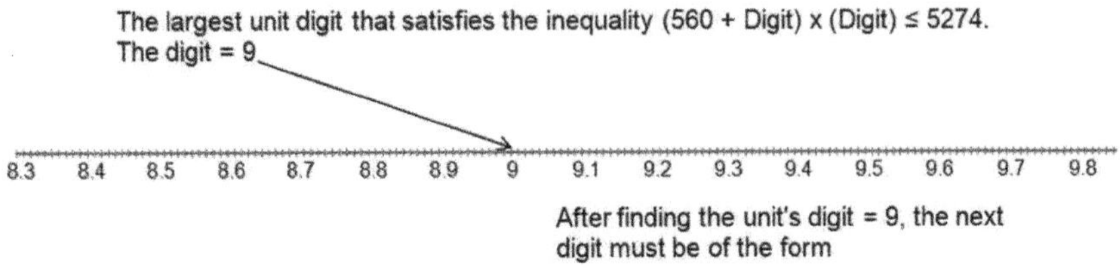

Figure 13.14. The digit 9 of the units. The next digit will be a decimal graphed in a number line.

In the ancient procedure the following steps were taken to present the best results (for irrational roots): subtract the 5121 from the 5274 in the Shi. This produces 5274–5121 = 153 (see Figure 13.15).

10000	1000	100	10	1.	
		2	8	9	*Shang*
		1	5	3	*Shi*
		5	6		*Fang fa*
				9	*Yu Fa*
				1	*Xia fa*

The The Yu fa ("corner divisor") is placed under the Fang fa. It is used to find the 153 from the Shi as described in the text

Figure 13.15. Calculation of decimals of the root. Results after the first subtraction.

The subtraction was done in two steps: First 560 x 9 = 5,040 was subtracted from the Shi and recorded on the board. The result was 234. Then 9 x 9 = 81 was subtracted from the first result of 234. The result of 153 was the new Shi and again recorded. Figure 13 shows these results on the board: The Shang is now at 289, the Shi is at 153, the Fang fa is at 56 (an intermediate number for the calculation process) and the Yu fa is at 9 (the Yu fa is the "corner divisor" and is used to calculate the last Fang fa before recording the result obtained so far.) We then proceed with the calculation by doubling the 9 from the Yu fa and adding the result to the 56

(= 560). The outcome is 578 in the Fang fa. In ancient China the calculation would often end here, so the root of the calculation would then be given as $289\frac{153}{578}$. In decimal form this number represents a root of 289.26470.... The actual root in five decimals is 289.26458.

10000	1000	100	10	1.	
		2	8	9	Shang
		1	5	3	Shi
		5	7	8	Fang fa
				1	Xia Fa

Figure 13.16. The counting board before extending the calculations to decimals.

In Figure 13.14 we see the counting board with the results at this stage before extending the calculations to decimals. We can continue the decimal calculations by extending the system into decimals hereby following the same procedures. For example: If we probe the digits n from 9–0 to find the largest digit such that $(578 + \frac{n}{10}) \times \frac{n}{10} \leq 153$, we find (starting from the digit 9):

Probe 1: 578 + 0.9 = 578.9; 578.9 x 0.9 = 521.01 which is very high compared to 153!

Probe 2: 578.5 x 0.5 = 289.25, still too high!

Probe 3: 578.2 x 0.2 = 115.64. So the selected digit is n = 2 and the root is 289.2...

What do we need to subtract from 153 in the Shi after this digit? That number turns out to be 115.64 resulting in 37.36. The next digit is calculated with the same procedure.

THE GEOMETRIC ORIGIN OF THE ROOT EXTRACTION METHOD

The Geometric Framework

Li Yan and Du Shiran (1987) and Yong and Se (2004) concluded that the root extraction method grew out of the arithmetic methods for multiplication and division using rod numerals that in turn were based on geometric reasoning. We would like to add that it is also a brilliant exploitation of the decimal notation methods to write any number with digits and powers of the number 10. We argue that by exploiting the integer form of 10, units, 10ths and so forth, the Chinese created a way to calculate the roots digit by digit.

Just like the division and multiplication, the root extraction method was originally derived using a diagram similar to the one shown in Figure 13.17. Using the counting board and rod numerals, the Chinese mathematicians were able to convert geometric quantities to numbers thereby creating a systematic method linking geometry to algebra and back. As shown in Figure 13.17, the big square ABCD has an area equal in numerical terms to the number from which to extract the root. The side of the yellow square area represents the first part, in our case the integer hundreds part. Here the numerical form comes in very handy. The hundred's part can be found easily if one knows the initial number. If we have 83,674 then this number is less than 1,000,000 and more than 10,000. This means that the root is less than 1,000 and more than 100 in other words it must have exactly three digits. There is a **100's** digit, a **10's** digit a **unit's** digit followed by **decimal** digits. In our example the hundreds digit cannot be more than 2, and it must be more than 1.

The big top left square represents the square of 200 which equals 40000. After subtracting 40000 from 83674 the remainder is 43674. In terms of our contemporary algebraic language this is what happens: If the root of 83674 is $200 + x$, then $(200 + x)^2$ must be equal to 83674. Expanding the square we get 40000 + 2 x 200 x (x^2). The 40000 and the 2 200 are clearly derived from the 2(hundred) determined in the first digit determination. The ancients found a working system to bring this to the counting board and formulate their algorithm. The x includes the 10s part of the root, the unit's part and the remaining decimal parts.

To find the 10's part of the numeral, the reasoning is as follows: Let $x = 10n + \boldsymbol{a}$, where $\boldsymbol{a} < 10$, and n is the largest digit, such that the equality $400 \times (10n + \boldsymbol{a})+(10n + \boldsymbol{a})^2 = 43674$ still holds. The idea is that there are only a limited number of choices for N, if we recall that "***a***" must be less than 10. The equality is transformed into this inequality: find the largest n such that $400 \times 10n + (10n)^2 \leq 43674$. We ignore the decimal parts of the root and determine the digit by probing:

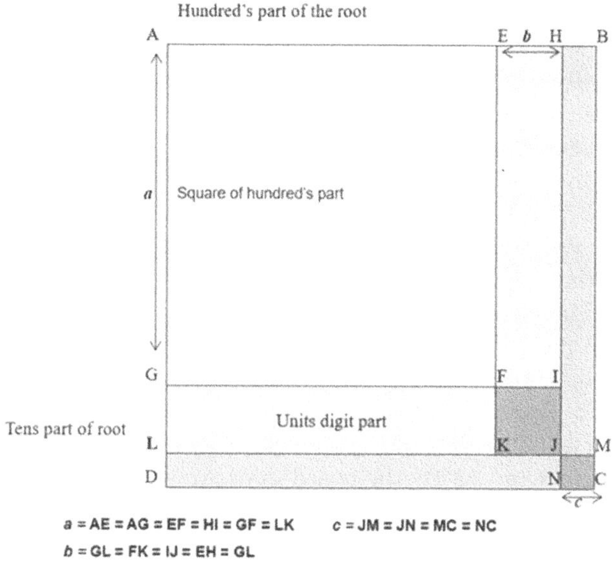

Figure 13.17. The geometric origin of the root method.

Probe 1: Choose $n = 9$. Then we have $400 \times 90 + 90^2 = 44100$. Clearly this outcome is too high.

Probe 2: Choose $n = 8$. Then $400 \times 80 + 80^2 = 38400 < 43674$, and we have found the largest digit N.

In the diagram in Figure 13.17 the gnomon-like shape EHJLGF represents (2 x 200 + 80) x 80 = 2 x 200 x 80 (rectangles EHIF and GFKL) + 80 x 80 (= small blue square FIJK). The determination of the digit $n = 8$ is the arithmetic part of the method. The proof that it is correct as a method is in the geometric part, where it is visually logical that we can subtract the gnomon—like shape from the remaining part beyond the yellow square and move closer to the actual root.

FROM THE ROOT EXTRACTION METHOD TO THE QUADRATIC EQUATION SOLVING METHOD

From an early stage—at least since the classic mathematical text dating from no later than 100 A.D., the *Nine Chapters of the Mathematical Art* (Yong & Se, 2004)—it was recognized in China that solving quadratic-type problems was similar to "completing the square" as was done in the process of calculating square roots. This suggested a solution to problems involving forms that we now write $\boldsymbol{x^2 + ax = b}$ as with a and b being positive numbers. For clarity, we will use some algebraic notation to connect with contemporary practice. The board configurations are given later. Suppose we need to solve a problem involving an equation such as $x^2 + 3x = 100$. An approximate number would be 8 because $8^2 + 3 * 8 = 88$, while $9^2 + 3 * 9 = 108$.

It became clear to the ancient Chinese mathematicians that such a step did not present any serious problem for solving this type of equations. From here the procedure for finding the root of the equation would be identical to the steps for root extraction. The number 88 would be subtracted from the 100 leaving 12. The algebraic formula would be: $(8 + a)^2 + 3(8 + a) = 100$, where a is the irrational decimal portion of the root. After expansion and simplification, we get $19a + a^2 = 12$. The next value for "a" must be between 0 and 1, because at 1 we get 20 and at 0 we get 0. In Figures 13.18 and 13.19, the board steps are given for the calculations. Setting up the numbers on board #1 we'll start from the left. Then the next coefficients are calculated step by step. Comparing ancient and modern steps, it is obvious that the ancient steps are on the counting board only, while the modern steps use algebraic symbolism with variables. We then proceed with the calculation of the second coefficient (see Figure 13.20).

To select the next digit we use the coefficients calculated in the last board: –12; 19; and 1. From here the next digit of the root, a decimal, is selected: Find the largest digit n, such that $\left(19 + \frac{n}{10}\right) \times \frac{n}{10} \leq 12$. The selected digit turns out to be $n = 6$. $\left(19 + \frac{6}{10}\right) \times \frac{6}{10} = 11.76$. We will proceed with the counting boards showing more columns since our calculations go into decimals (see Figures 13.21 and 13.22).

To find those coefficients we take the 0.6 and multiply by 1 (the first coefficient). This is then added to the 19 of the next coefficient giving 19.6. Now we multiply again 0.6 x 19.6 and add the result to – 12 to find 0.6 x 19.6 – 12 = – 0.24, our new third coefficient for the next board. The same conditions apply again to the selection of the next decimal and digit. Find the largest digit n such that $(20.2 + \frac{n}{100}) \times \frac{n}{100} \leq 0.24$. A quick check shows that $n = 1$ is the only option for n. The next coefficients will be 1, 20.22 and – 0 (see Figure 13.23).

To find the next coefficient we first take 0.011 x 1, add to 20.2, giving 20.21. Multiply again 0.01 x 20.21, add to – 0.24 and find – 0.0379. This is the third coefficient. We then multiply 0.01 again by 1 and add that to 20.21 to find 20.22. This is our next coefficient. Set the same conditions and use the expression to get the next decimal: "Find the largest such n that $\left(20.22 + \frac{n}{1000}\right) \times \frac{n}{1000} \leq 0.0379$." The selected digit is $n = 1$. On the basis of the ancient calculations the solution of the quadratic equation should be $x = 8.611\ldots$

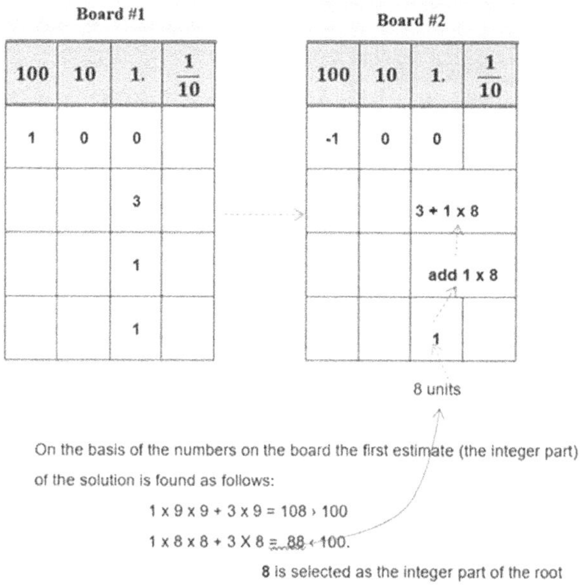

Figure 13.18. First action of the coefficients calculation for the quadratic form.

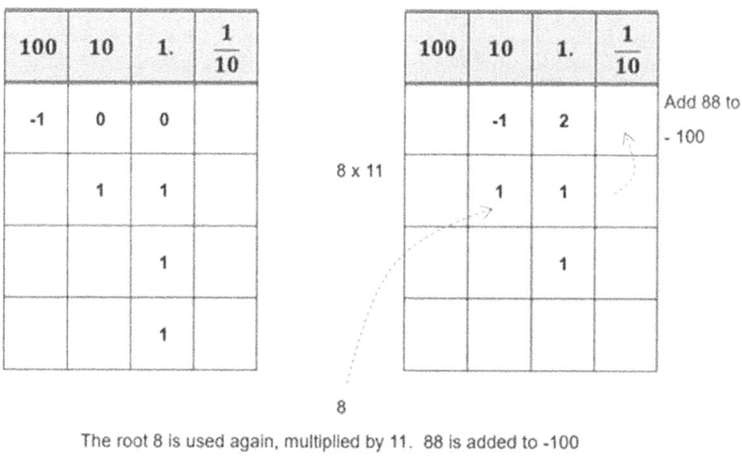

Figure 13.19. The formation of the second coefficient is complete. The formation of the third (constant) term is not yet done.

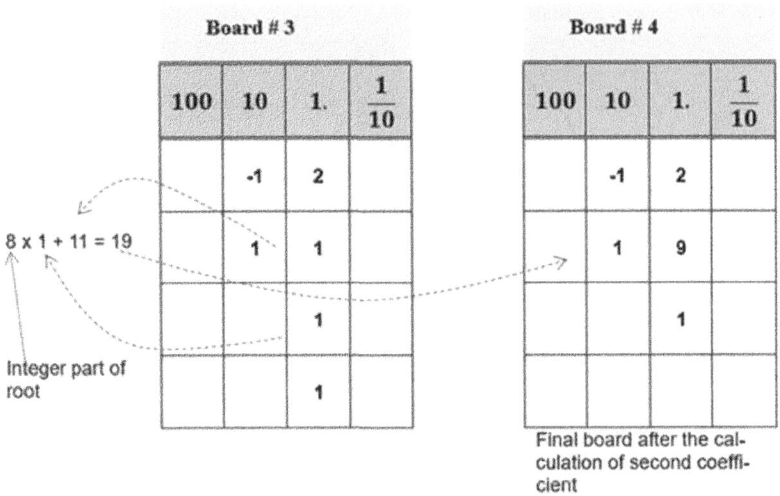

Figure 13.20. Last action of the coefficients calculation for the quadratic form.

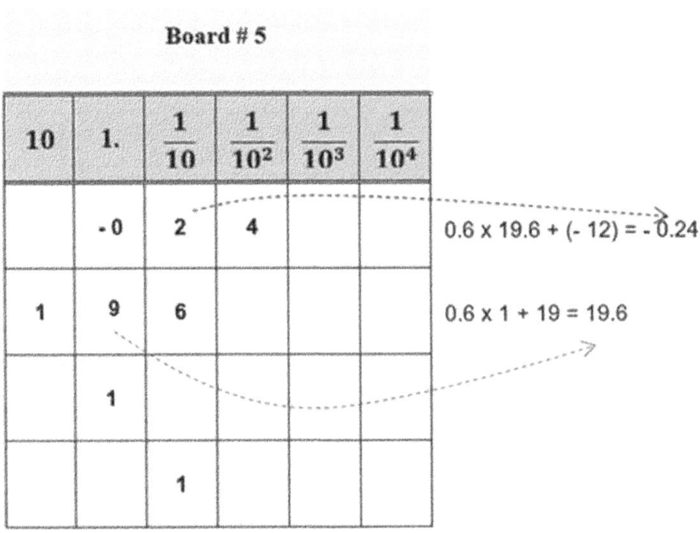

Figure 13.21. Calculation of third (constant) coefficient.

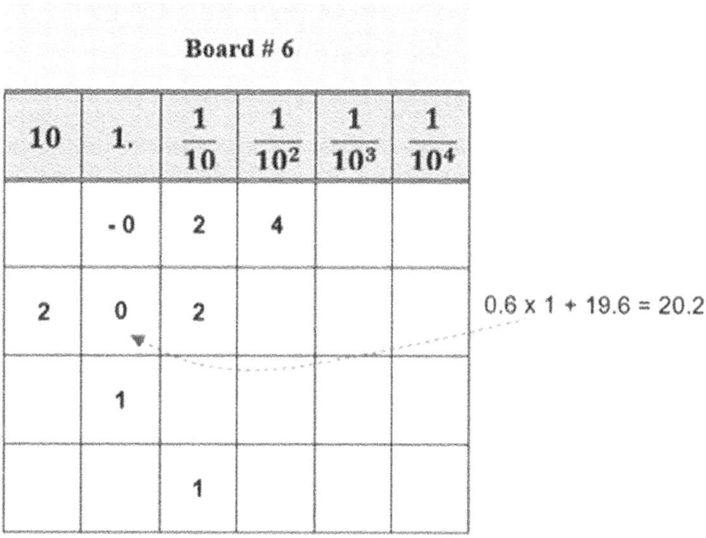

Figure 13.22. Calculation of the second coefficient for the first decimal.

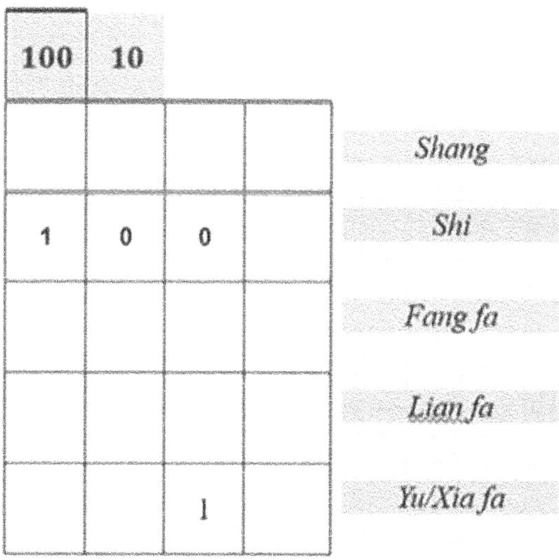

Figure 13.23. Calculation of the second coefficient for the second decimal.

THE CUBE ROOT EXTRACTION METHOD BY "ITERATED MULTIPLICATION"

According to Li Yan and Du Shiran (1987), the methods of solving cubic equations was called "the corollary to taking cube roots." This method was developed from the method to extract cube roots of any number using the rod numerals and the counting board. The methods for extracting square and cube roots was already mentioned in the *Nine Chapters on the Mathematical Art* dating from not later than 100 A.D. It also contains the method for solving quadratic equations called "corollary to taking square roots" (Yong & Se, 2004). The book, *The Mathematical Manual of Zhang Qiujian,* also contains the square root extraction method (Li Yan & Du Shiran, 1987).

We will explain the cube root extraction method with rod numerals using the number 100 from which the cube root will be extracted in two decimals. That cube root is an irrational number. There is no clear indication that the concept of irrationality existed in ancient Chinese mathematics. However, there is evidence that the Chinese mathematicians were well aware of the fact that certain roots of numbers (square or otherwise) could not be calculated exactly no matter how many digits were calculated. The third century mathematician Liu Hui commented on this when he stated:

> One cannot determine its value (i.e., the value of the root). Therefore it is only when one names it (i.e., the number) with "side" that one does not make any mistake (or that there is an error). This is analogous to the fact that, when one divides 10 by 3, to take its rest as being $\frac{1}{3}$, one is hence again able to restore (fu) its value. (Liu Hui, as cited in Chemla, 2012, p. 454)

Most of the rod numeral calculations in the literature are with integers, but it is our intention to show that if we follow the algorithm strictly, it works perfectly with decimal fractions. The cube root will be calculated to two decimals. The method we use is from the 11th century Chinese mathematician Jia Xian (Li Yan & Du Shiran, 1987). According to Li Yan & Du Shiran (1987), the method was presented by Yang Hui in the book titled *Reclassification of the Mathematical Methods in the Nine Chapters*, published in the 13th century. We will first present this method and explain it in modern algebraic language and symbols. We want to emphasize that in ancient China, symbols like "x" and exponents such as in "x^3," were unknown. The ancient Chinese were aware that roots had something that we now call "degrees." In their terminology it was called the "root extraction number" (Li Yan & Du Shiran, 1987). We will then present a method that follows more closely the actual steps taken by mathematicians in the past.

Methods Used in the 11th Century

Early in their history the ancient Chinese mathematicians had deduced that the coefficients of any binomial expansion $(a + b)^n$ could be found through what is now called the "Pascal Triangle." Therefore, from a technical standpoint the name of the diagram is incorrect, as Pascal did not invent this triangle. In China it was called "The Source of the Method of Extracting Roots" (Li & Du, 1987).

The cube root calculation. Figure 13.24 represents the initial board for the cube root calculation. We will start with a counting board with the Shi = 100 (our given number). The Shang is again our row for the results of the calculation. The "Fang" represents the (coefficient of the) square of the numbers we find through calculations; the "Lian," is the (coefficient of the) linear term. The "Yu" is the borrowing rod or corner. It reflects the place value of the result that is being calculated.

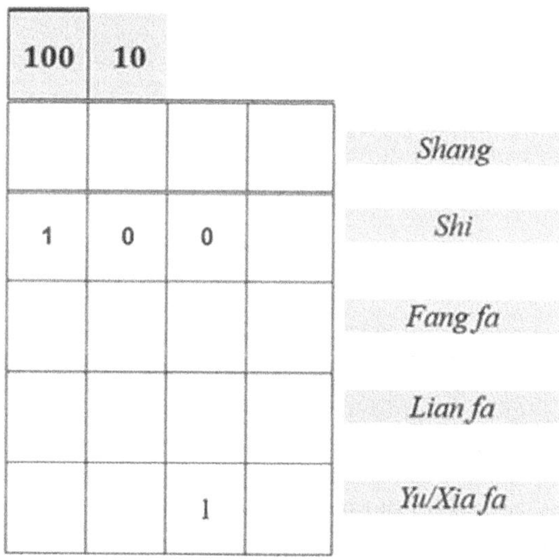

Figure 13.24. The initial board in the cube root method.

If x is the cube root of 100, then $x^3 = 100$. The integer part of the root must be 4 so we replace x by $4 + y$, with $0 < y < 1$. The "y" is the irrational part of the cube root of 100. Expanding this results in: $y^3 + 12y^2 + 48y = 100 - 64 = 36$. This is exactly our expression derived from the coefficients on the counting board below.

Actions on the board. The first digit in our case is simply found by reasoning that the cube of 5 is 125, and the cube of 4 is 64, so the real root must begin with digit 4. Place the 4 in the Lian row, then follow the steps.

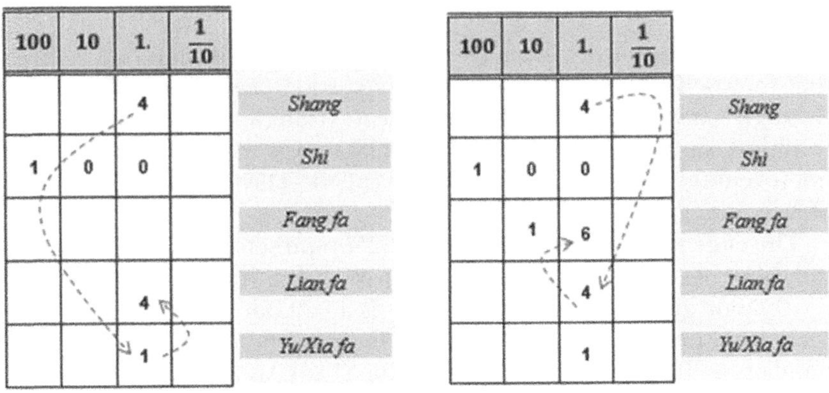

Figure 13.25. (a) **First action** on the board: Multiply 4 x 1 = 4. Add this to the empty Lian fa, (b) **Second action** on the board: Multiply 4 x 4 = 16 and add to Fang.

Connecting with modern algebraic symbolism we notice that the rows can be seen as representing the coefficients of cubic expressions with the yu/xia fa as the coefficient of the cubic term; the lian as the coefficient of the quadratic term; the fang as the coefficient of the linear term and the shi as the opposite of the constant term. In modern symbolic notations we have:

$$1 * y^3 + 12*y^2 + 48 * y - 36 = 0 \textit{ or } 1 * y^3 + 12* y^2 + 48 * y = \mathbf{36}$$

It should be noted that all the calculations we do with the machinery of variables and symbols for operations were done by the Chinese through their counting boards and their rod numerals. We argue that the Chinese had a powerful insight into the structure of their algorithms for them to be able to expand their machinery. History has shown that the evolution of these ancient tools did not end but continued well into the field of what we now call Linear Algebra, including the Gaussian methods and the early use of determinants. Hart (2011) estimated that the visual arrangement on the counting board with rectangular arrays of rod numerals inspired the notion of the matrix and the determinant, a unique method only found in China and probably later transmitted to Europe.

To find the next digit of the root, a decimal $\frac{n}{10}$, we must find the largest digit n such that

$$48 \times \left(\frac{n}{10}\right) + 12 \times \left(\frac{n}{10}\right)^2 + \left(\frac{n}{10}\right)^3 \leq 36$$

In the literature on the extraction of roots in China or the determination of roots of polynomials, not many books (with the exception of *Fleeting*

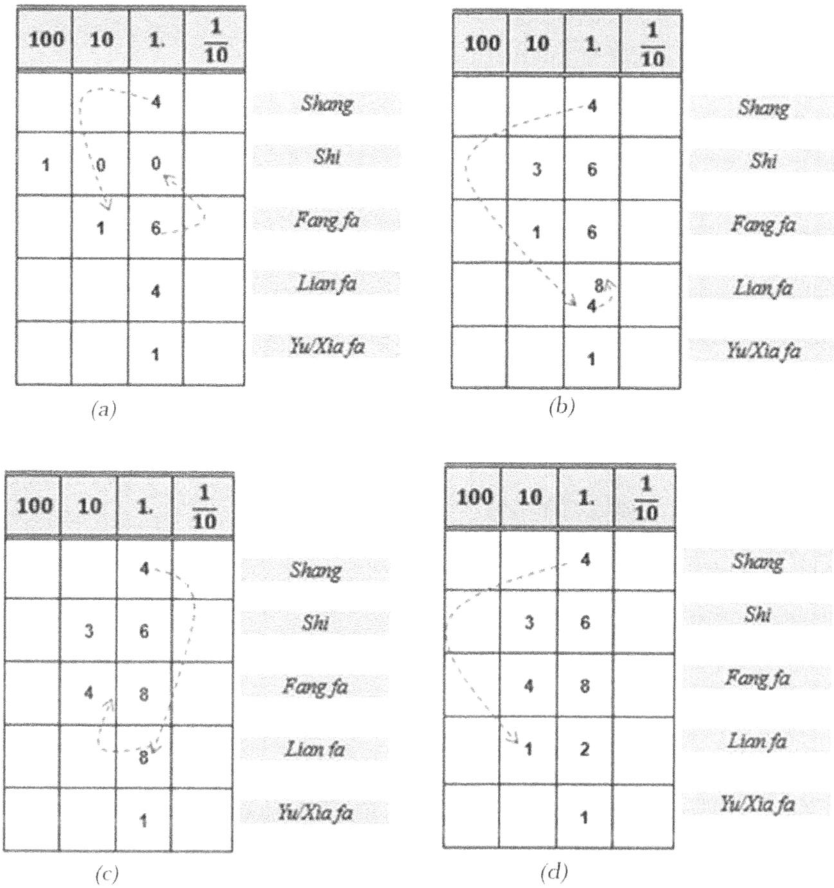

Figure 13.26. **(a) Third action:** Multiply 4 x 16 = 64 and subtract from the Shi, 100 – 64 = 36. **(b) Fourth action:** The Shi is 36 and the root digit 4 is added again to the 4 in the Lian, doubling the value to 2 x 4 = 8. The 8 stays in the Lian row. **(c) Fifth action:** Multiply the digit 4 by the 8 (= 2 × 4) to get 32 (= 2 ×4 × 4) and add the 32 to 16 from the Fang to get 48 (= 3 × 4^2). **(d)** The counting board after the sixth action: Add root digit 4 to the Lian of 8 to get a new Lian of 12 (= 2 × 4 + 4 = 3 × 4).

Footsteps by Young and Se (2004) clearly explain how this particular step is actually executed. We offer the following explanation: by probing we should first determine the exact largest digit n that will satisfy the inequality:

$$48 \times \left(\frac{n}{10}\right) + 12 \times \left(\frac{n}{10}\right)^2 + \left(\frac{n}{10}\right)^3 \leq 36 \quad \text{.................................} \quad \textbf{(1)}$$

Probe # 1: Let $n = 9$. Then $48 \times \left(\frac{9}{10}\right) + 12 \times \left(\frac{9}{10}\right)^2 + \left(\frac{9}{10}\right)^3 = 53.649 > 36.$.

This means that $n = 9$ cannot be selected as our next digit.

Probe # 2: Let $n = 8$. 46.592>36

Probe #3: Let $n = 7$. We will find in the same way for the expression (1): 39.823 > 36.

Probe# 4: Let = 6. We calculate the expression and find: 33.336 < 36!

Clearly = 6 is the largest integer that produces a value still less than or equal to 36. The $n = 6$ is selected. This result means that in one decimal, the approximation of the cube root of 100 must be 4.6 ...

100	10	1.	$\frac{1}{10}$	
		4	6	*Shang*
	3	6		*Shi*
	4	8		*Fang fa*
	1	2		*Lian fa*
			1	*Yu/Xia fa*

Figure 13.27. The counting board just before the calculation of the second decimal: the last column to the right is a decimal column with values between 0.0 and 0.9. The selected digit is 6, meaning that the actual value is 0.6.

The Concept of Fractions

The concept of fractions was mastered at a very early stage in the evolution of the Chinese Arithmetic system. Liu Hui seems to have used decimal fractions in his time to convey his calculation of π. His concept for decimal fractions developed from measures that convey units of length (Yong & Se, 2004; Li Yan & Du Shiran, 1987).

After finding the digit 6, place value 0.6, placed to the right of the digit 4 on the board, we proceed as follows: take the 6, and multiply it by the "yu" or "xia fa", with place value 0.1. Then add this 0.6 to the 12 in the "lian" row, to produce 12 + 0.6. This becomes our new lian. Again take the 6, multiply by the 1 (=0.1) in the "yu" and multiply that by (12 + 0.6) from the new lian to produce (12 + 0.6) x 0.6. We add (12 + 0.6) x 0.6 to the old fang and obtain a new fang, equal to (48 + (12 + 0.6) x 0.6). The ancients then multiplied this new fang by the 0.6 that we used twice already. The result is:

$$(48 + (12 + 0.6) \times 0.6) \times 0.6 = 3.336 \ldots\ldots (1)$$

This is subtracted from the 36 which represents the last "*shi.*" The difference is 2.664. This will be our new "*shi.*" In modern algebraic notation: The initial *shi* = 100. The first digit must be 4 units because $4^3 = 64$ and $5^3 = 125$. Therefore the cube root of 100 is between 4 and 5.

Let $\sqrt[3]{100} = 4 + \boldsymbol{a}$. Then $(4 + \boldsymbol{a})^3 = 100$. After expansion of the cubic form we find

$$(4 + \boldsymbol{a})^3 = 4^3 + 3 \times 4^2 \times \boldsymbol{a} + 3 \times 4 \times \boldsymbol{a}^2 + \boldsymbol{a}^3 = 100$$

We know that $\boldsymbol{a}$ will be a decimal of the form $\boldsymbol{a} = \frac{n}{10} + b$, where n takes on the digit values of the Base-10 system. This corresponds completely with the Chinese number words and numerals. So n is an integer from the set {0, 1, 2, 3, 4, 5, 6, 7, 8, 9}, and $0 \leq b < \frac{1}{10}$. In other words, $\boldsymbol{b}$ is the remaining fractional part of the irrational portion of the cube root of 100. After subtracting 64 from 100 (on the board) 36 remains. Now we try to find the largest digit N such that the value of

$$3 \times 4^2 \times \left(\frac{n}{10}\right) + 3 \times 4 \times \left(\frac{n}{10}\right)^2 + \left(\frac{n}{10}\right)^3 \leq 100 - 64 = 36$$

Algebraically this means that we do not calculate the exact and complete fractional part of the root, but only the digit that corresponds to the 10th place value. The equality has become an inequality because we should not expect the rational fraction to be the exact solution. The form allows us to do "probes" using the digits from 0 – 9, to determine which one is the largest digit n that still gives us a value less than or equal to the 36. $n = 6$ is our choice. In symbolic language we have the root now as $4.6 + y$. We try to find the next digit through the discrete numbers $\frac{m}{100}$ with m between 0 and 9.

The contemporary algebra of the procedure. First we found the unit's digit $\boldsymbol{a}$ of the root. Then we found the 10th digit $\boldsymbol{b}$. To move to the next

digit we assume $(\boldsymbol{a} + \boldsymbol{b} + y)^3 = \text{N} = 100$, where $y = \frac{m}{100}$ + (irrational decimals), where m is a digit for the second decimal.
Expanding the expression above we find:

$$(\boldsymbol{a} + \boldsymbol{b})^3 + 3 * (\boldsymbol{a} + \boldsymbol{b})^{2*} y + 3 * (\boldsymbol{a} + \boldsymbol{b})^* y^2 + y^3 = n$$

After subtracting $(\boldsymbol{a} + \boldsymbol{b})^3$ from n we have

$$3 * (\boldsymbol{a} + \boldsymbol{b})^2 * y + 3 * (\boldsymbol{a} + \boldsymbol{b}) * y^2 + y^3 = n - (\boldsymbol{a} + \boldsymbol{b})^3$$

The fraction $y = \frac{m}{100}$ + (irrational decimal) is such m that is the largest digit that fits into the y part of the cube root of 100. If we ignore the irrational part lower than $\frac{1}{100}$ in the fraction we can search for the largest digit (between 0 – 9) such that:

$$3 * (\boldsymbol{a} + \boldsymbol{b})^2 \times \left(\frac{m}{100}\right) + 3 \times (\boldsymbol{a} + \boldsymbol{b}) \times \left(\frac{m}{100}\right)^2 + \left(\frac{m}{100}\right)^3 \leq n - (\boldsymbol{a} + \boldsymbol{b})^3$$

After substituting $\boldsymbol{a} = 4$, $\boldsymbol{b} = 0.6$, and $100 - (4 + 0.6)^3 = 2.664$, we do our probes:

Probe #1: $m = 9$: $3 * (4.6)^2 * 0.09 {}_{+} 3 * 4.6 * 0.09^2 + 0.09^3 = 5.825709.$

This is more than 2.664. The 9 is rejected.

Probe # 2: $m = 8$. $3 * 4.6^2 * 0.08 + 3 * 4.6 * 0.08^2 + 0.08^3 = 5.167232.$

Again over the limit of 2.664. The 8 is rejected.

Probe # 6: $m = 4$. $3 * 4.6^2 * 0.04 + 3 * 4.6 * 0.04^2 + 0.04^3 = 2.56128.$

With this we have the digit closest to the 2.664 we calculated so far. In two decimals the cube root of 100 must be 4.64...

THE CALCULATION OF ROOTS OF "POLYNOMIAL EQUATIONS OF HIGHER DEGREE"

After developing and understanding the cube roots and higher roots the scholars in China found a way to extend the method to finding roots of any problem that involves, in contemporary terms, equations of higher degrees than two. In a sense we can speak freely of "equations" and "polynomials" because the system created by the scholars may not look like what is current today, but it functioned just as if they had a code that represented similar structures and allowed mathematicians to do anything they wanted to do.

Ulrich Libbrecht writes in his study "Chinese Mathematics in the thirteenth Century" in relation to the notion of equations in China that "the equality was not used; for although it was very important in the development of algebra in Europe, it was of no use when the operations were performed on the counting board. The equal sign did not exist, but this lack was compensated for by the method itself" (Libbrecht, 1973, p. 203).

Suppose we want to solve numerically the following equation:

$$2x^5 + x^4 + 3x^3 + 5x^2 + 3x = 20$$

The system requires that we find a first (integer) part of the root by simple substitution: It is obvious that at $x = 1$ the value of the left part of the equation is 14 (below 20), while for $x = 2$ we have 124, greater than 20, so it is reasonable to assume that the root's integer part is 1. So let the root be $x = 1 + a$, where is the decimal part of the root, a fraction less than 1. We will first describe the algebra of the initial step and then proceed with the methods of solving the polynomial (Lam Lay Young, 1980; Li & Du,1987).

Our general fifth degree equation is:

$$a_0x^5 + a_1x^4 + a_2x^3 + a_3x^2 + a_4x + a_5 = 0$$

So $a_0 = 2$; $a_1 = 1$; $a_2 = 3$; $a_3 = 5$; $a_4 = 3$; and $a_5 = -20$ (negative numbers were quite well known and accepted in China centuries before other cultures.) The first integer part of the root will be designated as $\boldsymbol{h}$ in our algebraic translation. So the root is $x = \boldsymbol{h} + y$. Plug this back into the polynomial. This substitution will result in new coefficients of a new fifth degree polynomial. The whole operation after this is aimed at finding those new coefficients by step-by-step calculations. The separate substitution terms are stated and expanded as shown in Figure 13.28.

After we collect the coefficients of like terms we can find per term the new coefficient as follows: First, we put all constant terms together and do the same for all linear terms, for all quadratic terms, and so forth. We then order the terms starting with the constant term, then the linear term, then the quadratic term, and so forth, just like the ancient Chinese did.

We start with the constant term. This term is in fact the value of the polynomial when we use the first integer part h for the unknown. We will get the "triangle" shown in Figure 13.29.

Figure 13.29 shows how new coefficients are generated from a given set of coefficients that the Ancient Chinese calculated with their methods, going from one digit of the solution to the next digit. The new equation is essentially a statement about the remaining (irrational) part (= $\underline{\mathbf{y}}$) of the root of the polynomial form. Its coefficients are determined by the value of $\underline{\mathbf{h}}$. The Chinese system calculated these values in its own way and every

$$a_0(h+y)^5 = a_0h^5 + 5a_0h^4y + 10a_0h^3y^2 + 10a_0h^2y^3 + 5a_0\,hy^4 + a_0y^5$$

$$a_1(h+y)^4 = a_1h^4 + 4a_1h^3y + 6a_1h^2y^2 + 4a_1hy^3 + a_1y^4$$

$$a_2(h+y)^3 = a_2h^3 + 3a_2h^2y + 3a_2hy^2 + a_2y^3$$

$$a_3(h+y)^2 = a_3h^2 + 2a_3hy + a_3y^2$$

$$a_4(h+y)^1 = a_4h + a_4y$$

$$a_5 = a_5$$

Figure 13.28. The algebra associated with the Chinese system: the expansion of each term per row before collecting the similar terms.

Constant term:	$a_0h^5 + a_1\,h^4 + a_2\,h^3 + a_3\,h^2 + a_4h + a_5$
Coefficient of y^1 :	$5a_0h^4 + 4a_1h^3 + 3a_2h^2 + 2a_3h + a_4$
Coefficient of y^2 :	$10a_0h^3 + 6a_1h^2 + 3a_2h^1 + a_3$
Coefficient of y^3 :	$10a_0h^2 + 4a_1h + a_2$
Coefficient of y^4 :	$5a_0h + 1a_1$
Coefficient of y^5 :	$1a_0$

Figure 13.29. The algebra of the calculation of new coefficients.

stage laid the foundation for finding the coefficients for the next equation. The rows are then ready to be used to find the next digit of the root. The numbers of the "Pascal Triangle" are visible. This shows that, after iterated multiplication to find the roots of high degree polynomial, the rows on the counting board could be found quickly if the person computing the root knew the "source." Even the addition of the coefficients had a clear order that one could memorize easily to proceed quickly. The constant term is the polynomial calculated for first root part **h**.

If one takes a close look at the coefficients in the diagram in Figure 13.27 one may recognize the beginning of "the source of extraction of roots" or "Pascal Triangle." Those coefficients are listed in Figure 13.30.

1	1	1	1	1	1	**Constant term**
5	4	3	2	1		**Coefficient of y^1**
10	6	3	1			**Coefficient of y^2**
10	4	1				**Coefficient of y^3**

Figure 13.30. If we drop all variables (that did not exist in China) and all plus-signs we will find the "Source of Extraction of Roots" or Pascal's Triangle.

Solving a fifth degree equation. We now return to the ancient way of solving a fifth degree polynomial equation using rod numerals. Our equation is still

$$2x^5 + x^4 + 3x^3 + 5x^2 + 3x = 20$$

Step 1. Place all coefficients on the board (see Figure 13.31): Take the first integer digit (1) of the root we found above and multiply this by the 2 in the *Xia Fa* row in Figure 13.31. Add the result to the 1 in the next row (*Lian fa*). Then multiply the (2 + 1) again by digit 1 and add to the 3 in the next *Fang fa* row, then we have 6 there. Again take 1 and multiply by 6 and add to 5 from the next row, find 11. Multiply 11 by 1 again and add to 3 in the next higher row, find 14. Multiply 14 x 1 and add to –20 and find –6. If this number was zero, we would stop and x = 1 would be the solution. Because it is not zero we continue the process of calculation.

Step 2. We continue the process with Step 2 as shown in 13.32. Remember if the last number (= 14 – 20 = – 6) is not zero then the calculation continues. The next step is the calculation of the coefficients in the second, third, fourth and fifth rows. We multiply h = 1 by coefficient 2. The result is added to the next higher row to get 2 x 1 + 1 = 3. This multiplication and addition process is continued. Row six (–6) is not changed in the process (see Figure 13.33).

This is also visible in the algebraic expressions from the new polynomial (see Figure 13.33).

Coefficient	
-20	
3	*Shang*
5	*Shi*
3	*Fang fa*
1	*Lian fa*
2	*Yu/Xia fa*

Figure 13.31. Solving a fifth degree equation with the counting board.

	Coeff.	
	-6	
- 20 + 14 = - 6		
	14	*Shang*
11 x 1 + 3 = 14		
	11	*Shi*
6 x 1 + 5 = 11		
	6	*Fang fa*
3 x 1 + 3 = 6		
	3	*Lian fa*
2 x 1 + 1 = 3		
	2	*Yu/Xia fa*

Figure 13.32. Step 2 on the counting board.

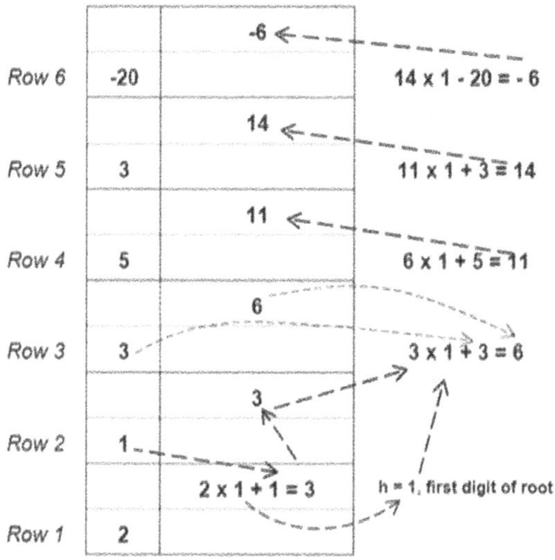

Figure 13.33. The column and row calculation process.

Figure 13.34 shows how the ancient Chinese mathematicians executed their row calculations when solving a fifth degree polynomial equation. The first column on the left contains the coefficients of the original problem. We use the word "problem" because our equations were not yet part of the ancient world of Chinese mathematics. In Figure 13.35 we show the diagrams with the numbers for all the steps in the calculation of the next set of coefficients for finding the next digit of the root of the fifth degree equation.

Step 3. The next step is usually not explained in detail. Through probes we find the next (decimal) digit. Algebraically we need to find a number $\frac{n}{10}$ with $0 < n \leq 9$, such that

$$2\left(\frac{n}{10}\right)^5 + 11\left(\frac{n}{10}\right)^4 + 27\left(\frac{n}{10}\right)^3 + 40\left(\frac{n}{10}\right)^2 + 36\left(\frac{n}{10}\right)^1 \leq 6$$

Probe # 1: $n = 9$ ....98.881 much too high. $n = 9$ is rejected.

Probe # 2: $n = 8$..........................73.385. Rejected..........

We probe and find that $n = 1$ gives the only value under 6:

Probe # 9: $n = 1$.......................... 4.02812. This means that the root is 1.1 in one decimal accuracy.

	- 6	No change to this number
New coeff →	36	22 x 1 + 14
	22	= 11 x 1 + 11
	11	= 5 x 1 + 6
	5	= 2 x 1 + 3
	2	

	- 6	No change to this number
	36	No change to his number
New coeff →	40	= 18 x 1 + 22
	18	= 7 x 1 + 11
	7	= 2 x 1 + 5
	2	

	- 6	No change to this number
	36	No change to this number
	40	No change to this number
New coeff →	27	= 9 x 1 + 18
	9	= 2 x 1 + 7
	2	

	=6	No change to this number
	36	No change to this number
	40	No change to this number
	27	No change to this number
New coeff →	11	= 2 x 1 + 9
	2	

Figure 13.34. Steps in the calculation of the next set of coefficients for finding the next digit of the root of the fifth degree equation.

For the rows and columns, it means that we take this value of 4.081 and add it to the –6 in the first row of the counting board to find –1.97188. The next board configurations are shown in 13.35. Use the digit 0.1 and multiply by 2 then add to 11.2 to find 11.4. Repeat again: 0.1 x 11.4 + 28.12 = 29.26. Repeat again, then stop and start again at 0.1 x 2 ... and stop when the last column has been reached. Now we can move to find the next digit.

-1.97188	No change to this number
36	
22	
11	
5	
2	

-1.97188	No change to this number
40.2812	42.812 x 0.1 + 36
42.812	28.12 x 0.1 + 40
28.12	= 11.2 x 0.1 + 27
11.2	= 2 x 0.1 +11
2	

- 1.97188	No change to this number
40.2812	No change to this number
45.738	Multiply and add
29.26	Multiply and add
11.4	Similar method
2	

- 1.97188
40.2812
45.738
30.42
11.6
2

- 1.91882
40.2812
45.738
30.42
11.6
2

Figure 13.35.

Based on the last column, the next digit is the largest (digit) number m, such that

$$2 \times \left(\frac{m}{100}\right)^5 + 11.8 \times \left(\frac{m}{100}\right)^4 + 30.42 \times \left(\frac{m}{100}\right)^3 + 45.738 \times \left(\frac{m}{100}\right)^2 + 40.2812 \times \left(\frac{m}{100}\right) \leq 1.91882$$

Probe # 1: $m = 9$.... Value = 4.0187 too high, so $m = 9$ is rejected.

Probe # 2: $m = 8$.... Value = 3.5313 too high, $m = 8$ is rejected................

Probe # 5: $m = 5$..... Value = 2.1323, still too high, $m = 5$ is rejected.

Probe # 6: $m = 4$..... Value = 1.6864, $m = 4$ is selected as the next digit.

In two decimals the solution to the 5th degree problem is $x = 1.14$...

When we plug in $x = 1.14$ in the fifth degree polynomial we find

$$2(1.14)^5 + (1.14)^4 + 3(1.14)^3 + 5(1.14)^2 + 3(1.14) - 20 = -0.097\ldots$$

Development of Chinese Solutions of Polynomial Equations

The diagram in Figure 13.36 shows the conceptual connections between various aspects of the Chinese development of solution methods for polynomial equations, including the linear algebra branch in relation to tools and systems indigenous to China. The method invented by the ancient Chinese was duplicated centuries later by the mathematicians Ruffini and Horner in the 19th century in Italy and Britain. There is a continuous development from the written language to the counting board with rods to represent numerals. The system was based on powers of 10. Since the classic, *Nine Chapters on the mathematical Art*, we know for sure that all the algorithms for arithmetic were in place, including extractions of square and cube roots. For the Chinese mathematicians there was no division between "algebra" and "geometry." The geometric insights into areas and volumes seemed to have inspired the reasoning around root extractions. The study of counting board configurations also seemed to have inspired extensions of root calculations for any quadratic equation. Pascal's triangle was another fruit from the study of counting board solving of "equations." The same counting board studies also influenced the creation of determinant like methods for solving systems of linear equations.

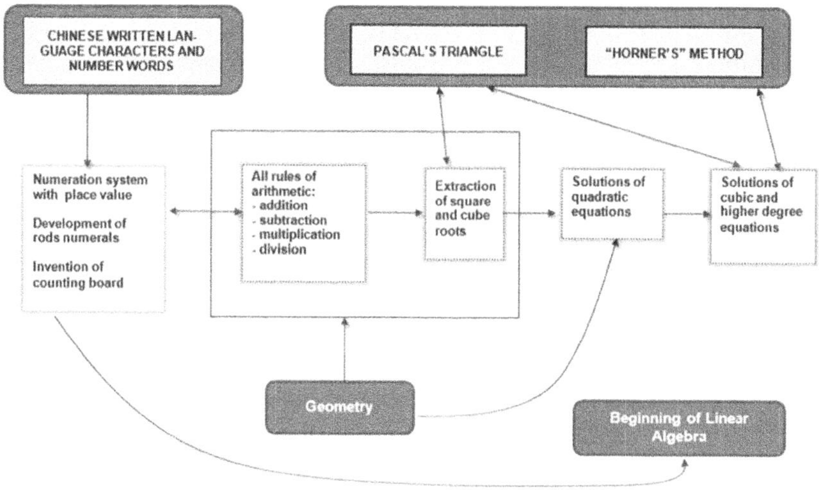

Figure 13.36. Evolution of Chinese concepts and tools for solving polynomial equations.

ASSESSMENT OF THE ANCIENT CHINESE SOLUTION OF HIGHER DEGREE PROBLEMS

Wang and Needham (1955) proved that the method invented by the ancient Chinese was duplicated centuries later by the mathematicians Ruffini and Horner in the 19th century in Italy and Britain. For a good description of the method we refer to Horner's (1819) publication and Alex Pathan and Tony Collyer's (2003) modern expose. Since the 18th century, calculus was the standard technique to analyze polynomials. The ancients did the same without derivatives or algebra. As with Pascal, the method still bears the name of the late creators. What makes the original system special is the extraordinary power of generalization that was demonstrated by the ancient Chinese in the formulation of the method. Not only did they invent the Triangle, broke through the barrier of quadratic forms, expanded their method to cover all types of high degree equations, all types of coefficients, positive and negative, they also proposed and named the technique of conceptualizing the unknown as a separate mathematical object in their system (Li Yan & Du Shiran, 1987; Pathan & Collyer, 2003). The same extraordinary power of generalization was again demonstrated when determinant-like methods were invented centuries before other cultures (Hart, 2011). Regarding a possible transmission of Chinese knowledge to Europe, Joseph (2011) points out that when Jesuit scholar Matteo Ricci was sent to China by the Roman Catholic Church in

1582 to extend its influence to Asia, one of his explicit instructions was to "gather information on scientific matters from the East." Hart (2011) suggests in his recent study, *The Chinese Roots of Linear Algebra*, that a book that was presented to the Chinese court by the Jesuits as a Chinese translation of the European "Epitome Arithmeticae Practicae" dating from 1583, by the German Jesuit scholar Christopher Clavius contains text that cannot be found in the European book. The translated book contained solutions to the problem of solving systems of linear equations in unknowns, which were completely unknown in Europe at that time. Hart suggests that the Clavius book does not contain any of those problems. In his quest to find the Chinese sources of the methods that were quite unjustly presented as knowledge developed in Europe, he eventually uncovered the determinant—like methods that were used in China in those days.

According to Li Yan and Du Shiran. (1987), the ancient mathematician known as Master Chen, commented in *Zhoubi Suanji* (a book on astronomy and mathematics from 100 B.C.–100 A.D. and still available) on how to learn mathematics and grasp the principal idea.

Master Chen said:

> The method of learning: after you have learnt something beware that what you have learnt is not wide and after you have learnt widely, beware that you have not specialized enough. After specializing you should worry lest you do not have the ability to generalize.

These words describe the power of mathematical thinking and the specific Chinese insistence on generalization, which according to Chemla (2003) comes in the Chinese mathematical world before abstraction.

REFERENCES

Chemla, K. (2003). Generality before abstraction: The general expressed in terms of the paradigmatic in mathematics in ancient China. *Science in Context, 16*, 413–458.

Chemla, K. (2012). Reading proofs in Chinese commentaries. In. K. Chemla (Ed.), *The history of mathematical proof in ancient traditions.* (pp. 423–486). Cambridge University Press.

Hart, R. (2011). *The Chines roots of linear algebra.* The John Hopkins University Press.

Horner, W. G. (1819). A new method of solving numerical equations of all orders, by continuous approximation, *Philosophical Transactions of the Royal Society of London, 109*, 308–335.

Joseph, G. G. (2011). *The crest of the peacock. Non-european roots of mathematics.* Princeton University Press.

Yong, L. L. (1980). The Chinese connection between the Pascal Triangle and the solution of numerical equations of any degree. *Historia Mathematica, 7*, 407–424.

Yong, L. L., & Se, A. T. (2004). *Fleeting footsteps* (Revised ed.). World.

Libbrecht, U. (/2005). *Chinese mathematics in the thirteenth century. The Shu-shu chiu-chang of Ch'in Chiu-shao.* Dover. (Original work published 1973)

Li Yan, & Du Shiran. (1987). *Chinese mathematics: A concise history* (J. Crosley & A. Lu, Trans.). Clarendon Press.

Merzbach, U., & Boyer, C. (2011): *A history of mathematics* (3rd ed.). John Wiley & Sons.

Netz, R. (2004). *The transformation of mathematics in the early Mediterranean world: From problems to equations.* Cambridge University Press.

Pathan, A., & Collyer, T. (2003). The wonder of Horner's method. *The Mathematical Gazette, 87*, 230–242.

Plofker, K. (2009). *Mathematics in India.* Princeton University Press.

Wang, L., & Needham, J. (1955). Horner's method in Chinese mathematics: Its origins in the root extraction procedures of the Han dynasty. *T'oung Pao, 43*, 345–401.

CHAPTER 14

MATHEMATICS TEACHERS' PROFESSIONAL DEVELOPMENT

The Case of the Indigenous Game of NCUVA and the Common Fraction Concept

Marthie Van der Walt, Erika Potgieter, and Divan Jagals

ABSTRACT

This chapter offers two phases of research to narrow the gap between teachers' professional development in South Africa and the affordances of an indigenous game that demonstrates the common fraction concept. First, the conceptualization and development of a case of the indigenous game of Ncuva is introduced to teach the concept of fractions in a rural school as a professional development tool. Second, this case is implemented within a short learning program (SLP) in ethnomathematics to describe and interpret teachers' perceptions of the ethnomathematical approach. We applied cultural historical activity theory (CHAT) as overarching theoretical framework of the conceptualization and development of the case study in the first

Evidence-Based Inquiries in Ethno-STEM Research: Investigations in Knowledge Systems Across Disciplines and Transcultural Settings, pp. 249–278

phase of the research. Data from previous research projects were used in the development of the case. The data collection techniques for the second phase of the research included observations, field notes and a focus group interview. Saldana's (2009) coding technique was used to identify codes, categories and emerging meaningful themes from the data.

The results include a case on infusing the game of Ncuva in the teaching-learning of fractions and the teachers' perceptions and ideas of the ethnomathematical approach in the mathematics classroom. Findings of the case within a SLP indicate that some of the participants have resistance to this new "model" of teaching mathematics, maybe because they have fixed ideas about mathematics and its teaching. Furthermore, an aspect highlighted by some of the participants was the opportunity they had to relate an ethnomathematical approach to their own teaching (mostly teacher centered) practices. These results confirm that an ethnomathematical approach as methodology for continuing professional development education enables teachers to look at themselves and at their professional performances, with a view to more effective actions.

INTRODUCTION

South Africa's education system has numerous gaps that need to be addressed. Students and teachers in South Africa face several constraints that affect good performance in the STEM subjects (Centre for Development and Enterprise [CDE], 2014). One of the factors listed in the CDE report is inadequate teacher education resulting in underqualified teachers. CDE, in their 2011 report, *Value in the Classroom: The Quantity and Quality of South Africa's Teachers*, makes it clear that our teachers are often poorly trained. Teachers' underdeveloped pedagogical content knowledge is one of the major reasons for underperformance in the science arena.

International research shows that teachers are the main driver of the variation in student achievement in schools (Barber & Mourshed, 2007). With regard to "teacher quality" that is an important component in the pursuit of improved student performance, both the South African Department of Basic Education (DBE) and Department of Higher Education and Training (DHET) confirmed that one of the challenges they face, is a relative shortage of qualified teachers competent enough to teach specific subjects or learning areas including mathematics (CDE, 2014). In partnership with higher education institutions, the DBE developed a series of 16-credit short learning programs offered in a variety of topics of school subjects/curriculum to train teachers on best practices for teaching.

The South African Mathematics Curriculum

The South African education system has been undergoing extensive restructuring since 1994 to improve the poor performance of South African students, particularly in STEM education. It was also envisaged that this new curriculum based on principles that incorporate student centeredness, formative assessment, integration and critical thinking, would reflect the values and principles of the new democracy in South Africa (Chisholm, 2003).

Although the Curriculum Assessment Policy Statement (CAPS) of South Africa clearly states that indigenous knowledge systems (IKS) should be acknowledged with regards to the rich history and heritage it encompasses in the country (South Africa. Department of Basic Education [SADBE], 2012), teachers are often un/(under)prepared with regards to teaching-learning practices that are informed by the students' cultural practices, beliefs, values and attitudes. While the value of indigenous knowledge (IK) in education has been recognized, this recognition is yet to translate into practical mathematics curriculum processes (Luitel, 2013).

While the CAPS Grades R–12 is sensitive to issues of diversity such as valuing IKS (SADBE, 2012), "real world" problem solving is often neglected by Westernized curricula. The Westernized curriculum materialized due to the needs posed by a globalized world and the need to equip 21st century students with workplace skills for jobs that does not even exist yet in order to prepare them for demanding futures (Davidson, 2012; Smith & Morgan, 2016). Nonetheless, studies have shown that mathematics should be made more relevant, accessible and meaningful for students and enable students to apply their mathematical knowledge in the "real world" solving "real world problems." Teaching mathematics only in a western, decontextualized and abstract way neglects the need students have to bridge the gap between content and real life applications, in order to make mathematics meaningfully understandable.

Current trends in education include an ethnomathematics approach to the teaching and learning in mathematics that offers academic knowledge and skills contextualized "within the lived experiences and frames of reference" of students (Gay, 2000). It leads to more personally meaningful learning experiences and more readily and thoroughly learned mathematics (Maxwell & Chahine, 2013) and reflects the current movement of holistic pedagogy. Arguably, culturally relevant activities engage students in a "hands-on, minds-on" manner in order for meaningful learning to take place, which enriches students' learning experiences. However, teachers need to be equipped with knowledge and skills to address needs posed by a globalized world (Rhodes, 2015).

In what follows, we discuss an ethnomathematics approach to the teaching and learning of mathematics. The aim of the first phase of this study is to conceptualize and develop a case (contextualizing the mathematics concept of common fractions within the indigenous game of Ncuva, for Grade 6 students at a rural school) in Phase 1. The aim of the second phase is to present this case within a professional development intervention, a SLP in order to negotiate a broader understanding of the role of contextualizing of mathematics for South African students. Furthermore, we describe and interpret mathematics teachers' perceptions using an ethnomathematics approach. This leads to the affordances of using cases as a professional development tool. The research methodology firstly includes the conceptualizing and development of a case as a professional development tool and second, the implementation of the case within a SLP in ethnomathematics to describe and interpret teachers' perceptions of contextualizing the fraction concept in the indigenous game of Ncuva.

THE NEED FOR PROFESSIONAL DEVELOPMENT THROUGH CASE PEDAGOGY

In the developing world, it is estimated that approximately 50% of primary school students are educated in rural schools. As is also the case in South Africa, rural schools, often scattered and remote, require specific pedagogical approaches offering real-life and inclusive teaching and learning opportunities for its students (OECD, 2012). These specific pedagogical approaches are often acquired by teachers through professional development programs to improve quality of instruction.

Professional Development

Hunzicker (2011) argues that effective professional development engages teachers in teaching and learning activities, supports their pedagogical content knowledge, embraces the nature of teaching as professional career, is focussed on increasing instructional effectiveness, and is a collaborative and ongoing process. Therefore, professional development is holistic and improves the overall teaching and learning practice.

In South Africa, teachers and students from rural backgrounds are often exposed to a Westernized curriculum, restricting them from making connections between school mathematics curriculum and their everyday experiences. Elmore (1996) asserts that the teaching praxis depends on how teachers understand the nature of mathematical knowledge, how they understand the students' role in learning (mathematics) and how these understandings manifest in their classrooms. However, many South African teachers view mathematics as a discrete set of abstract facts that are disconnected and fragmented. These teachers are also accustomed to trying to address questions from students like "Why do I need to know this?" or "When am I going to use this?" Because students are faced with a Westernized curriculum, they are often restricted in making connections between school mathematics curriculum and their everyday experiences.

An ethnomathematical approach provides opportunity for indigenous games in the mathematics classroom, particularly if the game is inspired by the cultural context (Nkopodi & Mosimege, 2009). In South Africa, indigenous games like *Ncuva* illustrates the role of ethnomathematics in engaging students in mathematical ideas that are inherent in cultural practices. The nature of the construct of indigenous games suggests that while playing the games, students will engage in hands-on activities that reveal related mathematical concepts and processes. As facilitators, teachers can then use relevant mathematical knowledge to translate the students' ideas into meaningful mathematical explorations. Such a problem-solving endeavor would only be possible when a platform is created in which students' experiences outside the classroom is connected to the mathematical knowledge introduced in the classroom.

Contextualizing mathematics learning experiences is a means of presenting mathematics in a way that builds on students out of school experiences—students develop their mathematical language and strengthen their skills and knowledge in Western mathematics (Frigo, 1999). The implementation of a contextualized approach in the mathematics classroom would make mathematics more relevant and accessible for South African students. Furthermore, it provides the opportunity to give all South African students epistemological access to good education and provides affordances to learning high level mathematics (SADBE, 2012).

Gavarette (2015) found in her study that incorporating a sociocultural component in the curriculum, acknowledges the importance of the affective dimensions in mathematics teaching. Mathematical ideas are deliberately embedded in indigenous knowledge/real life contexts that provide a sense of purpose and relevance for learning mathematics in terms of the development of Western mathematical understanding (Ainley et al., 2006). Mathematization is the reverse process in which real contextual problems are expressed mathematically. In this sense, mathematics becomes the vehicle for solving real problems and it shows how mathematics can be applied in other disciplines (Fosnot & Dolk, 2001).

Professional development programs that employ material resources and innovative approaches designed to encourage the development of skilful practitioners are critical to the success of mathematics education (Merseth, 2003). The use of cases and case-based discussions provide an important yet often missing model of learning in many teacher professional development settings. The use of cases as a professional development tool is relatively unknown to the South African education sector and necessitates a case-based pedagogy that informs new instructional approaches to facilitate meaningful learning.

Case Pedagogy

Teaching with the case method bridges the theory-practice divide (Kinsella & Pitman 2012). According to Merseth (2007), cases are a natural mode of learning about reform-oriented teaching. In case pedagogy realistic examples are drawn upon and often provided in narrative format. These are examples taken from everyday practice, and developed into a case to assist in similar or new problems that may arise in practice (Lampert, 2010). The complexities of the classroom are brought to life by highlighting the multiple dimensions of teaching (Merseth, 2007). A complex yet interesting context invites analysis due to the integration of situations in which decisions must be made or problems solved, or reconsideration of practices that will lead to meaningful consequences for teaching and learning. In addition to the main character (e.g., the mathematics teacher) there are other(s) (e.g., the principal of the school) who offer their points of view. They not only see teaching mathematics differently (e.g., teaching the concept of fractions to students from rural settings) but they may present completely different accounts of the same event (e.g., traditional teacher-centred classroom versus active learning by infusing an indigenous game and use the context to teach fractions) (Merseth, 2007).

Through the description of actual schools and classrooms, these cases also offer a window that illuminates the contextual influences on teaching and learning. Challenging instructional situations in schools often present "not one well-defined issue but many ill-defined issues, intertwined like the fibres of a thick rope" (Kleinfeld, 1988, p. 9). The purpose of cases reflects the objective of enhancing the understanding of teaching praxis by teachers or administrators (such as principals at schools) (Merseth & Lacey, 1993).

In order to develop pedagogical skills, case materials need to both stimulate and encourage the imagination of the case reader to examine and create various pedagogical responses to the situation presented in the case. These characteristics of cases require that the cases provide ample opportunity for interpretation and alternative thinking by the case reader.

By covering mathematical topics /concepts considered "hard to teach" and/or "hard to learn"—such as the fraction concept—cases encourage an in-depth exploration of challenging mathematical concepts. The cases offer opportunities to closely examine both "old" and "new" pedagogical methods, giving teachers a chance to understand for themselves the possible benefits and pitfalls of new teaching approaches. After reading a "perfect lesson," teachers have been known to comment, "Well that's great but she could never do that with my kids in my school!"

By studying a case, participants can grapple and associate with instructional challenges in classrooms particularly in a small rural community school. They can examine the instructional practices of one teacher with entirely different beliefs about the most productive ways to teach mathematics and debate the merits of each. In such contexts, participants are left with the need to make a decision, to take action or to take strong position in a typical school situation where tension needs to be resolved (e.g., between the principal and the teacher) (Merseth, 2007) and seek to encourage deeper reflection about teaching praxis (Merseth, 2003). Participants who read the cases can learn to diagnose problems, to recognize multiple influences and perspectives, and to engage in the exercise of suggesting and analyzing possible solutions and courses of action. Thus, case participants can "try out" ideas and approaches, often building new understanding by listening to the interpretations and suggestions of others in the discussion. Also, the situation described in each case is bounded—it is specific, often delineating a particular dilemma or tension that demands exploration and resolution.

Cases such as the one conceptualized and developed for professional development in this study can help teachers develop new pedagogical techniques. By design, cases invite teachers to examine and discuss alternative instructional strategies and to invent new ones. Discussions can range from a consideration of the techniques and advisability, to the use of indigenous games in classrooms to help overcome the language barrier which English second language (ESL) students face whose first language is not English. Case participants are invited to explore and critique lesson plans and pedagogical actions. In this way, the cases and their discussion provide a vehicle to analyze practice and pedagogy.

Cases foster a constructivist approach to the development of knowledge of mathematics teaching praxis because the pedagogy fosters the ability to think with what one knows and to do something with that knowledge. Social constructivism shows strong epistemological elements that emphasizes the importance of culture and context in understanding what occurs in society and constructing (mathematical) knowledge based on this understanding (Prawat, 1995). Contextualizing mathematics in IK and/or real-life contexts, acknowledges that knowledge are constructed by people in a social context (Dubinsky et al., 2013).

THE INTERVENTION

Contextualizing mathematics activities need increased teacher understanding and appreciation of different pedagogical models that aim to build maths confidence and competence (Howard et al., 2003) In this sense, a SLP on ethnomathematics was developed and presented in a three-day workshop in one province in South Africa in 2016. This SLP was part of an outcome of a National Research Foundation funded project on the affordances of indigenous knowledge in education. The training aimed at introducing teachers to viable ways to contextualize mathematics in indigenous knowledge or real-life contexts and at the same time develop teachers' pedagogical content knowledge.

RESEARCH METHODOLOGY

This paper presents the conceptualization and development of "The case of the indigenous game of Ncuva to introduce the concept of fractions in a rural school" as a professional development tool (Phase 1). This case was implemented within a SLP ethnomathematics workshop (Phase 2) to describe and interpret teachers' perceptions of an ethnomathematical approach in the mathematics classroom during implementation of the case at a SLP workshop.

In the **first phase** of the research, we applied CHAT (Engeström, 1987) as overarching theoretical framework of conceptualizing and developing the case of Ncuva. The data collection techniques for **the second phase** of the research included observations, field notes and a focus group interview. Saldana's (2009) coding technique was used to identify codes, categories and emerging meaningful themes from the data.

PHASE 1 of the Research

The first phase entails the conceptualizing and development of the case of the indigenous game of Ncuva for the teaching and learning of the concept of common fractions in a rural school setting. This phase is considered more conceptual in nature and affirms to a nonempirical (or theoretical) design. Past data from different teachers at rural schools were used to conceptualize and develop a case study on the tensions and affordances of an indigenous game like Ncuva in the teaching and learning of common fractions. The empirical work used to animate the framework developed here is from a Grade 6 mathematics classroom in a disadvantaged school in a rural area in one province in South Africa.

Based on the lesson observations and focus group interviews it was evident that each of the classrooms functioned as an activity system. Teaching the fraction concept is complex. Teachers might know how to do fraction calculations, but their students might find it difficult to understand the concept of a fraction. Therefore, the development of a case (the case of Ncuva and the fraction concept) was necessitated by this need and a second generation CHAT lens was used to disseminate the interactions that came forth in this developed case

Theoretical Orientation of CHAT

Engeström (1987) built on Vygotsky's activity theory model (1978) and generated the second generation activity theory. Because the first generation CHAT overlooks the important role of tensions, we decided to utilise the second generation CHAT as conceptual framework for the case study developed for the professional development intervention. The emphasis, then, was on the particular role that tensions between the case's components. In this sense, it is important to note that Phase 1 of the research is solely theoretical in nature. What follows is an outline of the case pedagogy. The character (Samantha) is a fictitious teacher representing, expressing and experiencing typical situations in a South African school context.

Vygotsky (1978) conceptualized pedagogy or teaching praxis as involving mediation by a more competent peer or teacher within a zone of proximal development, focusing on teaching as evolving by a more experienced other. Engeström's (1987) system thinking enables one to conceive praxis as playing out in a rule bound context in which power and control influence praxis.

Central to Vygotsky's (1978) thesis is the notion that the individual's interaction with objects in the world is mediated by cultural artefacts: signs, language, symbols, and practical tools. Artifacts carry with them a history of use and are themselves altered shaped and transformed when used in activities (Säljö, 1999).

The activity in this case is teaching praxis (action and reflection). Activity theory is a theory of object driven activity. Objects are concerns, they are generators and the foci of attention, effort and meaning. Through their activities, people constantly change and create new objects. The new objects are often not intentional products of a single activity but unintended consequences of multiple activities (Engeström, 2008). The creation of a new activity is approaches of reflective remediation. A mediating concept or device can open up an entirely new question and lead to the formation of a new object and a new activity (Engeström, 2008, p. 11).

Figure 14.1 graphically represents how a human interacts with the world by means of cultural artifacts.

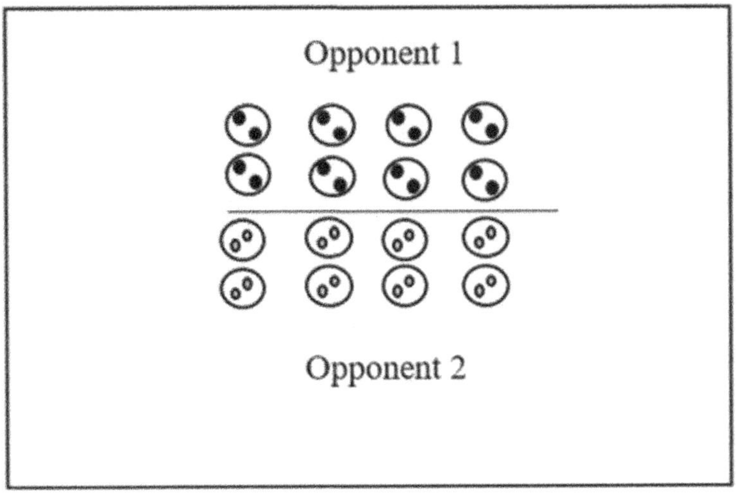

Figure 14.1. The Ncuva board game: *Boundary of game (known as the "river").*

The world is never approached directly in the course of the development of higher cognitive functions but is always mediated (Bateson, 1972; Wertsch, 1991). The subject, an individual or group, uses mediational means in order to act on the object of the activity.

We used CHAT in a conventional way, that is with the subject (Samantha) being a White novice mathematics teacher at a rural school (mostly Black students) and her teaching Grade 6 students for meaningful understanding of the common fraction concept the object of the teaching praxis (pedagogical activity system).

The epistemic assumptions the participant (Samantha) holds regarding learning will impact on how he/she infuses the indigenous game of Ncuva in the mathematics curriculum. Where a teacher believes that students learn through active engagement with the problem under discussion, s/he will use tools (game of Ncuva) in different ways to a teacher who believes that knowledge is innate and that students learn in a passive manner.

The **object** of an activity system represents the problem at which the activity is directed and represents that problem space that the teacher and students are working on (Hardman, 2008). In this case study we reflect on an activity system in which a novice teacher infused the indigenous game of Ncuva in teaching the concept of fractions. The object of the activity is the understanding of the fraction concept and included her pedagogical content knowledge (PCK) that become visible by her innovatively planned activity working with the official curriculum (CAPS).

For conceptual clarity, this chapter draws on Engeström's (1987) understanding of the object as "the 'raw material' or 'problem' space' at which the activity is directed and which is moulded and transformed into outcomes with the help of physical and symbolic, external and internal mediating instruments, including both tools and signs" (p.79).

Mediating artifact (indigenous game of Ncuva): one might view tools as resources mobilized by the teacher. Significantly, these tools mediate thought during the interaction between the subject and the context within an activity. These tools are both material (for example, the chalkboard) and psychological (for example language or symbolic systems such as mathematics symbols and language).

Tools are viewed as the resources available. Under tools this case study focuses on the infusing of an indigenous game (Ncuva) in the mathematics curriculum to encourage students to actively participate in constructing their own knowledge/understanding of the fraction concept. Resources such as textbooks, the pedagogies that unfold in the mathematics classroom such as using an indigenous game, its board and rules are equipment, signs (symbols) language (written and spoken) instruments and artifacts are all tools utilized in the classroom to support learning.

Samantha, a White novice mathematics teacher contextualized mathematics in students' indigenous game of Ncuva (as a tool) in the teaching and learning of fraction concept. Ncuva is also known as Tsoro, Tintjiwa, Mufuvha, Tshimayo, or Moruba. Ncuva is a board game

> *played by teams of one to six players a side per game. Four rows of four holes each are dug on flat ground or sculpted on to the board to make a total of 16 holes (see Figure 1 below). A line divides the board into two rows of eight holes each. To play, each player uses 16 "cows", or tokens, made of marula fruit kernels or small stones* (in an eight-hole board)*. The stones are placed in sets of four in the holes for the game to start. Ncuva is used to sharpen the mathematical and strategic acumen of its players.* (South Africa Department of Sport and Recreation, 2010)

Rules refer to norms, conventions and social interactions at play in the activity system. In the context of this study, we refer to the policy frameworks, the national curricula (CAPS SADBE, 2012), societal rules and academic rules of the school in question. Also, of significance, is the learning environment created in the classroom and the learning activity the teacher designed. The school principal's views (rules) on how one teaches and how students learn mathematics, can also be viewed as a rule.

The **community** (all influence the achievement of the object of the activity system) refers to all the stakeholders involved: teachers and students at the rural school, the principal, school governing body.

Division of labor refers to responsibilities, tasks and power relations among the different stakeholders. Stakeholders all have a shared responsibility, to provide quality mathematics education. The teacher wears a novice mathematics teacher's hat of continuing her journey of becoming a competent teacher, and second, that of a teacher, having to make pedagogical choices. The principal is overall responsible for what is happening in the school.

Tensions and Affordances Experienced in the Teaching of the Fraction Concept

As a reminder to the reader, this first research phase deals solely with a fictitious case pedagogy as portrayed in the Addendum. The teacher (Samantha), student's, and other contextual attributes should be seen as generalized expressions of a typical South African school setting. In these settings, teachers are sometimes underdeveloped in the necessary knowledge and skills to develop contextually based mathematics lessons (Howard et al., 2003). South Africa has a performance driven schooling system. Limited human resources impact negatively on both the teachers, students, school principal head of departments and school community. Samantha being a White woman with a Western culture and the students are from another culture.

The Students

Students from rural areas have different learning needs than their peers in urban and metropolitan schools. These students are mostly underschooled, have teachers that had insufficient training and relevant opportunities for professional development are scarce. Students at rural schools have not had exposure to many experiences their peers have in urban schools. Moreover, as a result of more un- or underqualified teachers being willing to teach at rural schools, students lack basic knowledge and skills in mathematics.

Professional development offered by governmental/departmental officials in many cases focuses on administration or issues related to the finishing/completion of the content of the CAPS. Therefore, little professional development are offered and time does not allow for innovative ideas/perspectives.

Most students come from low socioeconomic families/backgrounds. These students also do not have a proper support system at home when doing homework or learning for a test because parents have little schooling themselves. Therefore, the link between home and school are poor. Furthermore, the students at the rural school are ESL students with a restricted vocabulary. In part a lack of a code of the language can be one of the factors contributing to the poor performance.

Moreover, as a result of the limited resources, Samantha, the newly appointed well qualified mathematics teacher was required to be more resourceful.

Subject and Object

One of the major tensions (or, what Veresov, 2007, calls "dramatical collisions") that emerged in the mathematics teacher's teaching praxis (activity system), was that Samantha, the newly appointed white novice mathematics teacher failed with traditional teach-and-tell and drill-and-practice teaching strategy to teach Grade 6 students meaningful conceptual understanding of common fractions. Samantha, the mathematics teacher, had to reflect on previous experiences and use a combination of tacit and explicit knowledge in order to be effective in the classroom. She realized that drill and practice are not the answer to students' understanding/grasping of common fractions. Samantha, in her school, has one colleague she can use as a reflective partner.

The teacher's responses to these conflicts was to go ahead with her plan to infuse an indigenous game in her teaching of common fraction to enable students in a fun way to understand fractions through play. Samantha's first lessons reflected transmission-mode teaching, teacher-centered approaches with minimal student involvement, lessons that contained too much factual information, and that was not directed at developing skills, and far too little focus on meaningful understanding. These lessons were characterized by content-laden, teacher-centered lessons and transmission-mode pedagogies, yet during the reflection after such lessons, the teacher indicated that she did not feel pleased with the students understanding of the fraction concept and the principal did not feel impressed by the students' performance test on the fraction concept.

Subject, Community and Division of Labor

Samantha and the principal (also a previous mathematics teacher): The failing of Grade 6 students in a test on common fractions caused tensions between herself and the principal of the school who was a traditional (teach-and-tell and drill-and-practice) mathematics teacher himself. Samantha therefore does not have a support system at the school. Teachers were teaching to the test and finishing the curriculum.

A major tension emerged between the subject (novice mathematics teacher) and the administrator (principal) as part of the "community" in the activity system. For Samantha, it was extremely difficult to reason /discuss her teaching praxis with the principal. The principal still had a talk-and-chalk and drill-and-practice (repetition) approach to teaching and learning in mathematics. Samantha was upset and frustrated by the lack of the principal's willingness to discuss her plan to infuse contextualised mathematics in an indigenous game in the teaching and learning of the fraction concept. This showed us the tension between a reflective teacher and a nonreflective principal (practitioners). Their views of quality education were informed by the practices of the teachers that they had when they were school pupils.

Object and the available tools: The available textbooks in the school (provided by the DBE) is difficult for these rural students to learn and understand due to the abstract Westernized decontextualized mathematics therein.

Another aspect contributing to students learning of /in mathematics is language: the students at the rural school are all ESL students which implies an even greater challenge for students to understand the English technical language of mathematics.

Another tension was observed between the rules and the object. The current content overloaded curriculum (CAPS) do not allow time for innovative ways of teaching and learning. The curriculum is content and performance (in summative tests and examinations) driven.

PHASE 2 of the Research

Implementing "The case of Ncuva as a professional development tool within a SLP to describe and interpret teachers' perceptions of contextualizing mathematics in an indigenous game."

Methodology

Mainly, a qualitative approach research design was employed. This qualitative research study used an interpretivist approach (Creswell, 2013) that was used as overarching theoretical framework.

Context

The context in which this study had been conducted, was a three-day long workshop presented to 69 mathematics teachers in one province in South Africa in July 2016. This study posed to investigate mathematics

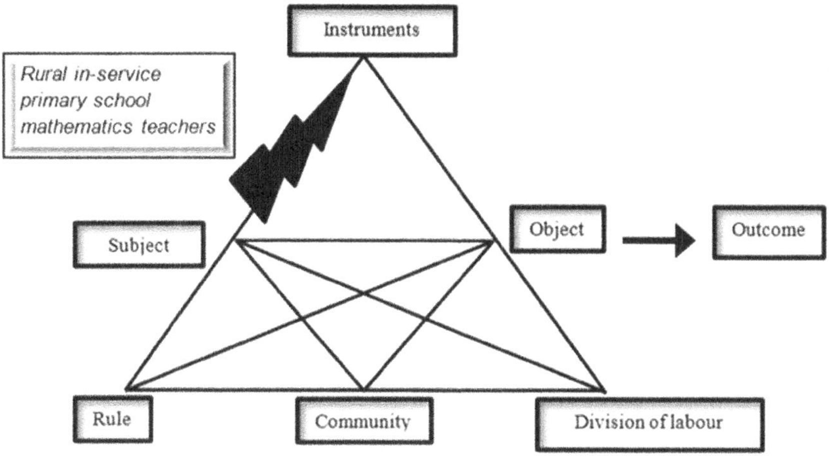

Source: Engeström (1987, p. 78)

Figure 14.2. Second generation CHAT theory.

teachers' perceptions of ethnomathematical ideas. This intervention (SLP) provided the 69 teachers with the opportunity to engage in developing contextualized activities, and developing skills to make and use mathematics-on-a-shoestring approaches in the underresourced classroom in order to negotiate a broader understanding of the role of contextualization of mathematics for South African students in rural schools.

Population and Sampling

Teachers were invited to attend the SLP workshop during the July school holidays in 2016. Mathematics teachers who agreed to participate in the study, were all teaching at rural schools and their ages differed between 33 and 55 years. All the teachers attending the SLP (N = 69) participated voluntarily in this SLP workshop and so did the 12 teachers participating in the focus group interview.

Data Collection Techniques in the Second Phase of the Research

The teachers at the SLP workshop (N = 69) had the opportunity to play the game of Ncuva themselves before the case of Ncuva was presented to them. After reading the case, they had to complete the assignment for the students, included in the case study. They engaged in discussions on the assignment and the technical and academic language which was used. Thereafter they reflected on the idea of an ethnomathematics approach in the teaching-learning of mathematical concepts using for example, an indigenous game (such as Ncuva) in order to negotiate a broader understanding of the role of contextualization of mathematics for South African students in rural schools

The data collection techniques included observations, field notes (over the three days, $N = 69$) and an hour long focus group ($N = 12$) interview. Observations were made and field notes were taken by the researchers during the presentation of the SLP ethnomathematics when these teachers engaged in the game of Ncuva. Videos were captured of these teachers when they were engaged in discussions during activities such as engaging in playing the game of Ncuva themselves. Thereafter "The case of Samantha's dilemma with teaching fraction concepts to grade 6 farm school (rural school) students and how she contextualized her lesson in the game of Ncuva," were read to them. They engaged in discussions about the different dilemma's Samantha faced as well as her and other possible solution(s) to the dilemma. Thereafter, 12 teachers participated in a focus group interview ($N = 12$) to capture their perceptions of this ethnomathematical approach to teaching the fraction concept.

Data Analysis in Phase 2 of the Research

Content analysis were used to analyze the data. Observations (field notes and video recordings) and the focus group interview were transcribed. Saldana's (2009) coding technique was used to identify codes, categories and emerging meaningful themes from the data. Content analysis, based on both the conceptual framework of this study, and multiple data collections ensured validity with regards to the findings. Trustworthiness were ensured by triangulating the data collected over the three days of the SLP as well as participant checking.

RESULTS

Results of the First Stage of the Research

The case of Ncuva and the teaching and learning of common fraction concept, as it was implemented within the SLP workshop in the second phase of the research, is available as an Addendum.

Results of the Second Stage of the Research

The results indicate that some of the participants have resistance to a new 'model' of teaching, mainly because they have set ideas about mathematics and its teaching. Two of the teachers seem to have the perception that mathematics is learned by continuous practicing which might indicate that they view mathematics as abstract facts.

Participants 2 and 6 were convinced that (P2) ... *every educator believes that when you have students that are having challenges in terms of, perhaps a concept, in most cases its always no, let there be constant practicing. Continue teaching them. Let them come to school in the morning, stay after school, ...*

Participant 4 who said agreed that more practice is the answer, but added and interesting response: *I would have given them more practice. ... so I would rather go the scientific way than the indigenous way, but that does not mean I defy indigenous knowledge.* It seems as if Participant 4 did not believe that indigenous knowledge is also scientific. Another participant agreed with Participant 4:

Also, an aspect highlighted by some of the group interviewees was the opportunity they had to relate ethnomathematical ideas to their own teaching (mostly teacher centred) practices. The results confirm that ethnomathematics and cases as methodology for continuing professional development education enables teachers to look at themselves and at their professional performances, with view to effective actions. This is evident from the following participants' responses:

Participant 12: *What I would do is just to facilitate to just make sure that everybody is involved. The students are not aware what playing this Moruba* (Ncuva)*, they were not aware that they are learning, so that I can introduce my concept and say "look, how many holes are empty"—they are counting, "how many holes ...," when they drop they pick, they drop, they subtract, they add. What I want to emphasize is that I have the interest of my students so that when my periods come, ... they love what they are doing.*

Participant 2 noticed the affordances the ethnomathematics approach holds for students' understanding in mathematics as well as their intention to also contextualize mathematics in indigenous knowledge: ... *So I should think for the educator to use the ethno-mathematics approach in that sense ... she used the indigenous resources around to bring the mathematics closer to the students, for them to see mathematics as something that is real, as something that is relevant, as something that is interesting and not abstract.... And I think it is welcome ... I believe I can also, ... apply such principle using different indigenous resources in my class.*

Participant 1 could not hide his passion for teaching and how he noticed the affordances of the new teaching praxis he could also implement in his own classroom: *Yesterday was the first time I saw the Ncuva game.... While the game was played yesterday, the only thing I have seen was my students. It doesn't matter where you come from, what matters most is you understand the rules of the game. It would tie your interest, to mathematics. It is not a barrier to indigenous knowledge.*

One participant appreciated Samantha's (in the case of Ncuva) innovative idea in her teaching praxis to assist her students to understand the fraction concept. It is important here to keep in mind that the game of Ncuva is not an indigenous game to her culture: *The teacher in the case did not know about this Ncuva, but she was driven by those students who was playing it. Then she started noticing "I can use this in a classroom situation." So, I don't think it would be a problem....The interest is that students should end up having interest in that part of fractions or subject and mastering that part.*

Results indicate that teachers were actively engaged in ethnomathematical activities. The teachers had a very positive attitude and pure mathematical reasoning flowed from their engagement in their discussions regarding the Ncuva-activities.

Figure 14.3 demonstrates how teachers work collaboratively in a community of professional development on the conceptualized case study (available in the Addendum).

Some of the participants responded differently on the question what they would have done in Samantha's circumstances.

Participant 6 seemed to be carefully optimistic: *Ok I would suggest that since the normal teaching is failing, we are not making any progress. Why don't we try something different? Maybe it will produce something.*

It seems as if participant 10 shared participant 6's optimism about Samantha's solution to rural students' understanding of the fraction concept *If I would find myself in a similar situation, my approach would be to find other methods of teaching that would engage the students to take be active part and in the case of the Ncuva, I think it is one way of being totally different approach to teaching, because some students will understand it better when I use this than the type of teaching I normally use to develop a concept for the students to understand it better. This way, students get into the activities, real-life activities, and the ones that are taking part are developing a concept.*

Although participant 10 was optimistic about Samntha's idea he/she showed his/her concerned about time limits set by an overloaded content driven CAPS (SADBE, 2012) *I am sceptical of time.*

Other participants were of the opinion that they will in the future, also contextualize mathematics concept(s), procedure(s) or relationship(s) in either indigenous knowledge or students' everyday life experiences: *she used the indigenous resources around to bring the mathematics closer to the students, for them to see mathematics as something that is real, as something that is relevant, as something that is interesting and not abstract.... And I believe I can ... apply such principle using different indigenous resources in my class* (Participant 2), and added that he learned a new pedagogy (teaching strategy) to implement in

Figure 14.3. Teachers working collaboratively.

his own classroom: … *one couldn't have known that one could actually devise a strategy using their real life experiences and making the mathematics interesting to them and making it relevant and meaningful, in ways whereby they will see it from another angle other than what they thought it was before.*

Participant 3's response "*I just realized that students can learn through play. If I want to introduce a concept, I will allow them first to play. To create a situation where they will be free and relaxed.*" Participant three sees to only view the play as a way to relax earners in order for them to feel free to ask questions in the classroom.

Participant 6: *Now it means that we develop mathematical understanding out of indigenous knowledge games or whatever. It will bring misconceptions right? Because of the different backgrounds that we come from.… I am saying that you have to be very cautious of the misconceptions that may arise. But simply saying that play, sing, draw this and all the other things, that will, when they are excited, then you bring in the mathematics understanding in it. But if you are creating it out of the game, then that becomes very interesting—If you develop out of the games the knowledge of playing the game, then that would be interesting. But if you are just simply trying to get their attention by singing, they get excited and all this, that is just not simply using that to the benefit of the students.*

Participant 11 was very clear about the principal's behavior: *I think the principal was very wrong.* He emphasized that contextualizing the concept of common fractions in an indigenous game benefits students' understanding of mathematics concepts: "*As teachers we ... have many strategies that we should use so that the students should understand.*"

DISCUSSION OF THE FINDINGS AND CONCLUSION

Infusing indigenous games in mathematics classrooms holds great potential, including its value for professional development of teachers. However, there are necessary imperatives that need to be in place to ensure that this is a success, as their use may convey negative connotations when these imperatives are not taken into account. Some examples of these necessary aspects include (1) understanding indigenous games to the extent that they serve useful to teach mathematics concepts at school level, (2) acknowledging that indigenous games hold rich historical, social and cultural meanings that should be taken into account when integrating them with your lessons. Failing to consider these implications can lead to misconceptions and misunderstandings of the mathematical ideas they convey. One way this chapter addresses these aspects is by focusing on the professional development of teachers and the affordances of an indigenous game to explain the common fraction concept. The authors first conceptualized and developed a model case (refer to the addendum) based on the indigenous game of Ncuva to teach the concept of fractions in a rural South Africa primary school. In this regard the case served as a professional development tool.

Second, an opportunity was created by means of a SLP in ethnomathematics in which the case could be implemented. Through implementing the case, we were able to describe and interpret teachers' views regarding the ethnomathematical approach that the case offers as this informed their impression of professional development. For a detailed account on the Short Learning Programme refer to the following link: (https://www.youtube.com/watch?v=hrA3_MpsA2Q&t=9s)

Discussion of the Findings of Phase 1: Conceptualizing the Case

We employed CHAT as theoretical orientation in the first phase of the research whereby data from previous research projects were used in the conduct of the case. The findings include a case on infusing the game of Ncuva in the teaching-learning of fractions and the teachers' perceptions and ideas of the ethnomathematical approach in the mathematics classroom. In its essence, the case represents the various components of CHAT whereby the object of the activity system, the mediating artifact (indigenous game of Ncuva), tools as available resources, rules as the norms, conventions and social interactions, the community (all influence the achievement of the object of the activity system) and division of labor as the responsibilities involving the tensions and affordances experienced in the teaching of the fraction concept.

The affordances of the indigenous game of Ncuva as mediating tool, is the fact that the students knew the game rules and were able to play the game without the teacher having to first teach them how to play the game. The teacher noticed the students are playing the game during break. She started to reflect on how she could utilize the board game to teach the fraction concept by playing a game. The game board is inexpensive as it can be drawn in students' workbooks, on rocks or in dirt. Students can use stones, corn or any other available pips to play the game with. The teacher demonstrated an innovative way to make the fraction concept understood by contextualizing the mathematics in an indigenous game they played during break.

Discussion of the Findings of Phase 2

Implementation of the case within a SLP revealed two notions of teachers' professional development through the affordances of the indigenous game case based pedagogy approach followed. On the one hand, some of the participants showed resistance to this new "model" of teaching mathematics, perhaps due to their fixed ideas about mathematics and its teaching of rational and abstract (Westernized) ideas. Chilisa (2012) explained that new perspectives (such as the model case) requires different and sometimes unique interpretations which could be gained from implementing the case in different settings. In doing so, the strategy assisted in participants contributing to our own cultural intuitions (Chilisa, 2012) and offered perspectives and interpretations with an in-depth narrative account of their notion of professional development. Such participation enhances our understanding of what this transformational resistance might hold.

On the other hand, an aspect highlighted by the participants was the opportunity they had to relate an ethnomathematical approach to their own teaching (mostly teacher centred) practices. These results confirm that an ethnomathematical approach as methodology for continuing professional development education enables teachers to look at themselves and at their professional performances, with a view to more effective actions.

Ethnomathematics thus provides opportunity for indigenous games in the mathematics classroom which to integrate and accommodate a valuable component of IKS (Nkopodi & Mosimege, 2009). The implementation of such a contextualized approach would make mathematics more relevant and accessible for South African students and including such indigenous games like *Ncuva* can form an integral part of the ethnomathematics and mathematics education of this country.

Furthermore, contextualized mathematics provides the opportunity to give all South African students epistemological access to good education and provides learning affordances in maths (SA DBE, 2012). It is important to note that incorporating a contextualised (sociocultural) component in the curriculum, acknowledges the importance of the affective dimensions in mathematics teaching (Gavarette, 2015). Based on this experience with these teachers, we conclude that an ethnomathematical approach is a sustainable and inclusive way to better the quality of mathematics education by developing teachers' skills to contextualize mathematics appropriately. This research project argues that contextualized mathematics could serve as a very good entry point into the abstract world of mathematics. More research is being done on this SLP-intervention.

Thus, the use of cases with multiple components of mathematics, pedagogy, student thinking, and context represents "a new instruction." Furthermore, because cases are often puzzle-like without specific approaches, rules, or answers, this can make them a useful teacher education and professional development tool. Teachers improve their knowledge of teaching strategies by experimenting with them and developing concurrently metacognitive knowledge about how, when and why these can be used to improve teaching (Borkowski et al., 1987). Through constructing appropriate strategies, the teacher has to think about similar experiences or problems and reflect critically on these to differentiate between the approaches. The experience of case pedagogy can assist in forging a new epistemological approach to the teaching and learning of mathematics, particularly if they are informed by a traditional African worldview.

Based on this experience with these rural schoolteachers, we conclude that ethnomathematics is a sustainable and inclusive way to better the quality of rural school education due to the assistance it provides teachers with to contextualize mathematics effectively.

ACKNOWLEDGMENTS

We hereby acknowledge the National Research Foundation (NRF)-funded project based on the affordances of indigenous knowledge in self-directed learning.

REFERENCES

Ainley, J., Pratt, D., & Hansen, A. 2006. Connecting engagement and focus in pedagogic task design. *British Educational Research Journal, 32*(1), 23–38.

Barber, M., & Mourshed, M. (2007). *How the world's best-performing school systems come out on top*. McKinsey and Company.

Bateson, G. (1972). *Steps to an ecology of mind: Collected essays in anthropology, psychiatry, evolution, and epistemology*. University of Chicago Press.

Borkowski, J. G., Johnston, M. B., & Reid, M. K. (1987). Metacognition, motivation, and controlled performance. In S. J. Ceci (Ed.), *Handbook of cognitive, social, and neuropsychological aspects of learning disabilities* (Vol. 2, pp. 147–173). Erlbaum.

Chilisa, B. (2010). *Indigenous knowledge methodologies*. SAGE.

Chisholm, L. (2003). *The state of curriculum reform in South Africa: The issue of Curriculum 2005*. In J. Daniel, A. Habib, & R Southall (Eds.), *State of the nation* (pp. 268–289). HSRC Press.

Creswell. J. (2013). *Qualitative inquiry & research design: Choosing among five approaches* (3rd ed.). SAGE.

Centre for Development and Enterprise (CDE). (2014). *From laggard to world class: Reforming science education in SA schools*. http://www.scienceinafrica.co.za

Davidson, C. N. (2012). *Now you see it: How technology and brain science will transform schools and business for the 21st century*. Penguin Books.

Dubinsky, J. M., Roehrig, G., & Varma, S. (2013). Infusing neuroscience into teacher professional development. *Educational Researcher, 42*(6) 317–329.

Elmore, R. (1996). Getting to scale with good educational practice. *Harvard Educational Review, 66*(1) 1–26.

Engeström, Y. (1987). The emergence of learning activity as a historical form of human learning. In *Learning by expanding: An activity-theoretical approach to developmental research*. Cambridge University Press.

Engeström, Y. (2008). *From teams to knots: Activity-theoretical studies of collaboration and learning at work*. Cambridge University Press.

Fosnot, C. T., & Dolk, M. L. (2001). *Young mathematicians at work*. Heinemann.

Frigo, T. (1999). *Resources and teaching strategies to support Aboriginal children's numeracy learning. Melbourne: Australian Council for Educational Research*. http://www.boardofstudies.nsw.edu.au/aboriginalresearch/index.html#numeracy

Gavarette, M. E. (2015). The challenges of mathematics education for indigenous teacher training. *Intercultural Education, 26*(4), 326–337.

Gay, G. (2000). *Culturally responsive teaching: Theory, research, & practice*. Teachers College Press.

Hardman, J. (2008). Researching pedagogy: an activity theory approach. *Journal of Education, 45*(1), 65–94.

Howard, P., Perry, B., Lowe, K., Ziems, S., & McKnight, A. (2003). Mathematics in indigenous contexts: A case study. *Mathematics Education Research: Innovation, Networking, Opportunity*, 436–443.

Hunzicker, J. (2011). Effective professional development for teachers: A checklist. *Professional Development in Education, 32*(2), 177–179.

Luitel, B.C. 2013. Mathematics as an im/pure knowledge system: Symbiosis, (w)holism and synergy in mathematics education. *International Journal of Science and Mathematics Education, 11*(1), 65–87. https://link.springer.com/article/10.1007/s10763-012-9366-8

Kinsella, E. A., & Pitman, A. (2012). Phronesis as professional knowledge. *Phronesis as Professional Knowledge*, 163–172.

Kleinfeld, J. (1988). *Learning to think like a teacher: The study of cases*. https://files.eric.ed.gov/fulltext/ED308039.pdf

Lampert, M. (2010). Learning teaching in, from, and for practice: What do we mean? *Journal of Teacher Education, 61*(1–2), 21–34.

Maxwell, K. L. H., & Chahine, I. C. (2013). Cultural immersion and mathematics teacher education: Explorations in morocco and South Africa. *Journal of Humanistic Mathematics, 3*(2), 62–75.

Merseth, K. K., & Lacey, C. A. (1993). Weaving stronger fabric: The pedagogical promise of hypermedia and case methods in teacher education. *Teaching and teacher education, 9*(3), 283–299.

Merseth, K. K. (Ed.). (2003). *Windows on teaching math: Cases of middle and secondary classrooms*. Teachers College Press.

Merseth, K. (2007). Mathematics and case-based instruction. An interview with senior lecturer Kay Merseth. *The Young Women's Online Journal of Teaching and Learning Mathematics*, https://www.gse.harvard.edu/news/04/08/mathematics-and-case-based-instruction

Nkopodi, N. & Mosimege, M. (2009). Incorporating the indigenous game of morabaraba in the learning of mathematics. *South African Journal of Education, 29*, 377–392.

OECD. (2012). Equity and quality in education: supporting disadvantaged students and schools. http://dx.doi.org/10.1787/9789264130852-en

Prawat, R. S. 1995. Misleading Dewey: Reform, projects, and the language game. *Educational Research, 24*(7), 13–27.

Rhodes, J. A. (Ed.). (2015). *Advancing teacher education and curriculum development through study abroad programs*. IGI Global.

South Africa Department of Sport and Recreation. (2010). *Indigenous games rule book*. www.srsa.gov.za

South Africa Department of Basic Education. (2012). *Curriculum and assessment policy statement (CAPS): Mathematics Grade 10–12*. http://www.thutongh.doe.gov.za

South Africa The Presidency. National Planning Commission. (n.d.). *Our future: Make it work. National Development Plan—2030 Executive Summary*. http://www.gov.za/sites/www.gov.za/files/Executive%20Summary-NDP%202030%20-%20Our%20future%20-%20make%20it%20work.pdf

Saldana, J. (2009). *The coding manual for qualitative researchers*. SAGE.

Säljö, R. (1999). Concepts, cognition and discourse: From mental structures to discursive tools. *New Perspectives on Conceptual Change*, 81–90.

Smith, C., & Morgan, C. (2016). Curricular orientations to real-world contexts in mathematics. *The Curriculum Journal, 27*(1), 24–45.

Vygotsky, L. (1978). *Mind in society: The development of higher psychological processes*. Harvard University Press.

Veresov, N. (2007, June). *Sign mediation: Magoc triangle; sogn-mediated action and behind*. Fourth Nordic Conference on Cultural and Activity Research, Oslo, Norway.

Wertsch, J. V. (1991). A sociocultural approach to socially shared cognition. In L. B. Resnick, J. M. Levine, & S. D. Teasley (Eds.), *Perspectives on socially shared cognition* (pp. 85–100). American Psychological Association.

ADDENDUM

The Case of Ncuva and Fractions

"You will make them practice mathematics! We do not 'play' in this subject! They are not in kindergarten anymore." These words rang in Samantha's head. Never before had she felt so accomplished and so affronted at the same time. She was adamant to use Ncuva as a means to teach the learners about fractions. She had been teaching this current topic for the past two weeks and she felt as if the progress made by the learners, simply went at snail's pace. She observed that the learners struggled with the identification of and arranging of fractions from smallest to largest and vice versa especially. "If these concepts were the basic fundamentals of fractions, how could learners in grade six still seem to struggle with this?"

Her heart raced as she entered the principal's office. "So Samantha, why did you need to see me so urgently?" "Sir, you may not agree, but please allow me to explain."

The Setting

Samantha was a newly appointed teacher at a rural school located outside of Potchefstroom in the Klipdrift district. Klipdrift Primary School was a small school compared to what she was used to. Five hundred learners she recalled from the interview she had with the principal. The principal governed the school with a firm hand. He was very serious of character she noted. He allocated three grade six mathematics classes to her comprised of 30 learners each.

"Ms Deacon, I would like to let you know that we manage these kids very strictly. They will take chances with you at first. Do not let them get away with it! They will obey you and you will make them practice their mathematics. Besides, that is the only way you can learn it right?!" As she entered the classroom for the very first time, she heard herself say "mathematics is everywhere!" in an excited voice. To this day, she still remembers the uninterested eyes that stared blankly back at her.

Teaching and Learning During the First Term

The first term of teaching was quite overwhelming for Samantha. At times she felt that the learners taught her more than she taught them! Of course, the other issue was that their language of learning and teaching differed from their preferred home language which was Setswana. When it came to teaching, Samantha felt as if the learners simply did not grasp what she was saying. Mathematics seemed like a foreign language to them. Her grade sixes struggled through their mathematics, not with it! How could these learners possibly be struggling? She made them practice their mathematics as Mr Downey suggested. She taught by the book and used her CAPS document as the ultimate guideline. Her first term average showed a dramatic decrease. This was unsettling to her.

The following day, the principal arranged for a meeting with her after school.

"Ms Deacon, I noticed a dramatic decrease in your term average. Can you explain this to me please? "Mr Downey, I am really trying my best. The issue was unsettling to me and to be honest, I cannot understand what I am doing wrong, I think that" "Please do not explain any further. I know where this is going. I suggested that you PRACTICE mathematics with these learners. You know how they are. They are simply not interested in school and do not care anyway. Most of their parents are heavy drinkers, their domestic circumstances are terrible, and they eat once a day which is the meal, we serve them during first break. We are here to help these learners to pass. That is all. Most of them never break this evil drinking cycle when they enter adulthood. From now on, I will be your mentor. In my day, I was a top mathematics teacher myself. I will give you two weeks to come up with a proper teaching strategy. We will arrange for a meeting as soon as you have something."

Time Is of the Essence

At break times, Samantha would sit and watch the learners engage in gameplay that was unfamiliar to her. She found herself wishing that the learners understood mathematics as well as they understood these games. After continuously watching the learners play day after day, Samantha soon realised why she grasped mathematics as a school learner. The mathematical context in which it was taught was familiar to her!

As soon as Samantha arrived home that afternoon, she raced to her computer. Klipdrift Primary School did not have computers for use by staff members. It was a luxury they simply could not afford. The Google home page emerged in front of her. The burning question on her mind was "how could learners understand mathematics in a certain context." The answer was plain and simple that stood out to her. By using indigenous knowledge practices! In mathematics specifically, this practice is called ethnomathematics.

Her mind raced through the images of learners playing at break times. One game immediately jumped to mind. The game of Ncuva. To her surprise, she found an indigenous games rulebook for games in South Africa online. As she read the rules of the game, she quickly understood. What seemed foreign to her as observer, rang with ideas for incorporation in the mathematics classroom.

Ncuva seemed to be time consuming but after thorough research, Samantha noted that the game's size can influence the time spent before a winner emerges. The game can be played in class by drawing 16 circles on a piece of card board. Each player "owns" 8 holes that is divided by a boundary known as the "river." There are 32 pips on the board that is divided between the two players. Opponent one receives 16 black pips and opponent two receives 16 white pips representing the "cows" in the game. Each circle is then filled with two black or two white pips depending which color "cows" are owned by which opponent. Samantha noted that the game posed a constant difference of eight. There are eight holes on each side, 16 in total. There are 16 "cows" on each side, 32 in total. This constant difference would assist the learners when it came to the identification of the numerator and/or denominator when doing fractions.

Figure 1. The Ncuva board game.

Day of Days

Samantha had to report on a new teaching strategy as Mr Downey requested.

"So Ms Deacon, do you have a plan of action to improve upon your current teaching strategy?" "I do indeed Mr Downey! In fact, I think that all of Klipdrift Primary's teachers can implement something like this. As I did my research I found that the incorporation of ethnomathematics will direct me not only to better my teaching practice, but it will enhance learning for the learners." "Samantha let me stop you right there. Approaches like these to teaching only work in research, not in actual classrooms. "I acknowledge that Mr Downey. I have been struggling to teach these learners about fractions. I observed them at break and found that they often played Ncuva. I propose to use the game of Ncuva as a means to help these learners to integrate their indigenous knowledge into mathematics." "Under no circumstances would that be an option! Learners do not need to 'play' games in this subject! They are not in kindergarten anymore! This approach simply will not work, and I can guarantee this! In mathematics, the learners are equipped with a set of skills by means of PRACTICING. I will give you one more week to find a better solution to your problem. Please go now, I am late for my next meeting."

Samantha left the meeting affronted. The principal did not even give her a chance to finish her proposal.

A colleague found her in the staff room and asked her about the meeting.

"There are so many kinds of approaches to teach mathematics today. Does he even know that Jane? He still fancies a very teacher-centered approach. I am going to implement this ethnomathematics approach even though he told me not to. What do I have to lose?" "Samantha, I know you are young and eager, but Mr Downey do not like it when you step out of the boundaries." "What have I got to lose Jane? I want to prove him wrong."

The Game of Ncuva and Mathematics

In terms of time available per period, Samantha decided to divide the learners into groups of two players per game. Being newly appointed had one disadvantage for her regarding the pairing of the two-player teams. She realized that she did not know her learner's abilities that well enough yet. Samantha discussed her idea with Jane and asked for advice regarding the pairing of learners. Jane advised her to bring about mixed ability teams in order for each learner to benefit even more from the lesson. Given the fact that the learners already know the game, she decided to give each group a worksheet with ten questions. After each turn the pair needs to answer a question based on the game, but regarding fractions. Pairing the learners helps them to work collaboratively in order to find an answer. The fact that the numerator and denominator value is constant, will aid Samantha when she assesses these worksheets. The questions on the worksheet is as follows:

1. How many holes are **on your side** of the Ncuva board?
2. How many holes **are there in total** on the Ncuva board?
3. Write the **number of filled holes as a fraction of all the holes on your side** of the Ncuva board.
4. Write the **number of unfilled holes as a fraction of all the holes on your side** of the Ncuva board.
5. Write the **number of filled holes with black cattle as a fraction of all the holes** on the Ncuva board.

6. Write the **number of filled holes with white cattle as a fraction of all the holes** on the Ncuva board.
7. Write the **number of filled holes as a fraction of all the holes** on the Ncuva board.
8. Look at the answers of questions 3, 4 and 5. **Arrange** these answers **from the smallest to the largest.** Carefully write down your answers in the spaces provided in the form of a **fraction**.

Figure 2.

9. Look at the answers of questions 5, 6 and 7. **Arrange** these answers **from the largest to the smallest.** Carefully write down your answers in the spaces provided in the form of a **fraction**.

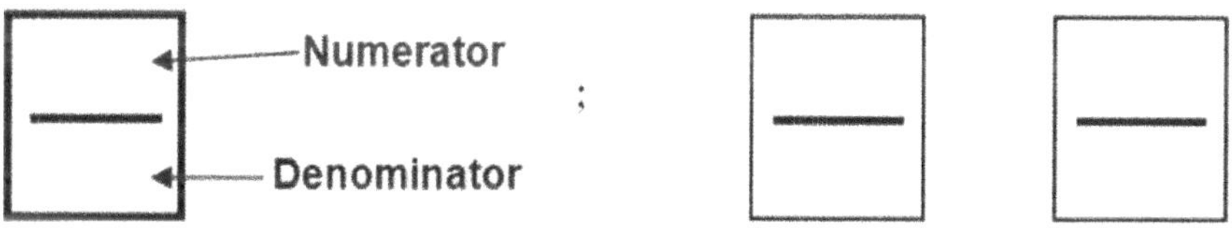

Figure 3.

10. What do you think is a **fraction**? Explain your answer.

The Aftermath of the Lesson

The time drew near for the lesson to commence. As soon as the learners came into Samantha's classroom, they bubbled with excitement—which never happened before. "Ma'am, this is the mathematics class. Why are we playing games today?" one of the learners asked. "Mathematics is everywhere Mariah, and today I am going to show you why I say this." As facilitator, Samantha watched every move as closely as she could in order to learn from her learners. Questions were asked frequently and at times, she felt quite overwhelmed with this approach. "Why did I try this?" she asked herself. As the session drew to a close, learners submitted their worksheets. Luckily the time spent on gameplay was just enough for a two period session. As she went through the worksheets, she found that almost every question was answered. This too was something she was not used to at all!

The learners had to demonstrate conceptual and procedural knowledge in order to complete this lesson successfully. While assessing the worksheets, Samantha realized that the learners were more willing to do mathematics. Although not all the answers were correct, she could definitely see improvement. This was her ultimate goal that she set for herself as a mathematics teacher. A sense of achievement came over her. She realised that context can be a powerful strategy to have when it comes to teaching.

Despite of this accomplishment, Samantha still had the following dilemma: Mr Downey, a teacher with more experience was in opposition to her ethnomathematical approach. He had to be convinced that this could work. Her worksheets served as a basis for improvement among her learners. She immediately arranged for a meeting with the principal. Improvement was what he wanted was it not?

CHAPTER 15

INDIGENOUS KNOWLEDGE IN THE LIFE SCIENCES CLASSROOM

Science, Pseudoscience, or a Missing Link? A Chat Lens

Josef de Beer

ABSTRACT

Although the school natural sciences curricula in South Africa ask for the inclusion of indigenous knowledge systems in the classroom, it is either done very superficially by providing an example or two, or ignored completely by teachers (Cronje, De Beer & Ankiewicz, 2015). This mixed-methods study (with emphasis on the qualitative inquiry) into the status of indigenous knowledge in the natural- and life sciences classroom in Gauteng, North-West, Northern Cape, Mpumalanga, and Limpopo provinces in South Africa, once again echoes what Rogan and Grayson (2003) reported: in South Africa the curriculum process focuses too much on the *what* (the curriculum itself), at the expense of the *how* (the implementation of the curriculum). Although the progressive curriculum makes it clear that indigenous knowledge should

Evidence-Based Inquiries in Ethno-STEM Research: Investigations in Knowledge Systems Across Disciplines and Transcultural Settings, pp. 279–307

be addressed, it provides very little guidance to teachers on how this should be done. Two problems are specifically highlighted in this chapter: teachers' lack of pedagogical and content knowledge in addressing indigenous knowledge systems, and their poor understanding of the nature of science (and how indigenous knowledge should be incorporated within this scientific framework). To effectively infuse indigenous knowledge in the school curriculum, effective epistemological border-crossing between indigenous knowledge and Western science is needed in terms of epistemological and ontological considerations, which places demands on teachers' knowledge and pedagogies. Too often "Western" science and indigenous knowledge are seen as conflicting epistemologies, disregarding the notion of border crossing, or Odora-Hoppers's (2002) idea of a "postmodern integrative paradigm shift" where indigenous knowledge and Western knowledge systems are seen as complimentary epistemologies. This chapter will elicit the value of indigenous knowledge in the life or natural sciences curricula from the perspective of embodied, situated, and distributed cognition (ESDC). Cultural-historical activity theory (CHAT) will be used as a research lens (linked to the theoretical framework of social constructivism), in looking at a funded research project, in which natural and life sciences teachers were trained in facilitating border-crossing between indigenous knowledge and "Western" science. This chapter also sheds light on interventions such as Ogunniyi's course based on contiguity argumentation theory (Ogunniyi & Hewson, 2008), and how such interventions can enhance teachers' understanding of how the nature of science and indigenous knowledge systems can be better integrated.

INTRODUCTION

Indigenous Knowledge and the Life Sciences Curriculum: A South African Context

Many of the new school science curricula in South Africa and elsewhere in the world contend for an approach to natural and life sciences which emphasizes the fact that there are different perspectives and world views to make sense of the world around us. Furthermore, indigenous knowledge is mentioned, which should, in conjunction with conventional modern science, enjoy attention in the life sciences classroom. Unfortunately, the new curricula are elusive as regards to the question of *how* indigenous knowledge can take its rightful place in the classroom (Onwu & Mosimege, 2004). While modern (Western) science and indigenous knowledge structures often share a common knowledge domain (Cronje, De Beer, & Ankiewicz, 2015), they often differ much in terms of methodology and epistemology (Onwu & Mosimege, 2004). The ways in which knowledge is either accepted or falsified differ much. This begs the question whether it is at all possible to include indigenous knowledge meaningfully in the life sciences curriculum. While some researchers (De Beer & Whitlock, 2009;

De Beer & Petersen, 2017) argue that indigenous knowledge, in some cases, is also the outcome of a scientific process of hypothesizing, experimentation, and careful observations, other researchers (Onwu & Mosimege, 2004) are of the opinion that it would be a mistake to subject indigenous knowledge to the same processes of verification as in the case of modern science. Onwu and Mosimege (2004) maintain that the two systems are different and require different forms of verification. Such a view does make sense, should one, for instance, think of the domain of spirituality and the metaphysical. Modern (Western) science does not negate spirituality and the metaphysical; its view is merely that, because proof for these cannot be provided and such metaphysical phenomena are not subjected to physical-causal laws, they fall outside the domain of natural science. In contrast, indigenous knowledge structures are holistic and include the metaphysical (De Beer & Van Wyk, 2016). Indigenous knowledge does not endeavor to explain reality by reducing it to subsections and immediate consequences; it looks at reality holistically and does not necessarily exclude the metaphysical. Cronje, De Beer, and Ankiewicz (2015) showed that the natural sciences and indigenous knowledge as knowledge systems share common characteristics, despite some differences.

Morrow (2009) coined the term "epistemological access" to describe the barriers to successful learning that are linked to worldview differences. He advocates for the acknowledgement that Africa has its own "alternative forms of knowledge." Often students need to make a transition between their own life-worlds and science, and we need to acknowledge that such "border crossing" (Aikenhead, 1996) may be accompanied by conflict. Ogunniyi's and Hewson (2008) contiguity argumentation theory (CAT) is traceable to Aristotelian contiguity theory, which asserted that one or two states of mind (or, in our case, science and IKS) tend to readily couple with, or recall each other to create an optimum cognitive state. CAT recognizes five categories describing scientific and/or IKS based conceptions within a learner's mind: dominant conceptions, suppressed conceptions, assimilated conceptions, emergent conceptions and equipollent conceptions (Ogunniyi & Hewson, 2008). A conception becomes dominant when it is the most adaptable to a given context.

It is probably this difference in methodology and epistemology which leads to the uncertainty among teachers, and the marginalization of indigenous knowledge in the natural and life sciences classroom. Odora-Hoppers (2004) points out that the inclusion of indigenous knowledge in the formal curriculum is often clinical and sterile. She uses the metaphor of structures preserved in bottles filled with formalin—a good description of the current "lip service" to indigenous knowledge: stereotyped examples presented by the teacher and repeated without more ado in the examination paper (De Beer & Whitlock, 2009). In this chapter I will refer to recent advances in this field, where indigenous knowledge is introduced in the natural and life sciences classroom using pedagogies that build on the tenets of the nature of science (De Beer, 2012; De Beer & Petersen, 2017a; De Beer & Petersen, 2017b). These pedagogies build on the tenets of science, and learners should use science process skills to explore indigenous knowledge phenomena.

Odora-Hoppers (2002) also speaks of the complementarity between indigenous and Western knowledge systems, and points out that there is a "postmodern integrative paradigm shift," which addresses "second generation indigenization." The exclusion of indigenous knowledge in past systems can therefore be addressed and corrected.

In this chapter, I wish to argue for a synergy between indigenous knowledge and modern science in the context of science education. This work has developed rhizomically over the past number of years, driven by funding from the South African National Research Foundation (NRF) and the Fuchs Foundation. I first briefly point out the important findings of a research project undertaken among teachers in Gauteng and Limpopo provinces in South Africa, and extensively reported in a research report (De Beer & Ramnarain, 2012). I then look at data emanating from several cycles of short learning programs (SLPs) that were presented to science teachers, the focus of which is how to facilitate the border-crossing between science and indigenous knowledge in the classroom.

Indigenous Knowledge in the South African Classroom: A Golden or Missed Opportunity?

Science education has undergone several waves of change over the past few decades. Wallace and Louden (1998) refer to three such waves of science curriculum reform in the international arena, namely (a) a focus on science as discipline knowledge in the 1950s and 1960s, (b) science as relevant knowledge in the 1970s (with an emphasis on scientific literacy and a science-technology-society approach), and (c) science as imperfect knowledge, with an epistemological shift towards cognitive science and more constructivist approaches in the 1980s and 1990s. The more constructivist approach to science education today asks of the life sciences teacher to acknowledge the indigenous knowledge (IK) that a learner might have as a catalyst for constructing new knowledge. In a true constructivist fashion, the teacher should build on this preknowledge of learners. This would assist in what Gibbons (2000) calls "mode 2 knowledge production," and teaching science that is context-sensitive: an important consideration in terms of the achievement of affective outcomes (Jackson, De Beer, & White, 2016). This poses a challenge to the natural or life sciences teacher as it not only places high demands on a teacher's pedagogical content knowledge (PCK) but also defy teachers' epistemological beliefs related to the nature of scientific knowledge. Abd-El-Khalick, Bell, and Lederman (1998) mention that a relationship exists between teachers' views on the nature of science, and how they teach.

South Africa with its rich cultural diversity and rich biodiversity offers wonderful learning opportunities to life sciences learners. As an emerging economy, South Africa is faced with a number of challenges such as unemployment, a lack of scientists (which prevents us from becoming a global player in the world economy), poverty and issues regarding food security, diseases, and pandemics such as HIV and AIDS, and a lack of proper healthcare for many South Africans, especially those in rural areas. These obstacles pose a threat to economic growth. Scientists such as Van Wyk and De Beer (2012) are of the opinion that the inclusion of indigenous knowledge can help to address many of these issues.

The rich indigenous knowledge (IK) and cultural practices in many areas in the country provide learners with a good entry point into the scientific world. Two examples of the application of IK:

1. The local people in the Giyani district in South Africa have been using the plant *Lippia javanica* (the fever tree) as an insect repellent for decades. The Council for Scientific and Industrial Research, in conjunction with the local community in Giyani in Limpopo, has registered a patent on the use of *Lippia javanica* as insect repellent. Clinical tests on citronella oil as a mosquito repellent showed a success rate of 40% while the oils in *Lippia javanica* seems to be 95% effective. The commercialisation of this application is also benefiting the community by creating jobs (Mothwa, 2011).
2. *Sutherlandia frutescens* (the so-called cancer bush) is an important medicinal plant in South Africa. Researchers have realised that this indigenous shrub common in South Africa has potent medical qualities that were known in early times by the Khoi, San and Zulu healers. Early people have observed that people suffering from cancer responded well to extracts made from this plant. They hypothesised that *Sutherlandia frutescens* may assist cancer patients, since there are active ingredients in this plant that assist the immune system to fight disease. Recent research has shown that the shrub contains an amino acid which fights depression, pinitol which helps patients to gain weight, and canavanine which is successful in treating retroviruses (Van Wyk et al., 2013). It is used to treat AIDS patients today. Although it does not cure AIDS, it definitely helps people with AIDS to enjoy a better quality of life. This is an interesting example of how modern science is giving status to the work of traditional healers.

Figure 15.1. *Sutherlandia frutescens*—the "cancer bush."

These developments also offer the opportunity to reflect on biotechnology and modern pharmacology in the life and natural sciences classroom. This provides teachers with golden opportunities to introduce learners to scientific processes and procedures that support indigenous knowledge claims—indigenous knowledge need not be pseudo-science: it can demonstrate rigorous scientific procedures. For example, botany can come alive if a teacher introduces simple chromatography techniques or organic chemistry. Simple chromatography can be done on a shoestring in the laboratory. As enrichment, teachers can also refer to modern techniques used in laboratories today, such as thin-layer chromatography (TLC), liquid chromatography coupled with mass spectrometry (LC-MS) or high-performance liquid chromatography (HPLC). These basic procedures are applied in most university laboratories, and they will give learners a better understanding of the nature of science, as well as future career opportunities.

Arguing for the Inclusion of Indigenous Knowledge in the Science Curriculum From a Neuroscience, and Embodied, Situated, and Distributed Cognition (ESDC) Perspectives

In a new democratic South Africa, all students have physical access to education. However, researchers like Le Grange (2016) rightfully ask if all students have epistemological access to education, though. Very often the incorporation of indigenous knowledge is substantiated from either a political or social justice perspective. In this chapter I do it from the learning psychology. Dubinsky et al. (2013) focus the attention on neuroscience learning concepts that directly inform teaching and learning. These authors also argue how such insights will directly inform and transform teacher preparation and teacher professional development. Some of the recent research findings emanating from neuroscience, that support the incorporation of indigenous knowledge in the science curriculum, are:

- An individual's regular and novel activities affect synaptic strength. The salience of an event or experience will determine its retention. Since learners come to the classroom with different prior knowledge based on their culturally learned experiences, the science teachers should use this as a point of departure of introducing abstract science concepts.
- The human brain endows us with a natural curiosity to understand how the world works. Harnessing natural curiosity of young learners engages and motivates them in the innate process of exploring their environment.
- Only experiences with an emotional stamp become committed to memory; decisions require operational emotional circuits (Dubinsky et al., 2013, p. 319).

Chahine (2013) provided a very insightful account of embodied, situated and distributed cognition (ESDC) from an educational perspective. She describes cognition as physiologically embodied, socioculturally situated, and ostensibly distributed among individuals (p. 434). The viewpoint of embodied cognition is that cognitive processes are deeply rooted in the body's interactions with the world (Wilson, 2002). Learners develop their worldviews based on their (situated) engagement with people and the environment. Such cultural knowledge is coconstructed (and therefore distributed in the community). This provides a very good entry into the abstract world of science.

Research Methodology

The research highlighted in this chapter spans almost a decade, and it employed an explanatory, sequential, mixed-methods design, as explained by Creswell (2009). The aim of this research was to determine teachers' perceptions of the inclusion of indigenous knowledge in the natural and life sciences classroom, as well as factors that prevent such epistemological border crossing, both before and after a short learning program. The following five research questions guided this study:

1. What are teachers' views of indigenous knowledge prior to in-service training opportunities (short learning programs)?
2. What role do short learning programs play in changing teachers' views of indigenous knowledge and its role in the science classroom?
3. Which factors prevent the inclusion of indigenous knowledge in the natural and life sciences classroom?
4. What pedagogies are effective in facilitating epistemological border-crossing between indigenous knowledge and the natural sciences?
5. What insights are derived from short learning programs on indigenous knowledge that were presented over a period of five years, by using Cultural-historical activity theory (CHAT) as a research lens?

I used the Rogan (2004) model as conceptual framework for this research. Rogan's work focuses mainly on capacity development of natural sciences teachers and professional development that enable teachers to implement the curriculum creatively. This links well with the conceptual prototype proposed by Chahine (2016), which was inspired by CHAT (see Tables 15.1 and 15.2).

The Intervention: A Short Learning Program (SLP)

Teachers participated in a three-day course[1] during which they engaged in cooperative learning, problem-based learning, and laboratory activities related to the infusion of indigenous knowledge in the natural and life sciences curriculum. The activities were designed in a way that would highlight the complimentary nature of both science and indigenous knowledge. The activities focused on doing tests on medicinal plants to determine its antimicrobial activities (an adapted Kirby-Bauer technique) in the laboratory (De Beer & Whitlock, 2009; De Beer, 2019), engaging in ethnobotanical surveys in the natural and life sciences classroom (De Beer & Van Wyk, 2011), exploring Australian (Noongar) indigenous knowledge related to the role of karrikins on seed germination (De Beer, 2012), Chinese and Egyptian indigenous knowledge related to the plant hormone ethylene (De Beer & Petersen, 2017a), and Dutch indigenous knowledge related to increasing the shelve life of cut flowers (De Beer & Petersen, 2017b). After the SLPs teachers had three months to compile and submit an evidence-based portfolio. In these portfolios teachers had to provide evidence that they could successfully do such epistemological border-crossing in the classroom.

Table 15.1.
Profile for Implementation

LEVEL	CLASSROOM PRACTICE	SCIENCE PRACTICAL WORK	SCIENCE IN SOCIETY	ASSESSMENT
1	**Teacher:** • Presents content in a well-organized way • Has a lesson plan • Provides adequate notes • Uses textbook effectively • Engages learners with questions	**Teacher:** • Uses demonstration to develop concepts • Uses specimens found in local environment for illustration	**Teacher:** • Uses examples and applications from everyday life—apply to science content	**Teacher:** • Uses written tests • Mostly recall type questions • Some questions are higher order thinking • Tests marked and returned promptly
	Learners: • Stay attentive and engaged • Respond to and ask questions	**Learners:** • Learners observe • Ask and answer questions	**Learners:** • Stay attentive and engaged • Ask and answer questions	**Learners** • Mostly apply rote learning • Sometimes apply higher order thinking
2	**Teacher:** • Textbooks used in conjunction with other resources • Engages learners with questions to encourage in depth thinking	**Teacher:** • Uses demonstration to promote a limited form of inquiry	**Teacher:** • Uses specific problem or issue faced by local community	**Teacher:** • Uses written tests • 50% of questions require higher order thinking • Some of the questions are based on practical work
	Learners: • Use additional resources to compile own notes • Engage in meaningful group work	**Learners:** • Some learners assist in planning and performing demonstration • Learners participate in cook-book practical work	**Learners:** • Teacher assists learners to explore the explanations of scientific phenomena by different cultural groups	**Learners** • Apply practical knowledge • Apply higher order thinking

(Table continues on next page)

Table 15.1.
Profile for Implementation (Continued)

LEVEL	CLASSROOM PRACTICE	SCIENCE PRACTICAL WORK	SCIENCE IN SOCIETY	ASSESSMENT
2	**Teacher:**	**Teacher:** • Learners communicate data using graphs and tables Ask and answer questions		
3	**Teacher:** • Probes learners' prior knowledge • Structures learning activities on relevant knowledge and problem solving techniques	**Teacher:** • Designs practical work to encourage learner discovery of information	**Teacher:** • Teacher facilitates investigation	**Teacher:** • Uses written tests • Tests include seen or unseen guided discovery type activities • Uses other forms of assessment as well
	Learners: • Engage in minds-on learning activities • Make own notes on the concepts learned from doing these activities	**Learners:** • Perform guided discovery type practical work in small groups—hands on • Write a scientific report	**Learners:** • Learners actively investigates science application in own • Gathers data, e.g., surveys	**Learners** • Apply practical knowledge • Apply higher order thinking • Includes open investigation of community project in assessment
	• Take major responsibility for own learning; partake in planning and assessment of own learning	• Can justify conclusions in terms of data collected • Design and do own open investigations	• Undertake long term community-based investigation • Apply science to specific need in community	• Create portfolio to present best work
		• Reflect on design and collected data • Interpret data		

(Table continues on next page)

Table 15.1.
Profile for Implementation (Continued)

LEVEL	CLASSROOM PRACTICE	SCIENCE PRACTICAL WORK	SCIENCE IN SOCIETY	ASSESSMENT
4	**Teacher:**	**Teacher:**	**Teacher:**	**Teacher:**
	• Facilitates learners as they design and undertake long-term investigations and projects	• Facilitates learning with design and data collection strategies	• Facilitates learning with the community project and identifying the need	• Creates opportunity for different types of assessment
	• Assist learners to weigh theories that attempt to explain the same phenomena	• Facilitates learning on data interpretation and conclusions		• Facilitates in compilation of portfolio
Level Behavior	**Learners:**	**Learners:**	**Learners:**	**Learners**
	• Takes major responsibility for own learning and undertake long-term investigations and projects.	• Design and do open investigations.	• Actively undertake a project in their local community.	• Develop portfolios.

Source: Adapted from Rogan and Grayson (2003, p. 1185)

Table 15.2.
Research Instruments Used in This Longitudinal Design-Based Research

Instrument	Description	Data Analysis	Number of Participants
Generic questionnaires on IK	Teachers' opinions on indigenous knowledge (Gauteng and Limpopo provinces)	Likert scale; percentages	255
Views-on-the-nature-of-indigenous-knowledge (VNOIK) questionnaire	Instrument developed by Cronje et al. (2015) to capture teachers' views on the nature of indigenous knowledge (in a pre- and posttest, following a short learning program)	The method described by Cronje et al. (2015) was used. A rubric was used, and responses were coded as informed-, partially informed or uninformed views. Averages were then determined.	300
Reformed Teaching Observation Protocol (RTOP)—Classroom observations (Sawada, Piburn & Judson, 2002).	Instrument used during classroom observations, to record teaching and learning activities	Likert-scale items as well as descriptions were analysed to determine how teachers implemented indigenous knowledge in their science lessons.	50
Teacher portfolios	Natural and Life Sciences teachers who attended the SLP's submitted evidence-based portfolios after the interventions.	Portfolios were analysed by using the coding method of Saldana (2009): categories were clustered to result in emerging themes.	120
Individual interviews	Personal interviews were conducted with teachers after classroom observations, to discuss their lived experiences of teaching indigenous knowledge.	Saldana's (2009) coding technique was used to distil emerging themes.	20

RESULTS

Teachers' Experiences of Infusing Indigenous Knowledge in the Natural—and Life Sciences Classroom

Quantitative Analysis: Teachers Views of the Infusion of Indigenous Knowledge Within Science Topics, and Their Views on the Nature of Indigenous Knowledge

The quantitative data obtained with a generic questionnaire that was completed by 255 teachers are summarized in Table 15.3. (These questionnaires were sent by post to all natural and life sciences teachers in the Gauteng and Limpopo provinces, and 255 completed questionnaires were received back.)

Table 15.3.
Teachers' Responses to Questions About the Implementation of IK in the Classroom (*n* = 255).

	Negate Vehemently	Do not Agree Fully	Agree	Agree Wholeheartedly
By referring to indigenous knowledge, learners perceive science as more relevant.	1.7%	11.7%	60.9%	25.7%
The natural sciences explain all natural phenomena perfectly.	7.6%	51.3%	33.5%	7.6%
The importance of indigenous knowledge is exaggerated.	4.5%	64.1%	25.6%	5.8%
There is no connection between indigenous knowledge and modern science.	9.4 %	63.4 %	22.8 %	4.4 %
Indigenous knowledge is often in conflict with modern science.	0.5 %	37.9%	51.4 %	10.2 %

The data show that life sciences teachers realize the importance of indigenous knowledge, and that they realize that it plays a role in making life sciences more relevant for the learners. A large percentage of the teachers (41.1 %) are of the opinion that natural sciences can fully explain natural phenomena—I shall address this subsequently when I discuss the nature of natural sciences. The majority of the teachers (61.6%) are of the opinion that indigenous knowledge and science are in conflict with each other. An important reason for this view is teachers' deficient knowledge regarding indigenous knowledge (Table 15.3). Another major problem is the fact that some teachers feel that indigenous knowledge is in conflict with their religious views (for instance, Christians who feel that indigenous knowledge also includes practices such as making contact with spirits of the ancestors). This highlights the naïve views that teachers sometimes hold of indigenous knowledge, as De Beer and Van Wyk (2016) emphasize that there are often very plausible scientific explanations for many practices in indigenous knowledge systems that seem to be metaphysical in nature.

In Table 15.4 a summary is provided of teachers' responses in the views-on-the-nature-of-indigenous-knowledge (VNOIK) instrument. This instrument, developed by Cronje, De Beer and Ankiewicz (2015) capture the views that science teachers hold of indigenous knowledge. The questions are based on the tenets of science and IK, for example, that IK is empirically based, inferential yet intuitive, creative, subjective, cultural, and holistic in nature (Cronje et al., 2015).

Table 15.4.
Average Percentage of Participants' VNOIK Before and After an Intervention[2]

Knowledge	VNOIK Before	VNOIK After	Percentage Change
Uninformed view	4%	0%	4% <
Partially informed view	83%	59%	24% <
Informed view	13%	41%	28% >

Source: (Cronje, De Beer, & Ankiewicz, 2015, p. 330).

The short courses seemed to assist teachers in their professional development, as there was an increase of 28% of teachers who held informed views on indigenous knowledge after the intervention (Cronje et al., 2015).

Qualitative Data Analysis

Several SLPs in Gauteng, Limpopo, Mpumalanga, North-West, and the Northern Cape provinces were presented. After these interventions, selected teachers were observed in their classrooms (and lessons analyzed using the Reformed Teaching Observation Protocol [RTOP] instrument). Individual interviews were conducted with these teachers, and transcribed, and the portfolios that teachers had to submit (in order to receive credits for attending the SLP) were analyzed. In Table 5.5, the themes, after analysis of the transcribed interviews, lessons, and portfolios, are indicated.

Table 15.5.
Themes That Emerged in the Qualitative Analysis

Data Sources	Categories	Emerging Themes
VNOIK, RTOP, portfolios, interviews	Deficient teacher knowledge- both content knowledge and indigenous knowledge- to address indigenous knowledge in the classroom.	**Theme 1** The challenges experienced with the infusion of IK in teaching Life Sciences mostly centres around a lack of teacher PCK
VNOIK, RTOP, portfolios, interviews	A lack of cooperative learning, problem-based learning and laboratory skills among teachers to infuse IK using 21st Century skills	
VNOIK, interviews	The view that indigenous knowledge does not belong in the life sciences classroom (different sphere of knowledge).	**Theme 2** Although the SLP provides teachers with more nuanced understanding of indigenous knowledge, a large percentage of teachers still have underdeveloped knowledge of the tenets of the nature of science and indigenous knowledge
Interviews	Indigenous knowledge is in conflict with religious views	
Interviews	Despite the fact that teachers see the value of IK in the science classroom, pace-setters and emphasis on examinations lead to tensions that negatively impact on delivery.	**Theme 3** There is a need for systemic support in order to facilitate such border-crossing
Interviews	A sense of disempowerment, constant curriculum change, but insufficient support.	
Interviews, RTOP	Lack of supporting material for teaching indigenous knowledge.	
Questionnaires, interviews	A sense of being overwhelmed as a result of the cultural diversity in the country whose indigenous knowledge should be highlighted.	

THEME 1: THE CHALLENGES EXPERIENCED WITH THE INFUSION OF IK IN TEACHING LIFE SCIENCES MOSTLY CENTERS AROUND A LACK OF TEACHER PEDAGOGICAL CONTENT KNOWLEDGE (PCK)

The data revealed that the participating teachers experienced several challenges in relation to the infusion of IK in teaching natural and life sciences, which mainly points to the lack of teacher PCK. A number of categories arose from the data.

Insufficient Training and Professional Development for the Implementation of the Curriculum

The participants acknowledged that the infusion of IK in life sciences posed challenges. The narratives also revealed that the teachers were concerned with the extent to which life sciences educators are trained to infuse their teaching with IK. Teachers acknowledged that they lack knowledge related to the contemporary curriculum and infusing IK in their teaching:

> My main problem is that of the lack of teaching methodology and it is difficult to teach the learners. I think this is teachers' greatest concern. Learners are adapted to western life and mostly their belief system makes them loose interest in their indigenous knowledge. This makes them relate IK to evil spirits (Satanism). The practical application of the indigenous knowledge in class is also difficult. As a teacher I am challenged to develop some practices that can convince them to be at least gravitated towards science.

Teachers need relevant pedagogical content knowledge in order to infuse their teaching with IK. A Life Sciences teacher needs *content knowledge*, that is, know-how of the indigenous knowledge claims of different cultural groups, for example, plants used for medicinal reasons; *pedagogical knowledge*, that is, how to effectively structure learning opportunities for learners when exploring IK claims, and *contextual knowledge*, that is, what the cultural background of the learners are. In order to facilitate the learning activities dealt with in the SLP, teachers need a good understanding of microbiology, plant physiology, and taxonomy. Due to the emphasis on transmission-mode teaching in schools, some teachers are also not familiar with innovative cooperative and problem-based teaching methods.

THEME 2: UNDERDEVELOPED KNOWLEDGE OF THE TENETS OF THE NATURE OF SCIENCE AND INDIGENOUS KNOWLEDGE

Although the SLP provides teachers with more nuanced understanding of indigenous knowledge, a large percentage of teachers still have underdeveloped knowledge of the tenets of the nature of science and indigenous knowledge. However, there has been an improvement in teachers' understanding of the tenets of science and indigenous knowledge. The comment below (from the questionnaire) summarizes the views of many teachers, prior to participating in the SLP:

> Indigenous knowledge brought fear to us as educators because we don't know how to approach it as we were never given courses or trained about it. We are just striving to understand it and feel that it is a challenge to us as we are having different belief systems. As teachers we are divided where some of us just brush it aside and continue with the other part of the syllabus.

However, many teachers in their portfolios and post-questionnaires indicated that they feel more empowered to facilitate such border-crossing after the SLP:

> The course made me to realise that there is not such a big difference between science and indigenous knowledge. However, I also realized that I know very little about the nature of science. I am determined to learn more about these aspects."

Another teacher commented as follows: "With my better understanding of the nature of science after this course, I think that I will teach differently in future."

Teachers' Own Negative Perceptions of IK

Some teachers are negative about IK and feel that IK cannot be scientifically proven and that it constitutes "pseudo-science"—a few even after the SLP. Many teachers do not have the know-how to incorporate IK in a scientific and rigorous way.

> Eish, ... some of the things cannot be scientifically proven and do not have a living evidence. Some cannot be brought to class so that they can be done practically or even show to learners when the teacher needs to give an explanation.

> Most of the learners have a negative attitude towards IK as they take it to be outdated and barbaric. Maybe it is because of their religion. To some people, anything that is not European is not just worth their time.

Religious Beliefs

The narratives revealed that most participants are bound by their religious beliefs in their infusion of IK. Belief systems discouraged some teachers to involve themselves in IK. Many Christian teachers thought that visiting a *sangoma* and making contact with their ancestral spirits is against their religious belief system.

> Most of the learners really give me a very negative attitude towards IK by saying it is against their religious beliefs. It is only few who seem to be interested in it. This makes some learners also to lose interest and become reluctant in doing the work.

THEME 3: THERE IS A NEED FOR SYSTEMIC SUPPORT IN ORDER TO FACILITATE SUCH BORDER-CROSSING

Lack of IK-Related Material Resources in School

The transcriptions of interviews revealed that it was difficult to include IK because of a lack of material resources. The participants indicated that it was difficult to obtain further information about indigenous knowledge. One participant mentioned that:

> We have the problem of not having enough material concerning IK. Teachers do not have places of reference or at times do not get necessary material to use in class. As I have already said that this type of knowledge is verbally transmitted, there is nowhere one can get the proof of its existence in the first place. Everybody comes up with his or her own statement. This leads us to the misconceptions and misrepresentations about indigenous knowledge.

Cultural Diversity and Differences

Teachers might be knowledgeable in terms of their own culture, but do not necessarily know much about IK in other cultures. This leads to a confusion of whose culture to consider as they do not match.

> Eish, ... teaching Life Sciences is very challenging, especially that you teach learners from different cultures and communities. As a teacher you cannot cover all their cultures, some of which you do not know. When you emphasise knowledge from one cultural group, others feel belittled while others feel excluded. At times you do not get the necessary or enough information or material you want to teach learners. Learners are from different cultures and environments and it is difficult to know and cater for their different cultural backgrounds in class. Our policy states clearly that we must include their indigenous knowledge in class, but it is difficult because as an educator I did not know them all. It is only a few that I am acquainted with.

Lack of Support Systems From the Department of Education During a Period of Constant Curriculum Change

The data (portfolios, questionnaires and interviews) indicate that the participants experienced negative emotions because of the perceived risk of not being able to convey the correct IK content to learners. The curriculum changes that were introduced, without thorough professional development support to accompany these changes, also create stress among teachers. Some participants indicated the following:

> There are changes in our education system that leads to the confusion and frustration on us. For instance, we are trained in an old method of teaching and they brought us OBE which was followed by NCS. It is difficult for us to apply them in class. When doing the NCS there are some outcomes that we need to accomplish with the inclusion or infusion of indigenous knowledge, but it is difficult because of the lack of material, different cultures from different communities and our own experiences as educators.

The participants indicated that there is lack of support from the Department of Education. As such, they try to work on their own by organising Saturday schools and inviting knowledgeable people from the community to come and share lessons on IK. The only available support comes from colleagues, educators from surrounding areas, parents and community elders. The participants mentioned that:

> We are not supported by anybody, instead we run around seeking help from others. We are just like a train moving without the head. You as a teacher will see how to come up or to solve problem.

The department needs to organize in-service education and training for the teachers in order to upgrade their skills and the knowledge they impart to learners. **Feeling of a lack of control over own practice** (e.g., changes in the education system and curriculum, and the inclusion of IK in the life sciences syllabus)

The narratives indicated that the participants felt they have no control over their teaching because of the changes in the education system and curriculum, and the inclusion of IK in the syllabus for life sciences. The following quote from Rogan and Aldous (2005) also applies to the inclusion of IK in Life Sciences:

> Teachers are at times described as being reluctant to changes, of being bound to tradition, or even active saboteurs of the intended curriculum. In our case studies, the low level of implementation, for the most part, cannot be described to "foot dragging or lack of effort." The interview revealed an overwhelming desire to faithfully implement the new curriculum, but tinged with an enormous perplexity about how to do so, and an uncertainty as to whether their efforts in the classroom were "the right way."

Utilizing Rogan's (2004) Profile of Implementation

When we use Rogan's (2004) model as conceptual lens, it is clear that the SLP facilitated teacher learning, and that limited movement was achieved in terms of the *profile of implementation*. Most teachers who functioned at Level 1 in Rogan's model (see Table 15.1) prior to the SLP, migrated to a Level 2. In this category teacher-centered practices were replaced by meaningful group work, and cookbook-type experiments replaced the occasional demonstrations. However, during the lesson observations after the SLP, the predominant trend was that indigenous knowledge was predominantly addressed by a number of examples, and relatively few cases of real inquiry learning. Some teachers did engage their learners in laboratory activities that were presented during the SLP. However, there was little evidence of innovation, and teachers engaging learners in contextualized investigations during the first cycles of this project.

In this project the researchers thus realized that such a relatively short-term intervention (the SLP) has limited value, if it is not accompanied by continued support. We therefore started to also introduce support mechanisms for teachers after the SLP, for instance the use of electronic platforms such as a Blackboard system, in which teachers could participate within online communities of practice. This paid off, and in the later SLPs (the more recent cycles), we started to see a migration to the higher levels in Rogan's (2004) profile of implementation, namely to Levels 3 (where some teachers engaged learners in contextualized problem solving) and even to Level 4 (In Calvinia, for example, one teacher engaged her learners in an ethnobotanical survey, where learners engaged in quite sophisticated research on the medicinal plants in the district). However, this asked of the facilitators of the SLP to continuously engage with the teachers online, and to provide support as needed.

It became clear that a focused, longitudinal, professional development program will be needed to empower teachers to function at Level 4. Although the SLPs definitely facilitated teacher professional development, it became evident that sustained support after the course is essential.

Studying the Intervention Through a Cultural-Historical Activity Theory (CHAT) Lens

CHAT has its origin in the work of Vygotsky (1978) and is a useful lens to study this project. The unit of analysis is the natural or life sciences teacher's professional development. CHAT provides a specific gaze into the activity system, and the professional learning that occurred as a result of the intervention. In Figure 15.2, I use a depiction of CHAT in the way described by Chahine (2016).

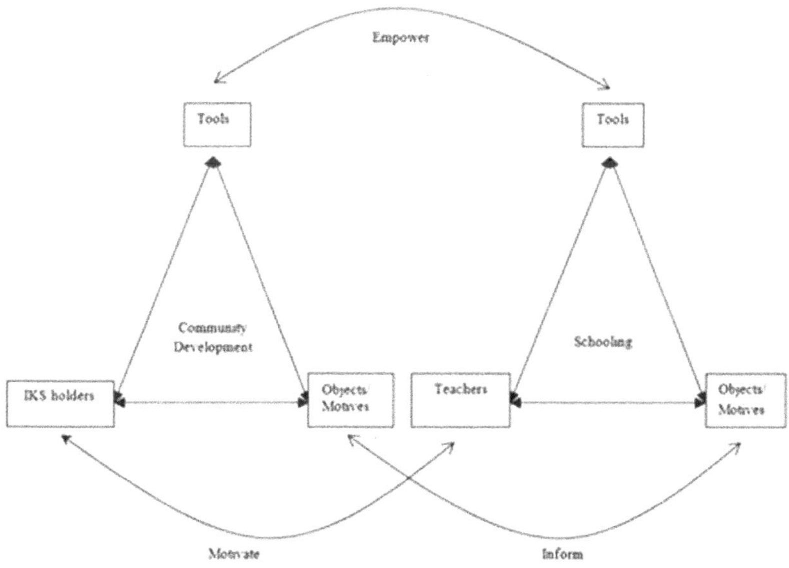

Source: Adopted from Chahine (2016).

Figure 15.2. A depiction of CHAT in the context of teacher professional development.

The activity system on the left focuses on the SLP. The natural sciences teacher (the subject) engages with the tools provided during the SLP (pedagogies such as problem-based learning, cooperative learning, as well as indigenous knowledge provided by indigenous knowledge holders) in order to obtain the object of the activity: more contextualized science education, where indigenous knowledge is infused with the curriculum themes. The rules that need to be considered are the tenets of science and indigenous knowledge, as well as guidelines for effective problem-based and cooperative learning. The supportive community includes the SLP facilitators, as well as the holders of indigenous knowledge with whom the teachers engage during the SLP. The division of labor in this context refers to the different "hats" that the teacher needs to put on: that of a facilitator of learning, a life-long learner, a critical reflective practitioner, and a researcher.

The second activity system in Figure 15.2 depicts the typical postintervention classroom. After the SLP researchers in the project visited some of the teachers who participated in the course in their classrooms. The purpose of these visits was to see how teachers implement indigenous knowledge in their teaching, and how they manage this epistemological border-crossing.

Engeström (2009) mentioned that an activity system is built around the object. McNeil (1999) referred to the "contradiction of control" which sheds light on the complexity of the object in an activity system. The second activity system shows this "contradiction of control." Whereas teachers generally realized the affordances of indigenous knowledge in the science classroom after the SLP, the incorporation thereof were often restricted by systemic tensions. Many of the classroom observations showed that little transfer took place to the school classroom after the SLP. During the interviews teachers were asked about this, and a number of factors were identified by the teachers for this unfortunate situation. One such factor is the Department of Education's "pace setters" (schedule), and the strong focus on preparing learners for the examination, which often result in teachers not changing their pedagogies. This then negatively impact on achieving the object of the activity system, namely contextualized science (through effective border crossing). Figure 15.2 highlights several tensions that emerges in the activity system. One such tension is that some teachers—even after a relatively successful SLP—still do not have nuanced views of the tenets of respectively science and indigenous knowledge, which negatively impact on border-crossing. It is for this reason that Chahine (2016) advocates a professional development approach (conceptual prototype) that entails training (the SLP), continued support (which could be online communities of practice), and eventually mastery (teachers building up experience). This is of course in line with the profile of implementation that was proposed by Rogan (2004) (see Table 15.1).

CONCLUSION: SUCH EPISTEMOLOGICAL BORDER-CROSSING IS INDEED POSSIBLE

The research done in this funded project clearly highlights the affordances of epistemological border-crossing, where indigenous knowledge is incorporated in the formal ("Western") science curriculum. Due to the shared tenets of respectively science and indigenous knowledge, is indeed possible to introduce indigenous knowledge in the natural science classroom, using the processes of science. Such an approach could result in what Gibbons (2000) called "mode 2 knowledge production," namely context-sensitive science. However, it is clear that more resources should be developed for use in the classroom, to highlight the complementarity of science and indigenous knowledge.

Major Problems That Negatively Impact on Such Epistemological Border-Crossing

The question posed at the beginning of the chapter, whether there can really be synergy between indigenous knowledge and modern science, has been answered. It is indeed possible to explain natural phenomena by using modern science and indigenous knowledge structures as lenses, and to show learners that these two interpretations can supplement and enrich each other. The above examples indicate that important scientific discoveries can indeed arise from such synergy.

There are two problems that may hamper the above integrated approach. First, teachers' deficient pedagogical content knowledge and second, teachers' deficient understanding of the nature of natural sciences.

Teachers' Expert Teaching Skills and Pedagogical Content Knowledge

Pedagogical content knowledge (or, as I prefer to refer to it, expert teaching skill), a concept formulated by Shulman (1986), refers to the integrated subject knowledge, didactic knowledge, as well as understanding of the learner, which in combination, enable the teacher to teach the concept so that learners understand the concept maximally. Teachers are not trained to address indigenous knowledge (De Beer & Ramnarain, 2012). In most classrooms where indigenous knowledge is indeed addressed, this is merely by means of an example or two. The same applies to the national Grade 12 examination papers (De Beer, 2011). However, to unlock indigenous knowledge heuristically for learners with the aid of investigative activities demands special expertise from teachers. De Beer and Van Wyk (2011) describe how learners may do an ethnobotanical survey in the life sciences classroom, while De Beer and Whitlock (2009) explain how medicinal claims relating to plants may be tested in the classroom by means of simple chromatographic techniques and simple tests for antimicrobial activity. Van Wyk and De Beer (2012) showed how the *matrix method* may be of value—a methodology which can be implemented very easily in the life sciences classroom.

Teachers' View of the Nature of Natural Sciences

There is no consensus in the literature about what the nature of natural sciences entails (Laugksch, 2000; National Research Council, 2001). For many researchers, the nature of aspects of natural sciences comprises aspects such as the philosophy, history and sociology of science. In this chapter, I concur with the interpretation of Dekkers and Mnisi (2003): What is the origin of natural-scientific knowledge and what is the state of scientific knowledge? Abd-El-Khalick, Bell, and Lederman (1998) are of the opinion that a person who has a sound concept of the nature of natural science will understand that this science is tentative (subject to change), empirically grounded (conclusions based on observations); subjective; partially the product of human conclusion, imagination and creativity and that it has a social and cultural context. However, research has shown that many teachers have naive and outdated views of the nature of science (De Beer & Ramnarain, 2012; Galagher, 1991; Lederman, 1992; McComas, 1998). If we expect the life sciences teacher to present various knowledge systems in juxtaposition to learners as possible explanations of natural-scientific phenomena, it is necessary to equip teachers with the understanding van values needed to compare the systems (Dekkers & Mnisi, 2003).

Unfortunately, there are many teachers who still regard indigenous knowledge (particularly concerning traditional healing) as magic, which is light years removed from science. This is sad—in a country that is suffering owing to the HIV/AIDS virus, very high levels of unemployment and a situation where South Africa is struggling to be internationally competitive owing to insufficient numbers of trained scientists, indigenous knowledge (and particularly ethnobotany) opens many doors to prospective scientists and entrepreneurs.

RECOMMENDATIONS

The funded research project adopted a design-based slant, and the following recommendations emerged from the findings:

1. Attention should be paid to expert teaching skill of life sciences teachers in professional development programs. Teachers should become acquainted with the latest developments in biotechnology and ethnobotany.
2. Teachers should get exposure to the nature of research, for instance by spending some time in laboratories where they work together with researches. This is happening, for example in the A-Team professional development program of the University of Johannesburg (De Beer & Ramnarain, 2012). Hopefully, such exposure will assist teachers to let go of the recipe book approaches to teaching life sciences and expose learners to the true nature of natural sciences.
3. A more systemic approach to teacher professional development is needed. As depicted in the conceptual prototype of Chahine (2016) short learning programs should be followed by continued support, and the opportunity for teachers to develop mastery.
4. Jautse, Thambe, and De Beer (2016) make the statement that a "third partner" is needed in the value chain between schools and universities in teacher professional development. This third partner can be the holders of indigenous knowledge, but also funders. The research has clearly highlighted the fact that learning materials are needed to effectively implement IK in school classrooms. Currently this project enjoys the support of the Fuchs Foundation,[3] which make it possible to provide teachers and schools with equipment and materials. This could enhance the transfer of knowledge and skills to the school classroom.

SUMMARY

Although the vast majority of teachers realize the value of the inclusion of indigenous knowledge in the life sciences curriculum, a large percentage of the teachers do not have the necessary knowledge or skills to facilitate such epistemological border-crossing. Teachers highlighted the following problems that render it difficult to address indigenous knowledge in the classroom: inadequate support from the Department of Education, deficient knowledge and skill, lack of supporting material (textbooks and other literature), the perceived conflict between indigenous knowledge and religious views, and the viewpoint that indigenous knowledge does not belong in natural sciences. In this chapter, I have presented examples which teachers may use to illustrate the close relationship between indigenous knowledge and modern science. A teacher's social accountability means, among other things, that he/she should make learners aware of the cultural and practical value of indigenous knowledge and stimulate their interest in disciplines such as ethnobotany and chemotaxonomy. Former President

Thabo Mbeki, in his African Renaissance Statement of 1998, held the following view:

> Africa's renewal demands that her intelligentsia must immerse itself in the titanic and all-round struggle to end poverty, ignorance, disease and backwardness, inspired by the fact that the Africans of Egypt were, in some instances, two thousand years ahead of the Europeans of Greece in the mastery of such subjects as geometry, trigonometry, algebra and chemistry. (Mbeki, 1998)

South Africa has proud Khoisan indigenous knowledge and traditions and the time has arrived for this knowledge to form a part of the formal school curriculum. Indigenous knowledge can make a contribution to efforts to solve the challenging problems with which our country and its people are struggling.

NOTES

1. A short video on the project can be seen by clicking on this link: https://www.youtube.com/watch?time_continue=548&v=1kK0VIHw1Tw
2. Table 15.4 reports on three interventions in respectively Gauteng, North-West Province and Mpumalanga, in which 46 teachers participated. This is ongoing research, and this Table 15.4 only reflects on already published results.
3. The following video link provide more information on the Fuchs Foundation project: http://education.nwu.ac.za/self-directed-learning/videos

REFERENCES

Abd-El-Khalick, F., Bell, R. L., & Lederman, N. G. (1998). The nature of science and instructional practice: Making the unnatural natural. *Science Education, 82*, 417–436.

Aikenhead, G. S. (1996). Science education: Border crossing into the subculture of science. *Studies in Science Education 27*, 1–52.

Chahine, I. C. (2013). The impact of using multiple modalities on students' acquisition of fractional knowledge: An international study in embodied mathematics across semiotic cultures. *Journal of Mathematical Behaviour, 32*, 434–449.

Chahine I. C. (2016). Investigating students' perceptions of indigenous mathematical knowledge systems through immersion in transcultural settings. *Journal of Mathematics and Culture, 10*(3), 50–69.

Creswell, J. W. (2009). *Research design: Qualitative, quantitative and mixed methods approaches* (3rd ed.). SAGE.

Cronje, A., De Beer, J. J., & Ankiewicz, P. (2015).The development and use of an instrument to investigate science teachers' views on indigenous knowledge.

African Journal of Research in Mathematics, Science and Technology Education, 19, 319–332.

De Beer, J. J. (2011). Evaluation of the Grade 12 national life sciences examination papers for 2011. Report for Higher Education South Africa.

De Beer, J. J. (2012). Investigating the influence of karrikins on seed germination. *The American Biology Teacher, 74*, 324–329.

De Beer, J. J., & Petersen, N. (2017a). A laboratory investigation on the role of ethylene in seed germination from an indigenous knowledge perspective. *The American Biology Teacher, 79*(1), 55–61.

De Beer, J. J., & Petersen, N. (2017b). Post-harvest physiology of cut flowers: A problem-based, cooperative learning activity for the Biology classroom. *The American Biology Teacher, 79*(7), 578–583.

De Beer, J. J., & Ramnarain, U. (2012). *The implementation of the FET Physical- and Life Sciences curricula: Opportunities and challenges* (Report prepared for the Gauteng Department of Education, ISBN 978-0-86970-721-0).

De Beer, J. J., & Van Wyk, B. (2016). Ethnobotanical and anthropological perspectives on the use of magic plants in traditional healing practices in South Africa. *Planta Medica, 82* (S01) S1–S381.

De Beer, J. J., & Whitlock, E. (2009). Indigenous knowledge in life science classroom: Put on your De Bono Hats. *The American Biology Teacher, 71*(4), 209–216.

De Beer, J. J., & Van Wyk, B.-E. (2011). Doing an ethnobotanical survey in the life sciences classroom. *The American Biology Teacher, 73*(2), 90–97.

Dekkers, P., & Mnisi, E. (2003). The nature of science—Do teachers have the understandings they are expected to teach? *African Journal of Research in Mathematics, Science and Technology Education, 7*, 21–34

Dubinsky, J. M., Roehrig, G., & Varma, S. (2013). Infusing neoroscience into teacher professional development. *Educational Researcher, 42*(6), 317–329.

Engeström, Y. (2009). From learning environments and implementation to activity systems and expansive learning. *Actio: International Journal of Human Activity Theory, 2*(1), 17–33.

Galagher, J. J. (1991). Prospective and practicing secondary science teachers' knowledge and beliefs about the philosophy of science. *Science Education, 25*, 97–137.

Gibbons. M. (2000). Mode 2 society and the emergence of context-sensitive science. *Science and Public Policy, 27*(3), 159–163.

Jackson, C., De Beer, J. J., & White, L. (2016, October 24–28). Teachers' affective development during an indigenous knowledge professional teacher intervention. *Proceedings of the ISTE Conference on Mathematics, Science and Technology Education, Kruger National Park.*

Jautse, P. P., Thambe, N., & De Beer, J. J. (2016, October 24–28). A day at the museum: A case study of how museums could partner with universities in teacher education. *Proceedings of the ISTE Conference on Mathematics, Science and Technology Education, Kruger National Park,* 438–445.

Laugksch, R. (2000). Scientific literacy: A conceptual overview. *Science Education, 84*, 71–94.

Lederman, N. G. (1992). Students' and teachers' conceptions of the nature of science: A review of the research. *Journal of Research in Science Teaching, 29*, 331–359.

Le Grange, L. (2016). Decolonizing the university curriculum. *South African Journal of Higher Education, 30*(2), 1–12.

Mbeki, T.(1998). *The African Renaissance Statement. Presentation at Gallagher Estate, 13 August 1998.* http://www.dfa.gov.za/docs/speeches/1998/mbek0813.htm

McComas, W. F. (1998). *The principle elements of the nature of science.* In W. F. McComas (Ed.), *The nature of science in science education—Rationales and strategies*. Kluwer.

McNeil, L. M. (2013). *Contradictions of control: School structure and school knowledge.* Routledge.

Morrow, W. (2009). *Bounds of democracy: epistemological access in higher education.* HSRC Press.

Mothwa, M. M. (2011). Teachers' experiences of incorporating indigenous knowledge in the life sciences classroom (Unpublished MEd dissertation). University of Johannesburg.

National Research Council. (2001). *Science Education Standards*. National Academy Press.

Odora-Hoppers, C. A. (2002). *Indigenous knowledge and the integration of knowledge systems: Towards a philosophy of articulation.* New Africa Books.

Odora-Hoppers, C. A. (2004, March). *Culture, indigenous knowledge and development: The role of the university.* Occasional Paper No. 5, Centre for Education Policy Development. Braamfontein, Johannesburg.

Odora-Hoppers, C. A. (2010). Transformative leadership and the humanization of the academy: Towards the integrative paradigm shift. *Keynote address during the launch of the Centre for the Advancement of Non-Racialism and Democracy, Nelson Mandela Metropolitan University.*

Ogunniyi, M. B., & Hewson, M. G. (2008). Effect of an argumentation-based course on teachers' disposition towards a science-indigenous knowledge curriculum. *International Journal of Environmental and Science Education, 3*(4), 159–177.

Onwu, G., & Mosimege, M. (2004). Indigenous knowledge systems and science and technology education: A dialogue. *African Journal for Research in SMT Education*, *8*(1), 1–12.

Rogan, J. (2004). Professional development: Implications for developing countries. In K. Osaki, K. Hosea, & W. Ottevanger (Eds.), *Reforming science and mathematics education in sub-Saharan Africa: Obstacles and opportunities* (pp. 155–170). Vrije Universteit Amsterdam.

Rogan, J., & Aldous, C. (2005, March). Relationships between the constructs of a theory of curriculum implementation. *Journal of Research in Science Teaching,* 1–25.

Rogan, J. M., & Grayson, D. (2003). Towards a theory of curriculum implementation with IHB reference to science education in developing countries. *International Journal of Science Education*, *25*(10), 1171–1204.

Saldana, J. (2009). *The coding manual for qualitative researchers*. SAGE.

Sawada, D., Piburn, M. D., & Judson, E. (2002). Measuring reform practices in science and mathematics classrooms: The Reformed Teaching Observation Protocol. *School Science and Mathematics*, *102*(3), 245–253.

Shulman, L. S. (1986). Those who understand: Knowledge growth in teaching. *Educational Researcher*, *15*(2), 4–14.

Van der Merwe, P. J. (1945). *Trek. Studies oor die mobiliteit van die pioniersbevolking aan die Kaap (1770–1842)*. Kaapstad: Nasionale Pers.

Van Wyk, B.-E., & De Beer, J. J. (2012). Sosiaal-aanspreeklike etnobotaniese opnames in die Kaapse flora: etiese beginsels, metodologie en kwantifisering van data [Socially responsible ethnobotanical surveys in Cape flora: Ethical principles, methodology and quantification of data]. *Die Suid-Afrikaanse Tydskrif vir Natuurwetenskap en Tegnologie* [The South African Journal of Natural Science and Technology], *31*(1).

Van Wyk, B.-E., Van Oudtshoorn, B., & Gericke, N. (2013). *Medicinal Plants of South Africa*. Briza Publishers.

Vygotsky, L. S. (1978). *Mind in society*. Harvard University Press.

Wallace, J., & Louden, W. (1998). Curriculum change in science: Riding the waves of reform. In B. J. Fraser & K. G. Tobin (Eds.), *International handbook of science education* (pp. 471–485). Kluwer.

Wilson, M. (2002). Six views of embodied cognition. *Psychonomic Bulletin & Review*, *9*(4), 625–636.

CHAPTER 16

EXPLORING EMBODIED, SITUATED, AND DISTRIBUTED COGNITION

Enhancing the Inquiry Learning Experiences of South African Science, Technology, and Mathematics Learners

Umesh Ramnarain

ABSTRACT

This chapter engages with the paradigm of embodied, situated, and distributed cognition (ESDC) in exploring how multimodal experiences (Chahine, 2013) from the physical and sociocultural worlds of the learners may support concept formation in science, technology, and mathematics (STEM). Indigenous knowledge of learners has often been marginalized in the STEM classroom due to teachers lacking in pedagogical knowledge on how indigenous knowledge can be infused into the STEM curriculum. Insights are gleaned from the overarching tenet of ESDC, namely that

Evidence-Based Inquiries in Ethno-STEM Research: Investigations in Knowledge Systems Across Disciplines and Transcultural Settings, pp. 309–321

"cognition is physiologically embodied, socio-culturally situated, and ostensibly distributed among individuals" (Chahine, 2013, p. 434). An extensive review of literature on the relationship between indigenous and scientific worldviews identifies key aspects that align with the ESDC and that have significance for classroom practice. The role of the teacher as "cultural broker" in facilitating the transition between the students' life-world and school science as advanced by Jegede (1997) in his theory of collateral learning is explored. George (1990) presents four conceptual categories on the relationship between the indigenous and scientific worldviews that can guide teachers in planning multimodal learning experiences in the science classroom. In this regard, an inquiry-based teaching approach in supporting learners investigate their ideas that emanate from an indigenous worldview, and the role of argumentation in this approach when learners straddling across the two worldviews are discussed.

INTRODUCTION

A strong force driving curriculum reform in schools, STEM in South Africa was the assertion that the previous curriculum of the apartheid era was both inaccessible and irrelevant to indigenous African students. This inaccessibility is related not only to the fact that many of these students were not offered the opportunity to do science, technology or mathematics as subjects, but also to students who did attempt the subjects performing poorly due to the apparent disconnect between the nature of these subjects and their worldview experiences. The research conducted by the Foundation for Research Development (1993) showed that in the mid-1980s, only 16% of indigenous African students passed science, compared to approximately 80% of mixed-race, Indian, and White students.

According to Naidoo and Lewin (1998),

> the apartheid policies inflicted on the African student a curriculum which many perceive as largely irrelevant to their needs, difficult to the point of impossible given the learning context, and taught by large number of unqualified science teachers in schools with few or no laboratories and science equipment. (p. 730)

Oluka (2007a, 2007b) highlights the "principle of relevance" in science learning by emphasizing that "Learning is contextualized within the local realities" (p. 9). This suggests that science learning needs to connect in some way to issues, problems and contexts that are familiar to students.

Integrating indigenous knowledge with STEM has been viewed as a means by which relevance in science learning can be reflected. Post-apartheid school curriculum reform in school science has acknowledged the role of indigenous knowledge in learning, and "Valuing indigenous knowledge systems" is one of the principles of the National Curriculum Statement (Department of Basic Education, 2011, p. 3). One of the primary reasons advanced for the integration of science with indigenous knowledge systems has been the need to fully recognize the wisdom and values that people living in South African have acquired over the centuries (Ogunniyi & Hewson, 2008). Another reason for the inclusion of indigenous knowledge in the science curriculum is that it can be a means by which issues on environmental sustainability in non-Western societies may be addressed (Hewson, 1988; Odora-Hoppers, 2002), and hence improve the quality of life for South African people. However, from a scholarly perspective, the role of indigenous knowledge in supporting meaningful and relevant science learning is terrain that needs to be more extensively navigated.

It is against this background that this chapter explores the role and value of indigenous knowledge in enhancing science teaching. However, the insights are also relevant to technology and mathematics education. The discussion is framed by the notion of ESDC (Chahine, 2013). By drawing on this framework, pedagogical guidelines and strategies on how indigenous knowledge can support science learning will be discussed.

VALUING INDIGENOUS KNOWLEDGE SYSTEMS

Indigenous knowledge systems (IKS) are embedded in the cultural milieu of all people (Maila & Loubsher, 2003). Seemingly, IKS is generally used in literature as a universal term to denote the worldviews of indigenous people. Odora-Hoppers (2001) defines IKS as knowledge that is characterised by its "embeddedness in the cultural web and history of a people including their civilization and forms the backbone of the social, economic, scientific and technological identity of such a people" (p. 4). Some detractors view indigenous knowledge as knowledge that is not conversant with the modern world, and it has been argued that indigenous knowledge is not considered as being scholarly because of the way in which it is developed and transmitted (Cobern & Loving, 2001; George, 1990). However, there is an overwhelming support for the integration of indigenous knowledge in the school science curriculum, and strides are being made to ensure the utilization of IKS by various academic institutions, individuals and nonprofit organizations throughout the world (Maila & Loubsher, 2003). But while IKS are being increasingly linked to global sustainability, there is still little consensus about *how* it should be introduced into the school science curriculum.

Situating Indigenous Knowledge Within an ESDC Framework

The paradigm of ESDC is now invoked in exploring the role of indigenous knowledge in science learning. This framework posits that cognitive processes are not limited to the symbolic processing of internal information structures, but are in fact embedded physiologically in action, situated in the sociocultural world and distributed among agents, artefacts and external structures (Hardy-Vallée & Payette, 2008). Drawing from this framework is the basic premise that cognitive processes are stimulated when students are immersed in multimodal experiences that are physical, cultural, social and linguistic (Chahine, 2013). Essentially, this premise entails that cognitive neural processes that become active during learning are stimulated through physiological experiences in the physical world of objects and anthropological experiences in the sociocultural world where these experiences are shared (distributed). This interdependence between the cognitive, physiological, and sociocultural systems that leads to learning is represented in Figure 16.1.

The above conception of cognition lends credence to the view espoused by Jegede (1999) that capitalizes on the significant role that the learner's

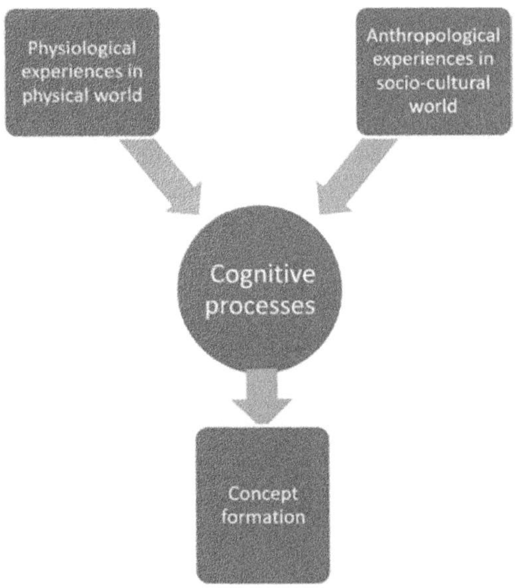

Figure 16.1. Author's depiction of the Integration of cognitive, physiological, and sociocultural systems.

cultural background and experiences plays in what and how to learn. Due to the cultural embeddedness of indigenous knowledge, and "the fact that indigenous knowledge systems reside among the majority of South Africans" (Le Grange, 2004, p. 578) it is clear that indigenous knowledge does encompass the anthropological experiences of learners in a sociocultural context, thereby contributing to cognitive processing and conceptualisation of natural phenomena.

It has been well documented in the science education literature that failure to acknowledge and address the indigenous knowledge of learners has led to difficulties in science learning (Jegede, 1996; Ogunniyi, 2007a; Okebukola & Jegede, 1990; Mavuru & Ramnarain, 2017; Msimanga & Shizha, 2014). In assuming an ESDC perspective, I argue that cognition is mediated by culture and takes place in a social context. This is also in accord with Jegede and Aikenhead (1999) who maintain that "the role of the social context is to scaffold the learner, and provide hints and help that foster co-construction of knowledge while interacting with other members of the society" (p. 1). Recognizing the social context of learning, as well as the effect of the learner's sociocultural background is of primary importance to ensuring success in science learning.

In South Africa, like other countries it is common practice for teachers to denigrate or discredit the knowledge systems of indigenous people, hereby dismissing a sociocultural perspective that could provide a trigger for learning science (Le Grange, 2007). When learners enter the science classroom, they encounter for the first time a science worldview that is disparate with their everyday sociocultural worldview. In school systems, learners are often forced to abandon their sociocultural worldviews and to replace with a new scientific way of viewing the world. This is referred to as "assimilation" and can alienate learners from science (Cobern & Aikenhead, 1998). In order to perform in school-based assessments they resort to rote memorization of key terms and processes that is counterproductive to meaningful learning (Tobin & McRobbie, 1997). Larson (1995) refers to these strategies as playing "Fatima's rules." If one subscribes to a "cognition beyond the brain" view then assimilation constraints learning because "cognitive processes are not limited to the symbolic processing of internal information structures," but embedded in the body, environment and culture (Hardy-Vallee & Payette, 2008). Jegede's collateral learning theory does provide some insight into how learners may be supported in achieving cognition that is ESDC. Jegede and Aikenhead (1999) opine that when instruction at school recognizes and is informed by the learners' worldview, there is smooth border crossings in school science. Success in school science depends on how effectively learners move between their life-world culture and the culture of science. Properly planned multimodal experiences can enable learners to smoothly transit across borders, and thereby acquire cognition in science. Opportunities need to be deliberately sought whereby the learning of concepts in the science classroom is mediated within a sociocultural framework. An example of mediation emerged in a study by Mavuru and Ramnarain (2017) when a science teacher in the teaching of the topic electrostatics recognized that learners had rich cultural beliefs on lightning. In this case, learners believed that they could be protected from being struck by lightning by exercising certain precautions they had learned from their elders. These precautions range from placing a tire on the chimney to covering mirrors to prevent lightning. Through prompting and guiding questions, the teacher mediated the learning space in getting learners to advance a scientific explanation for their cultural belief. In order to further explore the nature of this mediation, it is necessary to first consider the relationship between the learners' indigenous worldview and the worldview of science.

Categorizing the Relationship Between Indigenous and Scientific Worldviews

George (1999) presents four categories on the relationship between the two worldviews. In Category 1, indigenous (traditional) knowledge can be explained in conventional science terms. For example, the use of lime juice to remove some stains can be explained in terms of acid/oxide reactions. In Category 2, a scientific explanation of indigenous knowledge is not available, but is likely to be developed. For example, a brew made from the plant "vervine" (*Stachytarpheta sp.*) is used in the treatment of worms in children. This plant is considered in conventional science circles to have pharmacological properties, but appropriate usage has not been verified. Traditional knowledge in the Category 3 refers to cases where a conventional science link can be made to traditional knowledge, even though the underlying principles are different. For example, traditional wisdom warns that eating sweet foods causes diabetes. This is related to the conventional science principle that links diabetes with sugars. However, traditional knowledge states that sugars cause diabetes whereas conventional science claims that when one is diabetic, the ingestion of sugars can cause one's condition to worsen. In Category 4, some traditional knowledge cannot be explained in conventional science terms. For example, there is no conventional science explanation for the claim that the ingestion of certain foods can cause "heat" in the body. In Africa, the Venda people in Limpopo believe that lightning will strike anyone wearing red clothes.

Categories 1, 2, and 3 presents the best opportunity whereby the multimodal experiences of learners in the physical and sociocultural worlds can be exemplified for concept formation. In this regard an inquiry-based approach whereby learners are given the opportunity to investigate and construct ideas provide effective learning opportunities. The National Research Council (1996) in the U.S. defines inquiry as:

> a multifaceted activity that involves: making observations; posing questions; examining books and other sources of information to see what is already known; planning investigations; reviewing what is already known in light of experimental evidence; using tools to gather, analyze, and interpret data; proposing answers, explanations, and predictions; and communicating the results. Inquiry requires identification of assumptions, use of critical and logical thinking, and consideration of alternative explanations. (p. 23)

Using scientific tools, an inquiry-based approach will afford learners the experiences whereby ideas on phenomena derived from an indigenous worldview can be investigated using the scientific method. In inquiry-based environment, it is expected that learners will be able to merge their anthropological experiences of the sociocultural world with physiological experiences of the physical world that are afforded to them in the laboratory. De Beer and Whitlock (2009) discuss how learners engage in inquiry activities in investigating the traditional healing properties of certain plants. Learners are first asked to formulate a hypothesis. For example, they could hypothesize that garlic contains antimicrobial properties that could combat diseases. The antimicrobial activity of plant extracts is then tested by using a microbe-seeded agar plate and allowing the chemical substances in the plant to diffuse into the agar medium. An activity of this nature conducted in a group setting does provide learners with multimodal experiences (Chahine, 2013) that facilitate crossing the border between their life-world culture and the culture of science.

We argue that cases that fall within Category 4, may lead to cognitive conflict in teaching and learning contexts where learners take an intransigent position on their indigenous beliefs that conflict with a scientific worldview. According to Jegede and Aikenhead (1999), "These cultural clashes between pupils' life-worlds and the world of Western science can create hazards for many pupils who they must deal with to learn science effectively and meaningfully" (p. 6). They assert that in coping with such situations, learners either avoid constructing scientific ("foreign") knowledge, or they store the constructed scientific knowledge in their minds in a manner that does not align with their life-world experiences. In the later scenario, learners will access one schema or the other depending upon the context. Cobern (1996) describes the "segregation" of these two cognitive schemata within the minds of learners as "cognitive apartheid" (p. 580). Cajete (1999) maintains that "Negotiating meaning from one domain of meaning to another can be complicated" and the situation is exacerbated because "Students generally get very little help doing this kind of border crossing" (p. 97).

In addressing cases where there is a tension between the two worldviews, it is suggested by Aikenhead and Jegede (1999) that teachers play the role of "cultural broker" by helping learners "move between their everyday life-world and the world of school science" and "deal with cognitive conflicts between those two worlds" (p. 297). One way to do this is for teachers to invoke indigenous practices related to a phenomenon, and then support learners to engage with phenomenonfrom both scientific and indigenous worldviews. To facilitate students' border crossings between the two worldviews that teachers need to establish in their classroom, an environment where learners feel at ease in the less familiar culture (Aikenhead & Jegede, 1999; Msimanga & Shizha, 2014). This has to do with the social environment of the science classroom, the social interactions between a teacher and students, and the social interactions among students themselves. According to Jegede and Aikenhead (1999), teachers must promote discourse, so learners are:

> (a) talking in their own cultural interpretive framework as well as in the framework of Western science without cultural violence, (b) immersed in either the pupils' indigenous life-world culture or the culture of science, and (c) cognizant about which culture they are talking in at any given time. (p. 58)

This approach could involve learners being placed in groups and being asked to discuss questions such as "Does wearing red clothes improve the chances that you will be struck by lightning?" In acting as a "cultural broker," the teacher creates multimodal experiences (Chahine, 2013) that intersect the indigenous world view and the scientific world view, leading to cognition and concept formation.

The Role of Argumentation in ESDC

A discourse that supports embodied, situated, and distributed cognition may be characterised by argumentation whereby learners take a position on for example the lightning question posed above, and then justify their perspective from a particular worldview. Argumentation is described as the process where learners make claims to certain knowledge and then provide evidence to justify their claims (Osborne et al., 2001). Ogunniyi (2011) posits that "interactive classroom arguments and dialogues have tended to encourage teachers and students to externalise their viewpoints on any subject matter" (p. 968). There is, therefore, merit in exploring the role of argumentation in furthering the discourse that Jegede and Aikenhead (1999) consider necessary for learners to straddle across scientific and indigenous knowledge worldviews.

Many studies have recognised the value of argumentation in enhancing students' conceptual understanding (e.g., Driver, 1989; Niaz et al., 2002; Osborne et al., 2004; Ogunniyi, 2007a, 2007b). According to Ogunniyi and Hewson (2008), Toulmin's argumentation pattern (TAP) has been one of the most frequently used argumentation models by science teachers to enhance learners' understanding of the nature of science. However, the authors warn that TAP "is more applicable to a deductive-inductive classroom discourse than what is required when IKS is to be integrated with school science" (Ogunniyi & Hewson, 2008, p. 160). Furthermore, Ogunniyi and Hewson (2008) explored contiguity argumentation theory (CAT) which deals with both logical or scientifically valid arguments as well as nonlogical metaphysical discourses embraced by indigenous knowledge systems. CAT suggests that when a conflict arises in the mind of the learner when exposed to school science, there is a sort of "internal argument" or "conversation" that occurs at the micro-neuro-psychical level (Ogunniyi, 2007, p. 965). Ogunniyi and Hewson (2008) explored the use of CAT in an argumentation-based course (including dialogues and discussions) and its effect on teachers' ability to: express their views freely; develop a better appreciation for the new curriculum; and to select appropriate instructional methods to integrate indigenous knowledge into the science classrooms. They identify five categories in CAT that describe the way conceptions can manifest within a student's mind. These categories can also describe the movement of conceptions amongst learners involved in dialogues warranting the mobilization of scientific and/or indigenous knowledge-based conceptions. The five cognitive categories are: dominant conceptions, suppressed conceptions, assimilated conceptions, emergent conceptions, and equipollent conceptions. They state that "A conception becomes dominant when it is the most adaptable to a given context" but "in another context the same dominant conception can become suppressed by or assimilated into another more adaptable metal state" (p. 162).

Ogunniyi (2007a, 2007b) illustrates that a scientific claim that is supported by empirical evidence or a convincing explanation is more likely to be dominant in a science classroom than in a religious setting. An example of this is the theory of evolution that is strongly contested by certain religions. However, in a religious context the same dominant conception can become suppressed. An emergent conception is formed when an individual has no previous knowledge of a given phenomenon such as scientific concepts and theories. Examples of these include atoms, genes, entropy, theory of relativity and so on. Oguniniyi and Hewson (2008) explain that "An equipollent conception occurs when two competing ideas or worldviews exert comparably equal intellectual force on an individual" (p. 162). In that case, the ideas or worldviews tend to coexist in his/her mind without necessarily resulting in a conflict, for example, creation theory and evolution theory. Òtúlàjà, Cameron, and Msimanga (2011) caution that argumentation needs to be handled with care in the context of bringing indigenous into traditionally Western styles of education in science. They maintain that teacher beliefs about argumentation are critical for its acceptance and implementation as a teaching strategy.

In view of the above conception of CAT where both scientific and metaphysical discourses can be accommodated, I consider CAT to be a better fit than other argumentation models in promoting ESDC.

CONCLUSION

This chapter has drawn upon the paradigm of ESDC in establishing how multimodal experiences (Chahine, 2013) from the physical and sociocultural worlds of the learners may support concept formation in science. An extensive review of literature on the connection between indigenous and scientific worldviews, against the notion of ESDC has distilled key pedagogical principles that may be applied in the science classroom. First and foremost, the importance of indigenous knowledge in concept formation has to be acknowledged and recognized by the teacher. Second, the relationship between an indigenous worldview and a scientific worldview on a phenomenon has to be established. George's (1990) conceptualization of this in terms of the four categories does provide some insight into how teachers can plan multimodal experiences. Third, in order for learners to acquire a holistic understanding of a concept, they must be provided with opportunities whereby they border cross the scientific and indigenous knowledge worldviews. Fourth, teachers must promote a discourse that supports learner engagement from both worldview perspectives.

ESDC offers potential in driving a research agenda that explores how learning experiences that stem from the physical and sociocultural worlds of learners may lead to science concept formation. The chapter has discussed inquiry-based learning as approach that could be exploited in merging learners' anthropological experiences of the sociocultural world with physiological experiences of the physical world that are afforded to them in the laboratory. Inquiry-based learning is strongly promulgated by science curriculum worldwide, but its potential in facilitating science conceptual understanding with the paradigm of ESDC could invite future research.

REFERENCES

Aikenhead, G. S., & Jegede, O. J. (1999). Cross-cultural science education: A cognitive explanation of a cultural phenomenon. *Journal of Research in Science Teaching, 36*, 269–287.

Cajete, G. (Ed.). (1999). *A people's ecology: Explorations in sustainable living*. Clear Light Publishers.

Chahine, I. C. (2013). The impact of using multiple modalities on students' acquisition of fractional knowledge: An international study in embodied mathematics across semiotic cultures. *The Journal of Mathematical Behaviour, 32* (3), 434–449.

Cobern, W. W. (1996). Worldview theory and conceptual change in science education. *Science Education, 80*, 579–610.

Cobern, W. W., & Aikenhead, G.S. (1998). Cultural aspects of learning science. In K. Tobin & B. Fraser (Eds.), *International handbook of science education* (pp. 39–52). Kluwer.

Cobern, W. W., & Loving, C. C. (2001). Defining 'science' in a multicultural world: Implications for science education. *Science Education, 85*(1), 50–67.

De Beer, J. J., & Whitlock, E. (2009). Indigenous knowledge in the Life Sciences classroom: Put on your De Bono Hats. *The American Biology Teacher, 71*(4), 209–216.

Department of Basic Education. (2011). *Curriculum and assessment policy statement: Grades 10–12 physical sciences*. Government Printer.

Driver, R. (1989). Changing conceptions. In P. Adey, J. Bliss, J. Head, & M. Shayer (Eds.), *Adolescent development and school science* (pp. 79–104). The Falmer Press.

Foundation for research development. (1993). *South African science and technology indicators: Foundations for Research Development*. Government Printer.

Hewson, M. G. (1988). The ecological context of knowledge: Implications for learning science in developing countries. *Journal of Curriculum Studies, 20*, 317–326.

Hewson, M. G. (1990). Conceptualised science teaching in developing countries. Possibilities and dilemmas. *Proceedings of the 7th SAARDSE conference, Harare, Zimbabwe.*

George, J. M. (1990). Conceptualised science teaching in developing countries: Possibilities and dilemmas. *Proceedings of the 7th SAARDSE conference, Harare, Zimbabwe.*

Hardy-Vallée, B., & Payette, N. (Ed.) (2008). *Beyond the brain: Embodied, situated and distributed cognition*. Cambridge Scholars Publishing.

Jegede, O. J. (1996, August). *Fostering students' understanding of science concepts*. A special Keynote paper for the 37th Annual Conference of the Science Teachers' Association of Nigeria,Uyo, Akwam Ibom State.

Jegede, O. J. (1997). *Traditional cosmology and collateral learning in no-western science classrooms*. In *Report of an international scientific research programme* (joint research). Funded by the Grant-in-Aid for Scientific Research, Ibaraki.

Jegede, O. J., & Aikenhead, G. S. (1999). Transcending cultural borders: Implications for science teaching. *Journal for Science & Technology Education, 17*(1), 45–66.

Larson, J. O. (1995, April). *Fatima's rules and other elements of an unintended chemistry curriculum.* Paper presented to the American Educational Research Association Annual Meeting, San Francisco, CA.

Le Grange, L. (2004). Multicultural science in South Africa's national curriculum statement. *African Educational Review, 1*(2), 204–219.

Maila, M. W., & Loubser, C. P. (2003). Emancipatory indigenous knowledge systems: Implications for environmental education in South Africa. *South African Journal of Education, 23*(4), 276–280.

Mavuru, L., & Ramnarain, U. (2017). Teachers' knowledge and views on the use of learners' socio-cultural background in teaching natural sciences in Grade 9 township classes. *African Journal of Research in Mathematics, Science and Technology Education*. https://doi.org/10.1080/18117295.2017.1327239

Msimanga, A., & Shizha, E. (2014). Indigenous knowledge and science education in South Africa. What messages from the curriculum? In E. Shizha (Ed.), *Remapping Africa in the global space: Propositions for change* (pp. 137–150). Sense.

Naidoo, P., & Lewin, K. M. (1998). Policy and planning of physical sciences education in South Africa: Myths and realities. *Journal of Research in Science Teaching, 35*(7), 729–744.

National Research Council. (1996). *National science education standards.* National Academy Press.

Niaz, M., Agueilera, D., Maza, A., & Liendo, G. (2002). Arguments, contradictions, resistances, and conceptual change in students' understanding of atomic structure. *Science Education, 86,* 505–525.

Odora-Hoppers, C. (2001). *Indigenous knowledge and the integration of knowledge systems: Towards a conceptual and methodological framework*. HSRC.

Odora-Hoppers, C. (2002). (Ed.). *Indigenous knowledge and the integration of knowledge systems.* New Africa Books.

Ogunniyi, M. B. (2007a). Teachers' stances and practical arguments regarding a science-indigenous knowledge curriculum: Part 1. *International Journal of Science Education, 29*(8), 963–986.

Ogunniyi, M. B. (2007b). Teachers' stances and practical arguments regarding a science- indigenous knowledge curriculum Part 2. *International Journal of Science Education, 29*, 1189–1207.

Ogunniyi, M. B. (2011). The context of training teachers to implement a socially relevant science education in Africa. *African Journal of Research in Mathematics, Science and Technology Education*, *15*(3), Special Issue.

Ogunniyi, M. B., & Hewson, M. G. (2008). Effect of an argumentation-based course on teachers' disposition towards a science-indigenous knowledge course. *International Journal of Environmental Science Education, 3*(4), 159–177.

Okebukola, P. A. O., & Jegede, O. J. (1990). Eco-cultural influences upon students' concept attainment in science. *Journal of Research in Science Teaching, 27,* 661–669.

Oluka, S. (2007, March). *African condition and the changing goals of science and technology education.* Paper presented at the Distance Education and Teacher's Training in Africa (DETA) Conference, Makerere University, Kampala, Uganda.

Osborne, J., Erduran, S., Simon, S., & Monk, M. (2001). Enhancing the quality of argument in school science. *The School Science Review, 82*, 63–70.

Osborne, J., Erduran, S., & Simon, S. (2004). Enhancing the quality of argument in school science. *Journal of Research in Science Teaching, 4*(10), 994–1020.

Òtúlàjà, F. S., Cameron, A., & Msimanga, A. (2011). Rethinking argumentation-teaching strategies and indigenous knowledge in South African science classrooms. *Cultural Studies of Science Education*. https://doi.org/10.1007/s1142

Tobin, K., & McRobie, C. (1997). Beliefs about the nature of science and the enacted science curriculum. *Science Education, 6*, 355–371.

CHAPTER 17

INDIGENOUS TECHNOLOGY KNOWLEDGE SYSTEMS VIEWED THROUGH THE LENS OF A PHILOSOPHICAL FRAMEWORK FOR WESTERN TECHNOLOGY KNOWLEDGE SYSTEMS

Piet Ankiewicz

ABSTRACT

Odora Hoppers (2005) found that indigenous knowledge (IK) is often included in a clinical and sterile way in the formal curriculum, if addressed at all. She further mentioned the complementarity between indigenous and Western knowledge systems (Odora Hoppers, 2002). Technology curricula for schools seem to be based on the institutional and formal Western knowledge systems generated and manifested within disciplines (Ankiewicz, 2015b, 2016; Maurial, 1999; Maxwell & Chahine, 2013). Technology is embedded differently in different historical-cultural contexts (Ihde, 2006). There is currently a global movement to also include indigenous technological

Evidence-Based Inquiries in Ethno-STEM Research: Investigations in Knowledge Systems Across Disciplines and Transcultural Settings, pp. 323–343

knowledge in technology curricula for schools. However, the nature and tenets of indigenous technological knowledge have not been investigated and documented thoroughly, creating difficulties and challenges for curriculum designers, especially including and integrating it in curricula that mainly focus on Western technology. The purpose of this chapter is to investigate and conceptualise the complementary nature of indigenous technology knowledge systems (ITKS) and Western technology knowledge systems (WTKS) by using Mitcham's philosophical framework of technology,[1] which is based on WTKS, as point of departure. It is argued in the chapter that there is a complementarity between ITKS and WTKS from an ontological, epistemological, methodological and volitional perspective. Instead of separating the two knowledge systems curriculum designers should rather focus on the common tenets of ITKS and WTKS from an ontological, epistemological, methodological and volitional perspective.

INTRODUCTION

Technology is globally still a relatively new school subject that lacks a substantive research base and a well-established classroom pedagogy (Mawson, 2007; Rauscher, 2011). Unlike other school subjects where a well-established, subject-based philosophy exists at least for particular components, there is as yet no established philosophy for technology as subject—in fact, the dynamic nature of technology as such leaves its own philosophy in a tentative and flexible state. Curricula and learning programmes for technology education and their facilitation at chalk level thus often lack a scientifically founded, subject-based philosophical framework that may serve as a directive (Ankiewicz, 2013a, 2015a; Moreland & Jones, 2000; Van Niekerk et al., 2010).

Matters are complicated by the fact that technology at school level is globally a developing subject with no equivalent academic discipline, which may serve as a source upon which curriculum development and classroom pedagogy may rely in practice (Ankiewicz et al., 2006; De Vries, 2001, 2003a). In contrast, mathematics as school subject, for example, is based on the academic discipline of mathematics with an established, scientifically founded, subject-based, philosophical framework, which has been developed over centuries.

Mitcham's (1994) philosophical framework consists of four modes of the manifestation of technology, namely: object, knowledge, activity, and volition (Custer, 1995; De Vries, 2003a). The framework and its modes have affordances for science, technology, and society (STS) studies (Ankiewicz et al., 2006), classroom pedagogy (Ankiewicz, 2013b), teacher education (Ankiewicz, 2013c), curriculum development and evaluation (Ankiewicz, 2015c), as well as for the complementarity between indigenous technology knowledge systems (ITKS) and Western technology knowledge systems (WTKS) (Ankiewicz, 2015b). These affordances have been distilled into a fourfold set of criteria for curriculum development and evaluation (Ankiewicz, 2015c) that may be used to determine the practical implications of the framework for technology education. These criteria have been applied to the Curriculum Assessment Policy Statement (CAPS) (Ankiewicz, 2013a); the framework and learning guides for the academic majors of undergraduate technology student teachers (Ankiewicz, 2015a); and to indigenous knowledge (IK) in order to devise a heuristic to assist teachers to accommodate IK in technology lessons (Ankiewicz, 2016).

Odora Hoppers (2005) found that IK is often included in a clinical and sterile way in the formal curriculum, if addressed at all. Very often no IK is included, a practice that she calls "knowledge apartheid" (p. 17). It also seems that teachers generally present stereotyped examples, which are repeated without more ado in an examination paper (De Beer & Whitlock, 2009). Odora Hoppers (2002) further mentioned the complementarity between indigenous and Western knowledge systems: "The local contextual expertise and technologies that indigenous knowledge frames offer, can complement some of the mechanical and technical precision capabilities of Western knowledge systems to generate forms of creativity that benefit and empower everyone" (p. 16). Mitcham's (1994) philosophical framework was applied in previous research (Ankiewicz, 2015b) to both ITKS and WTKS. This research confirmed and strengthened the complementarity between the two systems. The complementary nature of the two knowledge systems allows for "border crossings" between them.

Technology curricula for schools seem to be based on the institutional and formal Western knowledge systems generated and manifested within disciplines (Ankiewicz, 2015b, 2016; Maurial, 1999; Maxwell & Chahine, 2013). Technology is embedded differently in different historical-cultural contexts (Ihde, 2006). Morrow (2009), for example, advocated for the acknowledgement that Africa has its own alternative forms of knowledge which requires epistemological access. According to Onwu and Mosimege (2004):

> Indigenous Knowledge Systems are local community-based systems of knowledge which are unique to a given culture or society and have developed as that culture has evolved over many generations of inhabiting particular ecosystem. Indigenous Knowledge Systems is a general term which refers broadly to the collective knowledge of an Indigenous People about relationships between people, habitat and nature. (p. 2)

There is currently a global movement to also include indigenous technological knowledge in technology curricula for schools. However, the nature and tenets of indigenous technological knowledge (unlike for the natural sciences) have not been investigated and documented thoroughly, creating difficulties and challenges for curriculum designers, especially including and integrating it in curricula that mainly focus on Western technology.

The purpose of this chapter is to investigate and conceptualise the complementary nature of ITKS and WTKS by using Mitcham's framework that is based on WTKS as point of departure. The following research question will be addressed: Based on Mitcham's four modes of the manifestation of technology— namely as object, knowledge, activity, and volition—to what extent are ITKS complementary to WTKS by identifying and comparing their common tenets?

A PHILOSOPHICAL FRAMEWORK FOR WESTERN TECHNOLOGY KNOWLEDGE SYSTEMS AS A POINT OF DEPARTURE

Ontology

From a Western point of view, technology as ontology is the first mode in which technology is manifested (Mitcham, 1994). Within their area of experience people are confronted with particular phenomena as discernible and identifiable entities or structures, which they may reflect on and may make the object of their analytical activity (Schoeman, 1983; Van Schalkwyk, 1996). Technology is such a phenomenon or onticity with the status of something that **is** or **exists** (Ankiewicz, 2013b, 2013c).

In people's conception of being or existence (Van der Walt et al., 1985) concerning technology, artefacts or material objects such as tools, machines and consumer products come to mind when the word "technology" is mentioned. "When people talk about technology today, they usually mean the products of modern engineering: computers, power plants, automobiles, nuclear weapons" (Bellington, 1986, as cited in Mitcham, 1994, p. 161; Custer, 1995). Mitcham (1994) argued that technology as object is the "most immediate, not to say the simplest, mode in which technology is found manifested, and it can include all human fabricated material artefacts whose function depends on a specific materiality as such" (p. 161).

From an ontological perspective, technology reflects the following fundamental aspects or universal characteristics (Ankiewicz, 2013b; Van der Walt & Dekker, 1982; Van Schalkwyk, 1996), namely that it is:

- a phenomenon unique to **humans**;
- employed by **using tools;**
- a way of human **form creation;**
- giving form to **nature;**
- for **human purposes;**

- to deliver a **product or process;**
- **being determined by world views.**

Any particular form of technology should possess the fundamental characteristics of technology as universal phenomenon, or else it would not be technology. Particular forms are characterized by particular participants, purposes, methods and directions; by the spirit of the times concerned and by a specific outlook on life and the world, characteristic of the particular circumstances (Van der Walt & Dekker, 1982; Van Schalkwyk, 1996). Considering the aforementioned fundamental characteristics, and according to Mitcham (1994), things that are produced by animals, such as bird nests, spider webs and beaver dams and which share certain characteristics with artefacts, are not classified as technology because one of the fundamental characteristics of technology is that it is a human phenomenon. Without humans there would have been no technology. By the same token, De Beer and Whitlock (2009) argued that IK is holistic and embedded in spirituality.

In order to determine whether a so-called "indigenous technology" knowledge system complies with the above-mentioned ontological views on technology (as embedded in a Western knowledge system) let us consider the traditional mat house (Figure 17.1) from Namaqualand that was mentioned in De Beer (this volume).

Source: Nortje (2011).

Figure 17.1. A mat house.

From the viewpoint of WTKS, one will agree that the mat house is a structure, more particularly a shell structure with a frame (structure). It has been uniquely constructed by humans through the manipulation of natural materials (long, light pieces of supple undressed wood and lightweight reed grass, usually the species of *Scirpus* or *Cyperus*). It represents human form creation in giving form to nature by using tools (light crowbar, a mat awl and a threading-needle) to deliver a product for shelter to satisfy a human need (shelter for human purposes) (Schapera, 1930; Van der Merwe, 1945). Thus, as a particular indigenous technology, the mat house possesses the fundamental characteristics of technology as a universal phenomenon and can therefore be regarded as technology.

Epistemology

According to Mitcham (1994), technology as knowledge, the second mode in which technology is manifested, has most frequently been the subject of analytical investigations in the epistemology or theory of knowledge. The epistemology of technology has its basis in theoretical reflections and more recently also in empirical studies (Broens & De Vries, 2003; Ropohl, 1997).

Technology as knowledge may be distinguished on the basis of various types of knowledge, for example maxims, rules and theories (Mitcham, 1994). Although a distinction is made between conceptual knowledge ("knowing that") and procedural knowledge ("knowing how") in technology (McCormick, 1997; Ropohl, 1997; Ryle, 1949), these two types of knowledge cannot be separated (McCormick, 1997). McCormick (1997) explained that conceptual or descriptive knowledge (the substantive nature of technology) refers to the links between knowledge items, to such an extent that when learners can identify these links, it is said that they show conceptual understanding. The author further noted: "Thus in the area of 'gearing' we hope that students will see the relationships among the 'direction of rotation,' 'change of speed,' and 'torque'" (p. 143). Furthermore, Gott (1988 as cited by McCormick, 1997) contended that conceptual knowledge that can be regarded as knowledge of devices or systems (as cited in McCormick, 1997) is clearly knowledge of technology as artefacts (Mitcham, 1994). Conceptual knowledge relevant to technology therefore includes "that drawn from other subjects, such as science, and that unique to technology" (McCormick, 1997, p. 153). In technology, conceptual knowledge is used as a resource for action rather than an end in itself (Assessment of Performance Unit [APU], 1994).

In Western view, technological conceptual knowledge is discipline based (e.g., civil technology, mechanical technology, electrical technology, engineering graphics and design (EGD). Odora Hoppers (2005) indicated that contemporary education is currently facing many challenges due to the ever-evolving landscape. To address these challenges, Maxwell and Chahine (2013) proposed a shift of focus from knowledge produced within disciplines to multidisciplinary research thus reaching beyond traditional limitations of disciplines. The authors further argued that indigenous technological conceptual knowledge is transferred orally in the form of storytelling and apprenticeships, and not through textbooks, standardized tests or a classroom setting. Therefore, despite the difference in transferring written discipline knowledge and the oral indigenous conceptual knowledge, people still "*know that.*"

From a philosophical perspective, conceptual knowledge is expressed in propositions while procedural knowledge ("*knowing how*") cannot be expressed completely in propositions. It is impossible, for example, to explain comprehensively to a person how to ride a bicycle in propositions. De Vries (2003a, 2003b) distinguished four types of conceptual knowledge ("*knowing that*") relevant to the development of technological artefacts. These are knowledge of its physical nature; knowledge of its functional nature; knowledge of the relationship between its physical and functional nature; and process knowledge (Ankiewicz et al., 2006; Broens & De Vries, 2003; Van Niekerk et al., 2010).

To illustrate, knowledge of a physical nature is reflected in the following proposition: X knows that an artefact has physical characteristics. For example, X knows that a corkscrew consists of a helix with a sharp point.

In a similar way, we represent knowledge of a functional nature as follows: X knows what the artefact, which may not as yet exist, is capable of doing. For example, X knows that a cork can be removed from a bottleneck by gripping the cork and pulling it out (and that a corkscrew is a device for doing this).

Furthermore, knowledge of the relationship between physical and functional nature can be depicted as follows: X knows that a particular physical characteristic or combination of characteristics of an artefact enables it to perform an action. For example, X knows that the sharp helix of a corkscrew enables it to get a grip on the cork (for it to be removed from the bottleneck).

To illustrate process knowledge, we can consider the following proposition: X knows that the cork can be removed from the neck of the bottle by first inserting the helix into the cork and then to pull at the handle of the corkscrew.

An epistemological aspect frequently coming to the fore is the nature of the relationship between the knowledge of the natural sciences and technological knowledge. In general, those who adhere to the positivist tradition in the philosophy of natural science would argue that technology is actually applied science (the accepted viewpoint of engineers and scientists and Mario Bunge, as their strongest philosophical representative). Proponents of this tradition would endeavour to build an epistemology of technology inspired by the encompassing model, which is based on scientific laws for its explanatory power (Mitcham, 1994). In contrast to Bunge's view of technology as applied science, one also finds the view according to which there is either no deducible link between one or more elements of technological knowledge and natural science, or that they are totally incompatible (Custer, 1995; Mitcham, 1994).

The complex relationship between natural science and technology is frequently clarified by making a distinction based on epistemological grounds (Custer, 1995), like the qualification of De Vries (2003b) according to which specific technological knowledge is normative, while natural science knowledge is not. Apart from the fact that knowledge about the relationship between the physical and functional nature is similar to knowledge in the natural sciences, the remaining three types of knowledge (knowledge of the physical and functional nature, and process knowledge) differ from natural science knowledge in that it has a normative nature which natural science knowledge does not have (Broens & De Vries, 2003). For instance, when we have knowledge of a computer, it frequently includes normative judgements: it functions well, or it does not function well. Hence, when technologists design artefacts their work is guided by very concrete objectives demanding them to first formulate design criteria and specifications. However, natural scientists are not guided by rigid preconceived objectives (Vincenti, 1990).

In other cases, technology and natural science are compared in terms of aspects like aims, research objects, methodology, characteristics of results and quality criteria (Custer, 1995; Mitcham, 1994; Ropohl, 1997). Examples would be that technological knowledge focuses on artefacts, while natural science knowledge is concerned with natural objects, and that natural science aims at cognition itself, technology only focuses on cognition inasmuch as it is useful to optimize the function and structure of technological artefacts.

To illustrate the extent to which IK can be categorized according to the abovementioned categories of conceptual knowledge, the example of the *Sutherlandia frutescens* (the so-called cancer bush) that was mentioned in De Beer (this volume), will be further explained:

Employing De Vries's (2003a, 2003b) types of conceptual knowledge, it is assumed that indigenous people know the physical characteristics of the extract made from cancer bush. It is also assumed that they *know that* a decoction (an extract made from cancer bush) may be used to treat people suffering from cancer (knowledge of a functional nature). It is further assumed that they *know that* (a decoction of) *Sutherlandia frutescens*

Source: Photograph taken by Josef de Beer.

Figure 17.2. Sutherlandia frutescens—the "cancer bush."

may assist cancer patients, since there are active ingredients in this plant that assist the immune system to fight disease (knowledge of the relationship between physical and functional nature). They may not necessarily *know that* the shrub contains an amino acid, which fights depression; pinitol, which helps patients to gain weight, and canavanine, which is successful in treating retroviruses. It is further assumed that they *know that* a specific method or procedure has to be followed when collecting and storing medicinal plant resources, as well as preparing and administering the decoction to treat patients (process knowledge) (Van Wyk et al., 2002). It is also assumed that their knowledge about the decoction or extract made from cancer bush includes normative judgements: it functions well, or it does not function well.

On the other hand, procedural knowledge (the syntactical nature of technology) is frequently referred to as tacit, personal or implicit knowledge. McCormick (1997) explained: "Design, modelling, problem solving, systems approaches, project planning, quality assurance and optimisation are all candidates for technological procedural knowledge" (p. 144). Unlike conceptual knowledge, procedural knowledge cannot be taught in the true sense of the word; "Technical know-how can be gained by thorough practice only" (Ropohl, 1997, p. 69).

As mentioned previously, indigenous communities transfer knowledge orally. Therefore, they learn primarily through apprenticeship, discipleship, and systematic strategies to support the memory (Maxwell & Chahine, 2013). It is specifically the notion of learning through apprenticeship by these communities that is conducive to the acquirement of procedural knowledge.

A theory of knowledge usually includes methodology (Van der Walt et al., 1985), but seeing that Mitcham (1994) classified methodology as one of the modes in which technology is manifested, methodology is dealt with separately. Technology as activity (methodology) in particular, provides insights into the procedural knowledge in technology.

Methodology

The third mode in which technology is manifested is technology as activity. Mitcham (1994) explained

> Technology includes more than material objects such as tools and machines and mental knowledge or cognition of the kind found in engineering sciences ... despite the quickness with which people think of physical objects ... when "technology" is mentioned ... activity is arguably its primary manifestation. Technology as activity is that pivotal event in which knowledge and volition unite to bring artefacts into existence or to use them; it is likewise the occasion for artefacts themselves to influence the mind and will. (p. 209)

Epistemology usually includes methodology (Van der Walt et al., 1985), which in particular provides insight into procedural knowledge in technology. Design processes constitute the object studied in the discipline of design methodology (De Vries, 2001), and two different paradigms, namely the rational problem solving, and the reflective practice paradigm form the basis of design methodology (Ankiewicz, 2013b, 2013c; Ankiewicz et al., 2006).

Various kinds of complex thinking processes (creative and critical thinking, decision making, problem solving, and design) underpin and form part of technological activities (Ankiewicz & De Swardt, 2002; De Swardt, 1998; De Swardt et al; 2010; Jakovljevic, 2002; Jakovljevic & Ankiewicz, 2016; Jakovljevic et al., 2004; Johnson, 1997; Reddy et al., 2003; Sharpe, 1996; Van Niekerk et al., 2010). McDonald (1998) coined this part of the technological process as the social (humanistic) elements, which are all aspects of human development. These elements involve not only the complex thinking skills, but also the development and practice of abstract personal and social attributes such as self-esteem, motivation, self-knowledge, knowledge of people, conflict resolution, communication, leadership, coordination and networking, the ability to give and accept orders, to accept responsibility and the ability to work effectively in a team.

It is common in technology to represent the procedural knowledge of technology in models using a stage-oriented format (De Swardt et al., 2010; Jakovljevic & Ankiewicz 2016; Jakovljevic et al. 2004; Johnsey, 1995; Jones, 1997; Van Niekerk et al., 2010). Most models of the technological process indicate a linear progress, assuming that the process is completed in a particular sequence (Johnsey, 1995; Mawson, 2003). One does, however, also find models with iterative design activities in the form of a loop (Garratt, 1998; Johnsey, 1995; Mawson, 2003; Todd, 1990).

Each stage of the technological process requires the application of some of the subprocesses of complex thinking (Ankiewicz & De Swardt, 2002). The technological process as a whole is not observable as it comprises various complex thinking processes that happen in the mind. The abovementioned thinking activities (minds-on) lead to doing activities (hands-on), which are observable and hence make the technological process partly observable (Ankiewicz & De Swardt, 2002). The hands-on activities manifest as procedural stages of the technological process. It is well known that procedural knowledge in technology can be described as a number of procedural stages (De Swardt et al., 2010; Jakovljevic & Ankiewicz, 2016; Jakovljevic et al., 2004; Johnsey, 1995; Jones, 1997; Van Niekerk et al., 2010) that can be briefly described as follows:

- *Statement of the problem* refers to the identification of the problem, need or want in the form of a short descriptive sentence;
- *Design brief* is the general outline of the intention how to solve the problem specifying special requirements and the context or the environment;
- *Investigation* is the process of gathering information related to the problem, needs or wants as identified in the first stage;
- *Proposal* implies a formal, written commitment of what needs to be made or designed in order to meet the need or want, or to solve the problem. The proposal consists of a written description of the specification for the product, as well as a time plan indicating how long the project will take to complete.
- *Initial ideas* imply generating ideas, discussion, analysis and choosing the best idea or combination of ideas;
- *Research:* During this stage an investigation is undertaken to solve problems and answer questions with regard to problematic aspects of the chosen idea, in order to develop it into a workable solution.

Research implies a more in depth and more specific search for information, than the initial investigation;

- *Development* refers to the refinement of the chosen idea by applying the research results to the problematic aspects;
- *Planning* refers to planning on how to make, assemble and finalize the product. The final idea is represented in a working drawing. The various drawings can be used to indicate different parts of the proposed system and the relationships between the components. A list of required resources is compiled and installed;
- *Realization/making* consists of the production of a product. It includes the realization of the planning and designing into a product;
- *Testing, evaluation, and improvement* involve the testing and a final evaluation of the product based on the criteria included in the design brief and proposal.

McDonald (1998) coined this part of the technological process as the mechanistic elements, which involve the practical hands-on activities. Each procedural stage of the technological process requires the application of some of the subprocesses of complex thinking (Ankiewicz & De Swardt, 2002).

Technology can therefore be regarded as both "minds-on" (complex thinking) and "hands-on" (practical activities) (McCormick & Davidson, 1996). The essence of technology is the interaction of mind and hand; thus inside and outside the head. It is dependent upon conceptual understanding but involves more. It also involves more than practical skill but is dependent upon it. Ideas conceived in the mind need to be expressed in concrete form (artefacts/products) (APU, 1994).

Figure 17.3 shows the APU model of the interaction behind the mind and the hand. The left-hand side of the figure focuses on the thinking and decision-making processes while the right-hand side focuses on the "hands-on" activities. It is crucially important not to divide/split the interactive process between the mind and hand to such an extent that the interactivity is ignored. This interaction supports the paradigm of *embodied, situated, and distributed cognition* (ESDC), which posits that cognitive processes are not limited to the symbolic processing of internal information structures, but are in fact "embedded physiologically in action, situated in the sociocultural world and distributed among agents, artifacts and external structures" (Chahine, 2013, p. 423).

The daily experiences of master craftspeople and knowledge holders of vital IK also exhibit the integration of mind and hand tools that indigenous communities continually employ to conceptualise, visualize, plan, and

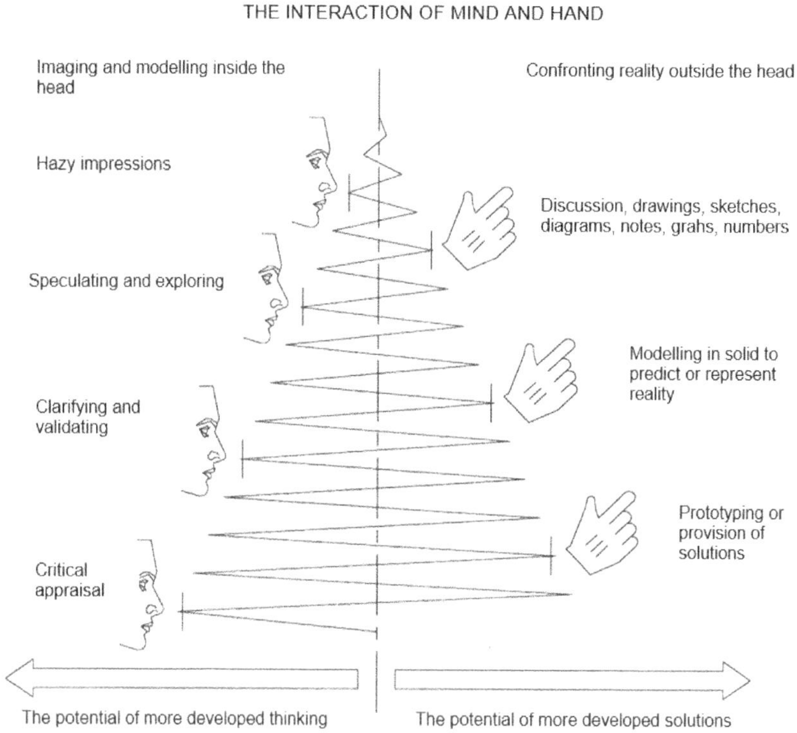

Source: APU (1994).

Figure 17.3. The APU model of the interaction behind the mind and the hand.

execute myriad activities as part of their daily practice (Maxwell & Chahine, 2013). For these craftspeople, the mind is the extension of the hand, so thinking is doing as they are immersing themselves in the creation of their arts (Maxwell & Chahine, 2013). It is obvious that the social/humanistic and mechanistic elements of the technological process are also involved in their technological activities.

Dorst (1997) contended that two radically different paradigms form the basis of the discipline of design methodology:

- The rational problem-solving paradigm (based on the work of Simon [1969]) where "objective" observation and logical analysis lead to general, formal design models and pave the way for objective interpretation—the structured approach generally associated with engineering.

- The reflective practice paradigm (as proposed by Schön (1983), which shifts the focus from general design models to the uniqueness of every design problem, where reflective communication with the situation takes place and where there is room for subjective interpretation—a less structured approach usually associated with architecture.

Employing Dorst's (1997) dual-mode model of design methodology, it is argued that the master crafts people and knowledge holders of vital indigenous knowledge follow the reflective practice paradigm rather than the rational problem-solving paradigm in the sense that they do not plan their design activities objectively in advance to a similar extent than engineers usually do. Prior to the execution of their final products, they may even design (conceptualize and visualize) and make their pilot products, also known as "prototypes," merely through trial and error.

Dorst (1997) distinguished three phases during the design activity, namely the conceptual phase, the information phase and the embodiment phase. All three phases may easily be related to the rational problem-solving paradigm, while the conceptual phase is the only one relevant to the reflective practice paradigm. The conceptual phase of the design process is a more subjective design activity and is therefore more adequately described by the reflective practice paradigm. During the information phase of such a design project most of the design activities involve objective interpretation and are therefore better described by the rational problem-solving paradigm (Dorst, 1997).

The claim proposed by Maxwell and Chahine (2013) that the indigenous communities in South Africa and Morocco were involved in activities by conceptualizing, visualizing, planning, and executing (making), may be related to the procedural stages of the technological process. The mere fact that they were conceptualizing relates to the conceptual phase of design activities, which is a more subjective design activity and is therefore more adequately described by the reflective practice paradigm. During the idea generation stage, Western technologists generally communicate visualized ideas of their products, mainly through freehand sketches. They also usually make working drawings of their final idea during the planning stage before they start making their products. It is highly unlikely for indigenous craftspeople to communicate their ideas through sketches and drawings. They would rather have mental pictures, like the craftswomen in Morocco who have mental pictures of their carpet designs without the aid of printed patterns or pictures. Indigenous craftspeople may also plan and execute their designs (conceptualizations and visualizations) in a structured manner as part of the rational problem-solving paradigm but to a lesser extent than engineers normally do. The fact that these communities may be unaware of their thinking processes or do not label them does not imply that they do not follow such processes.

Volition

The fourth mode in which technology is manifested is technology as volition (Mitcham, 1994), which is the most complex manifestation to grasp (Custer, 1995). Mitcham (1994) explained that technology as volition is the result of the volition of the practitioner involved.

Technologies are associated with a wide array of volitional activities, drives, motivation, aspiration, intentions and choice. The phrase "the will to" is found in various definitions of technology (Ankiewicz, 2013c; Mitcham, 1994). The underlying drive behind technological activity is one of improving some aspects of the "made world" for someone (APU, 1994). According to Custer (1995) it is more correct to refer to the volitional characteristics such as drive, impact, conscious choice and free will of technology rather than Mitcham's (1994) fourth mode that technology manifests itself as volitional behavior. The author further explained that there is a recurring tendency to think about technology in terms of its impact on entities outside of its essential nature, such as the impact of technology on the environment and society, and also the impact of human values and needs on technology.

Some of the complex thinking processes are also core to some aspects associated with technology as volition, such as innovation, entrepreneurial attitude and behavior. It is important, however, not to consider volition in isolation, but to integrate it with the previously discussed manifestations of technology, as technology has a human origin. It has already been mentioned that ITKS are holistic and embedded in spirituality from an ontological perspective. It must not be forgotten that it is humans who possess particular technological conceptual and procedural knowledge and who need to expand on and further develop their knowledge (Ankiewicz, 2013b). As noted by Maxwell and Chahine (2013), the quest for survival and self-development serves as motivation for the indigenous communities to execute myriad technological activities as part of their daily practice by applying their culturally embedded knowledge (conceptual technological knowledge) and competencies (procedural knowledge and skills). By observing the community dynamics in action, the authors highlighted their experiences as they observed the emergence of culturally embedded competencies, which empower the marginalized societies and grant them opportunities for survival and self-development. This also relates to some of the previously mentioned social (humanistic) elements of the technological process from a methodological perspective.

DISCUSSION

It follows from the discussion above that the complementarity between ITKS and WTKS can be strengthened through the lens of a philosophy of (Western) technology by identifying and comparing their common tenets. According to Mitcham's framework indigenous technology, as a particular form of technology, possesses the fundamental characteristics of technology as universal phenomenon, and therefore, also qualifies ontologically as genuine and true technology.

Epistemologically indigenous technology also consists of both conceptual and procedural knowledge. However, the indigenous technological conceptual knowledge is usually oral and not written like Western technological conceptual knowledge. The notion of learning through apprenticeship by indigenous communities is conducive to the acquirement of indigenous technological procedural knowledge.

Methodologically, indigenous technology is also activity based (hands-on activities with mechanistic elements), but unlike Western technology relies more on the reflective practice paradigm than on the rational problem-solving paradigm. However, there are also some instances where planned activities point towards the rational problem-solving paradigm, although to a lesser extent than is the case with Western technology. There are also complex thinking processes involved in ITKS even though these communities may not be aware of it.

In terms of volition the quest for survival and self-development serves as motivation (social/humanistic elements) for the daily technological activities of indigenous communities, as opposed to improving some aspects of the "made world" for someone in Western view. The aforementioned common tenets of indigenous and Western technology systems are summarized in Table 17.1.

Based on Mitcham's (1994) four modes of the manifestation of technology—namely as object, knowledge, activity, and volition, the common tenets of indigenous and Western technology systems indicate the large extent of complementarity between ITKS and WTKS. Instead of separating the two knowledge systems one should rather focus on the common tenets of them from an ontological, epistemological, methodological and volitional perspective. One cannot "do" (epistemology) technology (ontology) without knowing (epistemology) and having the desire to do so (volition) (Ankiewicz, 2013b; Ankiewicz et al., 2001).

NOTE

1. Parts of the theoretical framework that underpins this chapter have been published elsewhere in a different format and genre. The affordances of

Table 17.1.
The Common Tenets of Indigenous and Western Technology Systems

Philosophical Perspective	ITKS	WTKS
Ontology	• Possesses the fundamental characteristics of technology as universal phenomenon	• Possesses the fundamental characteristics of technology as universal phenomenon
Epistemology	• Consists of both conceptual and procedural knowledge • Indigenous technological conceptual knowledge is usually oral	• Consists of both conceptual and procedural knowledge • Western technological conceptual knowledge is usually written and discipline-based
Methodology	• Learning through apprenticeship • Activity based (hands-on activities/mechanistic elements) • Relies more on the reflective practice paradigm • Few instances where planned activities point towards the rational problem-solving paradigm • Complex thinking processes are subconsciously involved in ITKS	• Learning through thorough practice and repetition • Activity based (hands-on activities/mechanistic elements) • Relies on the rational problem-solving paradigm • Activities are usually planned and structured • Deliberately and consciously applying complex thinking processes
Volition	• The quest for survival and self-development serve as motivation (social/humanistic elements)	• The underlying drive behind technological activity is one of improving some aspects of the "made world" for someone

Mitcham's framework for science, technology and society (STS) studies (Ankiewicz et al., 2006), classroom pedagogy (Ankiewicz, 2013b), teacher education (Ankiewicz, 2013c), curriculum development and evaluation (Ankiewicz, 2015c), as well as for the complementarity between indigenous technology knowledge systems (ITKS) and Western technology knowledge systems (WTKS) (Ankiewicz, 2015b) have been distilled into a fourfold set of criteria for curriculum development and evaluation (Ankiewicz, 2015c) that may be used to determine the practical implications of the framework for technology education. These criteria have been applied to the Curriculum Assessment Policy Statement (CAPS) (Ankiewicz, 2013a); the framework and learning guides for the academic majors of undergraduate technology student teachers (Ankiewicz, 2015a); and to indigenous knowledge (IK) in order to devise a heuristic to assist teachers to accommodate IK in technology lessons (Ankiewicz, 2016).

REFERENCES

Ankiewicz, P. J. (October, 2013a). The alignment of the caps for technology in the senior phase with the philosophy of technology: A critical analysis. *Proceedings of the International Conference on Mathematics, Science and Technology Education (ISTE): "Towards effective teaching and meaningful learning in mathematics, science and technology," Mopani Camp, Kruger National Park*, 327–336.

Ankiewicz, P. J. (2013b). 'n Teoretiese besinning oor die implikasies van die filosofie van tegnologie vir klaskamerpraktyk [A theoretical reflection on the implications of the philosophy of technology for classroom pedagogy]. *Suid-Afrikaanse Tydskrif vir Natuurwetenskap en Tegnologie* [South African Journal of Natural Science and Technology], *32*(1), a386, 9 pages. http://doi.org/10.4102/satnt.v32i1.386

Ankiewicz, P. J. (2013c). 'n Teoretiese besinning oor die implikasies van die filosofie van tegnologie vir onderwyseropleiding [A theoretical reflection on the implications of the philosophy of technology for teacher education]. *Suid-Afrikaanse Tydskrif vir Natuurwetenskap en Tegnologie* [South African Journal of Natural Science and Technology], *32*(1), a387, 9 pages. http://doi.org/10.4102/satnt.v32i1.387

Ankiewicz, P. (2015a, April). The implications of the philosophy of technology for the academic majors of technology student teachers. *Proceedings of the PATT 29 conference, Marseille, France, Marseille: Presses Universitaires de Provence*, 13–25.

Ankiewicz, P. J. (2015b). Inheemse en Westerse tegnologiese kennisstelsels: Twee kante van dieselfde muntstuk? [Indigenous and Western technology knowledge systems: Two sides of the same coin?] *Suid-Afrikaanse Tydskrif vir Natuurwetenskap en Tegnologie* [South African Journal of Natural Science and Technology], *34*(1), a1309, 7 pages. http://doi.org/10.4102/satnt.v34i1.1309

Ankiewicz, P. J. (2015c). 'n Teoretiese besinning oor die implikasies van die filosofie van tegnologie vir kriteria vir vakkurrikulumontwikkeling en –evaluering [A theoretical reflection on the implications of the philosophy of technology for criteria for subject-curriculum development and evaluation]. *Suid- Afrikaanse Tydskrif vir Natuurwetenskap en Tegnologie* [South African Journal of Natural Science and Technology], *34*(1), a1170, 7 pages. http://doi.org/10.4102/satnt.v34i1.1170

Ankiewicz, P. (2016, August). The relevance of indigenous technology knowledge systems (ITKS) for the 21st century classroom. *Proceedings of the PATT 32 conference, Utrecht, The Netherlands*, 22–34.

Ankiewicz, P. J., & De Swardt, A. E. (2002, September-October). Aspects to be taken into account when compiling learning programme to support effective facilitation of Technology Education. *Proceedings of the National Conference for Technology Teachers, Port Natal School, Durban*, 76–81.

Ankiewicz, P. J., De Swardt, A. E., & De Vries, M. (2006). Some implications of the philosophy of technology for science, technology and society (STS) studies. *International Journal of Technology and Design Education, 16*(2), 117–141.

Ankiewicz, P., Van Rensburg, S., & Myburgh, C. (2001). Assessing the attitudinal technology profile of South African learners: A pilot study. *International Journal of Technology and Design Education, 11*(2), 93–109.

Assessment of Performance Unit (APU). (1994). Learning through design and technology. In F. Banks (Ed.), *Teaching technology* (pp. 59–67). Routledge.

Broens, C. J. A. M., & De Vries, M. J. (2003). Classifying technological knowledge for presentation to mechanical engineering designers. *Design Studies, 24*(5), 457–471.

Chahine, I. C. (2013). The impact of using multiple modalities on students' acquisition of fractional knowledge: An international study in embodied mathematics across semiotic cultures. *The Journal of Mathematical Behavior, 32*(3), 434–449.

Custer, R. L. (1995). Examining the dimensions of technology. *International Journal of Technology and Design Education, 5*(3), 219–244.

De Beer, J. J. J., & Whitlock, E. (2009). Indigenous Knowledge in Life Science classroom: Put on your De Bono hats! *The American Biology Teacher, 71*(4), 209–216.

De Swardt, A. E. (1998). *Technology education and the development of thinking skills: A case study* (MEd dissertation). Rand Afrikaans University, Johannesburg, South Africa.

De Swardt, E., Ankiewicz P., & Gross E. (2010). Implementing a technology learning programme in a school for learners with special educational needs: A case study, *Acta Academica, 42*(3), 230–248.

De Vries, M. J. (2001). The history of industrial research laboratories as a resource for teaching about science-technology relationships. *Research in Science Education, 31*, 15–28.

De Vries, M. J. (2002, June). Integration of knowledge in technological developments: Philosophical reflections on an empirical case study. *Proceedings of the Technological Knowledge: Philosophical Reflections Conference, Boxmeer, The Netherlands.*

De Vries, M. J. (2003a). The nature of technological knowledge: Extending empirically informed studies into what engineers know. *Techné: Research in Philosophy and Technology, 6*(3), 1–21.

De Vries, M. J. (2003b, July). The nature of technological knowledge: Philosophical reflections and educational consequences. *Proceedings of the PATT 13 Conference, Glasgow, Scotland*, 83–86.

Dorst, K. (1997). *Describing design* (PhD thesis).Technische Universiteit Delft, Delft.

Garratt, J. (1998). *Design and technology*. Cambridge University Press.

Ihde, D. (2006). The designer fallacy and technological imagination. In J. R. Dakers (Ed.), *Defining technological literacy: Towards an epistemological framework* (pp. 121–131). Palgrave Macmillan.

Jakovljevic, M. (2002). *An instructional model for teaching complex thinking through web page design* (Doctorandis Educationis thesis). Rand Afrikaans University, Johannesburg, South Africa.

Jakovljevic, M., & Ankiewicz, P. (2016). Project-based pedagogy for the facilitation of webpage design. *International Journal of Technology and Design Education, 26*(2), 225–242. http://doi.org/10.1007/s10798-015-9312-5

Jakovljevic, M., Ankiewicz, P., De Swardt, A. E., & Gross, E. (2004). A synergy between the technological process and a methodology for web design: Implications for technological problem solving. *International Journal of Technology and Design Education, 14*, 261–290.

Johnsey, R. (1995). The design process—does it exist? A critical review of published models for the design process in England and Wales. *International Journal of Technology and Design Education, 5*(3), 199–217.

Johnson, S. D. (1997). Learning technological concepts and developing intellectual skills. *International Journal of Technology and Design Education, 7*(1–2), 161–180.

Jones, A. (1997). Recent research in learning technological concepts and processes. *International Journal of Technology and Design Education, 7*, 83–96.

Maurial, M. (1999). Indigenous knowledge and schooling: A continuum between conflict and dialogue In L. Semali & J. Kincheloe (Eds.), *What is indigenous knowledge? Voices from the academy* (pp. 59–77). Falmer Press.

Mawson, B. (2003). Beyond "the design process": An alternative pedagogy for technology education. *International Journal of Technology and Design Education, 13*, 117–128.

Mawson, B. (2007). Factors affecting learning in technology in the early years at school. *International Journal of Technology and Design Education, 17*(3), 253–269.

Maxwell, K. L. H., & Chahine, I. C. (2013). Cultural immersion and mathematics teacher education: Explorations in Morocco and South Africa. *Journal of Humanistic Mathematics, 3*(2), 62–75.

McCormick, R. (1997). Conceptual and procedural knowledge. *International Journal of Technology and Design Education,* 7(1–2), 141–159.

McCormick, R., & Davidson, M. (1996). Problem solving and the tyranny of product outcomes. *Journal of Design and Technology Education, 1*(3), 230–241.

McDonald, M. A. (1998). *The social component of the technological process and interdisciplinary-integrated studies* (Magister Educationis dissertation). Rand Afrikaans University, Johannesburg, South Africa.

Mitcham, C. (1994). *Thinking through technology.* The University of Chicago Press.

Moreland, J., & Jones, A. (2000). Emerging assessment practices in an emergent curriculum: Implications for technology. *International Journal of Technology and Design Education, 10*(3), 283–305.

Morrow, W. (2009). *Bounds of democracy: Epistemological access in higher education.* http://www.hsrcpress.ac.za

Nortje, J. M. (2011). Medicinal ethnobotany of the Kamiesberg, Namaqualand, Northern Cape Province South Africa (MSc dissertation). University of Johannesburg.

Odora Hoppers, C. A. (2002). *Indigenous knowledge and the integration of knowledge systems: Towards a philosophy of articulation.* New Africa Books.

Odora Hoppers, C. A. (2005). *Culture, indigenous knowledge and development: The role of the university* (Occasional Paper no. 5:1–50). Centre for Education Policy Development (CEPD).

Onwu, G., & Mosimege, M. (2004). Indigenous knowledge systems and science and technology education: A dialogue. *African Journal of Research in SMT Education, 8*(1), 1–12.

Rauscher, W. J. (2011). The technological knowledge used by technology education students in capability tasks. *International Journal of Technology and Design Education, 21*(3), 291-305.

Reddy, K., Ankiewicz, P., & De Swardt, A.E., & Gross, E. (2003). The essential features of Technology and Technology Education: A conceptual framework for the development of OBE (Outcomes-Based Education) related programmes in Technology Education. *International Journal of Technology and Design Education, 13*, 27–45.

Ropohl, G. (1997). Knowledge types in technology. *International Journal of Technology and Design Education,* 7(1-2), 65–72.

Ryle, G. (1949). *The concept of mind.* Hammondsworth, UK: Penguin Books.

Schapera, I. (1930). *The Khoisan peoples of South Africa: Bushmen and Hottentots.* Routledge and Kegan Paul.

Schoeman, P.G. (1983). *Wysgerige pedagogiek* [Philosophy of education]. Nasionale Boekdrukkery.

Sharpe, D.B. (1996). Out with the old, in with the new. *Journal of Design and Technology Education, 1*(1), 24–36.

Todd, R. D. (1990). The teaching and learning environment: Designing instruction via the technological method, *The Technology Teacher, 50*(3), 3–7.

Van der Merwe, P. J. (1945). *TREK. Studies oor die mobiliteit van die pioniersbevolking aan die Kaap* [Trek. Studies about the mobility of the community of pioneers in the Cape]. Nasionale Pers Beperk.

Van der Walt, J. L., & Dekker, E. I. (1982). *Fundamentele opvoedkunde vir* onderwysstudente [Philosophy of education for education students]. Promedia Publikasies.

Van der Walt, J. L., Dekker, E. I., & Van der Walt, I. D. (1985). *Die opvoedingsgebeure: 'n Skrifmatige perspektief* [The education praxis: A scriptural perspective]. Instituut vir Reformatoriese Studies.

Van Niekerk, E., Ankiewicz, P., & De Swardt, E. (2010). A process-based assessment framework for technology education: A case study. *International Journal of Technology and Design Education, 20*(2), 191–215.

Van Schalkwyk, P. (1996). *Opvattinge oor tegnologie by leerlinge en onderwysers* (PhD thesis). Potchefstroomse Universiteit vir Christelike Hoër Onderwys, Potchefstroom.

Van Wyk, B.-E., Van Oudtshoorn, B., & Gericke, N. (2002). *Medicinal plants of South Africa.* Briza Publikasies.

Vincenti, W. G. (1990). *What engineers know and how they know it.* The Johns Hopkins University Press.

CHAPTER 18

THE *ISICHOLO* ZULU HAT

An Embodiment of South African Indigenous Technology

Erica Bass Flimmons

ABSTRACT

Technology is a form of practical action accompanied by practical reason; it is a means through which humans produce not only products and works, but also themselves as human in both their reflective and nonreflexive aspects (Pacey, 2001). The purpose of this chapter is to investigate indigenous technology while focusing on a specific indigenous technological object from the South African Zulu culture. I employ substantivism as the major theoretical framework underlying our views on technology in this study and I argue that indigenous technological objects are comprised of mathematical structures that are embodied in its design, formations, and technique.

Evidence-Based Inquiries in Ethno-STEM Research: Investigations in Knowledge Systems Across Disciplines and Transcultural Settings, pp. 345–359

INTRODUCTION

The introduction of this chapter begins with the development of technology, which is traced through a series of traditional periods. In the early nineteenth century, technology characterization followed cultural evolutionary theories of the times, which included religion, family and other cultural aspects (Spier, 1973). However, technology has long been recognized as playing a prominent role in man's efforts to survive (Lustig-Arecco, 1975). In etymology, the Greek root "techné" of the word "technology" means belonging to the arts, crafts, or skill. It was this sense, in which "technology" refers to a body of knowledge about the useful arts that prevailed from Renaissance times well into the industrial era (Misa et al., 2003). During the Renaissance times, technology embraced the means, which man used to modify his natural environment (Spier, 1970). During the nineteenth century "achievement, progress and purpose were all part of the public meaning of technology" (Vandeleur, 2010, p. 12). According to Misa et al. (2003), technology is a set of devices or an abstract form in itself and cannot be defined statically as its nature and meaning have changed over time and continues to change.

Broadly, technology is a form of practical action accompanied by practical reason; it is a means through which humans produce not only products and works, but also themselves as human in both their reflective and nonreflexive aspects (Vaughan, 2011). It is through various technologies that humans develop the habits, capacities, skills, identity, and knowledge that mark them out as individual members of a social and cultural group (Vaughan, 2011). Spier (1973) argues that all technology is composed of a series of coordinated techniques. These techniques are directed toward a particular goal as the situation is evaluated by the members of that culture.

Indigenous technology forms a body of knowledge, developed by a culture, that provides methods or means to control the environment, extract resources, produce goods and services, and improve the quality of life. "Indigenous knowledge" refers to the complex set of knowledge of these technologies existing and developed around specific conditions of populations and communities indigenous to a particular geographic area. This knowledge enables them to interact with and maneuver the environment to address their needs, wants or everyday challenges.

Indigenous technologies have been characterized as a branch of indigenous knowledge systems comprising the complex set of technologies developed and sustained by indigenous civilizations (Vandeleur & Schafer, 2011). Since indigenous technologies vary among ethnic communities, the inclusion of indigenous technology should be recognized as a knowledge system that coexists with the mainstream systems. There is no doubt that the inclusion of 'indigenous technology and culture' promotes the visibility of indigenous knowledge. Indigenous technologies have come into being over many (human) generations and have done so within the "ecosystem" of practices embedded in a particular culture (Spector, 2011). The Zulu people represent a Bantu group of South Africa and the largest ethnic group that embodies indigenous technology and oral culture. Afọlayan (2004) asserts that the transmission of nature into preserved technological frameworks is what makes the Zulu culture distinct. Despite not having a formal written text for communication purposes, the Zulu people fundamentally employ emblems to reflect on the world around them and to formulate profound explanations about life and existence. They developed a wide array of concepts and ideas inspired by life experiences and transmitted them from one generation to the other using carefully designed technological tools (Carton et al., 2009). Woven baskets and handicraft for tourist are ideas of indigenous technologies that are prevalent in the Zulu community

Regrettably, the technologies used in traditional societies are frequently viewed by modernists as noninnovative (Boyne & Rattansi, 1990) due to their subtle nature and low cost (Vandeleur, 2010). Local knowledge systems have often been described as "primitive," "unscientific" and "backward" compared to the Western system, which is seen to be 'uniquely scientific and universal' (McGovern, 1999). Modernism promoted the concept of Western science by adopting Western technologies. As a result, it has led to the devastation of nature and the destruction of cultures. Though the advent of European literary culture from the mid-17th century provided an alternative, the South African Zulu culture has prevailed by using indigenous technologies for sustainable use (Afọlayan, 2004). Such technologies forged novel cultural identity that is highly regarded by children and elders alike (Afọlayan, 2004). Zulu identities are showcased in rituals and social ceremonies that delineate the initiation and transition of youth to mature members of society. These traditions and customs are preserved through the practice of oral culture, which is the foundation of indigenous technologies. For instance, at the core of South African culture is the institution of marriage. Oral culture preserves traditions such as the making of a women's traditional *Isicholo* hat to wear after she has been married. This hat expresses the social status of a Zulu woman and illustrates the functional characterization of indigenous technology.

The purpose of this chapter is to investigate indigenous technology while focusing on a specific indigenous technological object from the South African Zulu culture. I argue that this object is comprised of mathematical structures that are embodied in the design, formations and technique. The following questions were investigated:

1. How do the colors, patterns and material employed in the design of the *Isicholo* hat reflect its social and cultural relevance?

2. What are the mathematical ideas embedded in the practice of the *Isicholo* hat?

THEORETICAL FRAMEWORK

Substantivism is the major theoretical framework underlying my stand on technology in this study. The substantivism theory views technology as "tools" that are connected to tradition, ideology and efficiency, which arise from sociotechnical change (Feenberg et al., 2006). There is no moment when technology is seen as separate from social context. Instead, technology and society are developed together. Technology is a part of culture and not something separate from it. The cultural feature includes goals, values, and moral rules that affect the creativity of designers. As such, technology is seen as a wide-scale idea with purpose and meaning to cultures (White, 1959).

The theory was initially coined by Ellul and Heidegger during the twentieth century. Instead Ellul and Heidegger asserted that technology constitutes a new type of cultural system that restructures the entire social world as an object of control. They maintained that technologies are more than independent "tools" waiting to serve as a function or to be transformed into whatever it is that human beings desire (Feenberg et al., 2006). This new type of cultural system is dependent on the technology tools that we decide to use as representation of our culture in order to transform our society. Furthermore, Ellul and Hideger argued that when the public is unaware of both the technical and social aspects of technology, then it will be seen as an alien entity which appears as an independent system in the face of existing cultural systems. Consequently, the main premise that underlies their theory is the fact that technology is not simply a means but basically an environment with "substantive impact." This theory capitalizes on cultural connections and endorses the notion that technology is a set of techniques and skills that are used in one's given environment to respond to everyday needs and challenges.

Broadly, substantivists claim that the standards embodied by technology must be seen as a culture of universal control that is a way of life (Pacey, 1992). Culture is man's compatibility with the environment around him and the relation he establishes with it. As a result, technology is considered as a means for change in the environment in order to make it compatible with necessary and inevitable human needs. By using the substantivism theory, we can describe the relationship between technology and culture in a more generalized fashion, namely, a part of society and changes with the environment. For example, within the Zulu culture, prototypes of indigenous technology affect the well-being of the majority of people in this community. The transformation of raw materials into meaningful, culturally connected, technological objects has to have community participation and approval to ensure that local technology is preserved and sustained. Such emblems are seen as valuable assets that empower the local community and support the preservation of their heritage and legacy. The *Isicholo* Zulu hat is a representation of a technological tool that connects ancestral traditions with purposeful daily practices in the Zulu community. It is the intent of this study to examine the cultural influences of this technological object, which is developed within the Zulu community, while observing the mathematical patterns involved in the construction of the item.

Brief Background

There are numerous terminologies in the literature that are closely tied to indigenous technology. These terms include "local technology," "native technology," and even "intrinsic technology." Different forms of technology related to local knowledge are employed and which constitute an important part of the cultural heritage of the Zulu culture in South Africa. These technologies include textile technology and cultural fabrics.

Indigenous Technology

Indigenous technology literature can be connected to the beginning of mankind through various accounts. In ancient times, man was relentlessly experimenting with his living environment using available raw materials such as plants, animal skins and brayed hides to make clothing (Afọlayan, 2004). By using goods that were natural resources prevalent in their environment, local inhabitants created earlier forms of indigenous technology. In this respect, technology is particularly viewed as the "systematic study and development of making and doing things" (Ogungbure, 2011). However, to indigenous people, technology comprises the cultural mindset within a social praxis for the sustenance of local wellbeing (Ogungbure, 2011). Customs, family, kinship, and community are major cultural pillars at the forefront of the Zulu peoples' belief system (Vaughan, 2011). There is a plethora of indigenous technological objects developed and adaptable in the material market ranging from beadwork to basket weaving and clothing material that narrate the ancestors' stories and traditions and delineate the history of the Zulu development (Vandeleur, 2010). The advent of woven and manufactured cotton and other fabrics brought about far reaching transformations in dressing materials and forms (Afọlayan, 2004). However, in the traditional clothing material sector, efforts have remained to use indigenous technology in order to keep traditions alive (Mabunda, 2010).

Local technologies in nations around the world consist of a series of techniques that local people know about and can acquire. In South Africa, local technology is defined from the perspective of culture (Vandeleur, 2010). Individuals convert local raw materials to be used for cultural purposes into technological objects. The technology in use is a subset of techniques that have been developed and mastered (Stewart, 1978). This would explain why a nation such as South Africa, does not possess the technology from Western nations because it does not have the knowledge of techniques, and therefore would not produce items that are foreign to their environment (Ogungbure, 2011). For example, textile material technology emerged many years ago in the Zulu culture by local women designing and making traditional dress attire using raw materials that are available to them in their immediate environment. For the Zulu people, dress makes and defines the person and does not exclude ornaments and adornments.

Textile Technologies

Textile technology is made up of various related products such as fiber, yarn, and fabric. These products are then processed into consumer goods. The consumer goods include the social and economic culture in which the designing process takes place while fulfilling a specific need or purpose (Ogungbure, 2011). Raw fibers are usually transformed into materials of meaning in most South African cultures. For example, Chahine and Kinuthia (2013) studied a project focused on basketry by the Zulu women. They state that "different basket designs reflect various constructional solutions synchronized to meet certain criteria, such as the material used in making the basket, specialty and purpose of the basket, and the size of the person who will be carrying the basket" (pp. 22–23).

There are several elements of Africa that have influences on cultural textile such as cultural practices, human and natural resources, and local knowledge. When looking at textile design from a broad perspective, it seems evident the importance of culture to understand the technological skills needed to produce such relevant objects (Mabunda, 2010). In South Africa, "technological advancement cannot be separate from culture; culture is the motivating force that drives technological success in their society" (Ogungbure, 2011 p. 89). In Zulu culture, each indigenous technology is a product of culture and textile technology includes versatile expressions of life and is influenced by the way people interact with nature. The Zulu people of South Africa value their culture and therefore understand that any textile will be connected to their life expereinces (Precious, 2010). The textile aspect in cultural textile is one's skilled talent and technique while creating cultural material that serves universal purposes for one's society.

THE STUDY

Context

The Zulu community is among four indigenous groups in South Africa including the Nguni, Tosonga, Venda and Sotho-Tswana. The Nguni group includes the Zulus, Xhosa, Ndebele and Swazi people. Amongst the entire Nguni group, the Zulus are the most well-known due to their advanced system of preserving indigenous technologies and cultural practices. In the 18th century they ascended from the hundreds of small clans occupying the northern regions of KwaZulu-Natal on the eastern seaboard of South Africa (Kapff, 2003). With an approximate population of 8 million, the Zulus are the largest ethnic group in South Africa (Kapff, 2003). The Zulus emerged toward the end of the 18th century, when they constituted themselves into a number of Nguni speaking chiefdoms north of the Tugela River. "The military efforts of King Shaka, led to the emergence of the Zulus, from a small insignificant group to the most dominant indigenous demographic group in South Africa" (Afọlayan, 2004, p. 214). Though the Zulu empire was weakened and fragmented after Shakas's death, the Zulu's sense of cultural and ethnic identity survived to become a major rallying point in South African politics in the early 1990s.

Indigenous technologies are considered a key element to the Zulu community in South Africa and customs and traditions are communicated in an oral manner. While the Zulu did not use written texts to communicate, they nonetheless developed a wide array of representations, including oral narrative, to transmit their life experiences from one generation to another.

With regards to textile materials, women play a major role in preserving the cultural heritage that is significant to the technology of the women's traditional attire and its value to society (James, 1995). In many African cultures, adorning one's head is a significant part of culture attire. *Isicholo* is the Zulu word for married woman's headdress; these hats are traditionally given to a bride at wedding ceremonies. After one has received the gift, the hat is worn to indicate their married status at important social and cultural events. Single Zulu women do not wear the hat, which identifies them as an unmarried woman (Arnoldi & Kreamer, 1995). The technology used to construct this hat expresses the social status of a Zulu woman and embodies underlying mathematical components.

Description of the Hat

In relation to the Zulu culture, the original *Isicholo* hat consists of using lamb wool, palm fiber, cord, ochre, hard board, and a small part of the human hair. Since the construction of the hat includes using one's human hair, the sized of the hat became more enlarged over time. As the community's resources expanded, the material for constructing the hat were transformed into a more versatile range of objects. The modern-day hat design includes using cotton fiber in a tightly woven basket like frame while employing traditional weaving techniques with yarn. A fine cotton cloth is then stretched over the framed and sewn into a concentric circular pattern that includes embroidery. The process of converting these raw fibers into finished apparel is a time-consuming task. Zulu women identify the hat as serving a major function in their culture.

The *Isicholo* hat presented in Figure 18.1 is an imitation of the authentic head piece. The hat is detachable and mainly worn during ceremonies. The foundation of the hat is made up of black cloth that is stretch to cover the outside and inside of the hat. Furthermore, several colors of string, which are prepared from a combination of red ochre and fat mixed with the dyed grass fiber. The colored yarns are threaded in symmetrical patterns to make up the design of the hat.

Methods for Data Collection

I employed four collection methods to examine the structure, mathematical and social relevance of the indigenous object, that is, *Isicholo* hat. The methods included: interviews, site visits and field observations, col-

Figure 18.1. The modern *Isicholo* hat purchased at the uShaka Marine beachfront in Durban, South Africa made in 2011

lection of artifacts, and informal conversations with sellers. I collected anecdotal narratives and prototype of the hat from several locations including the tourist site of Shakaland cultural village in the KwaZulu Natal province of South Africa.

I interviewed local Zulu tour guides in the guest office at Shakaland. After attending a traditional attire cultural workshop earlier that day, I noticed the traditional attire that is worn by the Zulu personnel at the cultural village especially the women's headdress. My main questions focused on the following:

1. "What is the attire worn for nightly programs?"
2. "How do you make the original *Isicholo* hat made?"
3. "Who is responsible for the material?"

When explaining the answers, the tour guides explained how each item is constructed by giving descriptions. They also described the roles of the men and women as reflected through their clothing. My conversations with the tour guides culminated in the most detailed and profound description of each individual piece of traditional Zulu clothing. For example, one tour guide explained,

> In present day, married Zulu women keep their heads covered in public. For ceremonial occasions, they wear elaborately woven hat with beadwork attached. The old hat beaded headbands (umqhawazi and isembozo) was made out of Dried grass, intertwined with red cotton and human hair and covered with red ochre, which refers to the living cow and also evokes the blood of the earth. The hat sends a message to the people, it conveys that she is no longer among the unmarried women amaghikiza; she belongs to the new stage of life, the married woman-umfazi.

This information helped me understand how oral traditions and knowledge are handed down to each generation. The Zulu women learn how to make the *Isicholo* hat by watching older married women engaging in the design and execution process. The tour guides insisted that note taking during our conversation would serve as a distraction from actually learning about Zulu culture. They went on to suggest that watching and listening intently were merely enough to grasp most concepts of Zulu practices (see Figure 18.2).

I also, visited the Phansi Museum, one of the biggest and most spectacular collections of African arts, crafts, and indigenous technologies in the world. The tour guide in the museum, before explaining the apparel items, asked me, "are you married, single, or engaged?" She expressed that she could not tell my marital status by the attire I had on. She indicated that the *Isicholo* hat is a means to identify the women's marital status. She further

Figure 18.2. The original *Isicholo* hat assembled by weaving grass in conjunction with hair with a flat top.

explained that the patterns and shape of the hat represent which tribe the women are from. It was obvious from her description that the *Isicholo* hat represented the social status of a woman. By seeing this indigenous technology on the life-size dolls, I then saw mathematical patterns and designs that needed further investigation. After understanding the social and cultural relevance of the hat to Zulu culture and traditions, I was ready to understand the mathematical representations.

While walking through the museum, I observed the hanging clothing displays. The clothing display is the most popular exhibit at the museum. This room includes thirty life-size puppets adorned with ceremonial dresses from all over Southern Africa. After asking permission, I took several photos of single, married, pregnant, engaged, and elderly Zulu women and men and asked specific questions related to the colors used in clothing, beadwork messages and pattern usage.

While at Shakaland cultural village, I purchased two distinct postcards that displayed women traditional clothing attire. While at dinner in Durban's WestPoint Mall, one tour guide glanced at my postcards and inquired about my purchase. I expressed to him that I was seeking information about the clothing style and the relationship to mathematics. He then initiated the conversation describing the clothing that was on my postcard and indicating that the women were from the Ndeble tribe. He explained the patterns in the clothing styles describing the concealed meanings behind the clothing. During our chat, the tour guide mentioned that he was raised as a traditional Zulu but received a Western education. While he understood my investigation, he was puzzled at how the *Isicholo* hat patterns were designed. He explained that people in the city of Durban do not see that weaving and beadwork are associated with mathematics. Instead South Africa mathematics education is focused on westernized ways of learning mathematics, so mathematics has no significance in these oral traditional practices.

During my field work, I visited kiosks of local vendors located on the beachfront in Durban and who sell items in the city made by their families in the rural areas. The vendors spoke little English, but they bargained with prices skillfully. To hold a conversation with the seller, you must make a purchase first before asking any questions about the items. I approached a woman at the beachfront, she was weaving a beaded skirt, so I inquired about the price of the *Isicholo* hat first then asked about the beaded skirt. With the purchase of the hat, the woman happily entertained my questions. My opening questions were, "what math is involved while making the *Isicholo* hat, how many lines were numbered and how was the pattern traced? The women looked disengaged from the conversation and was not particularly interested in what I was asking. When asking about who taught her how to make the *Isicholo* hat and about the meaning of the colors and lines, she replied She replied pointed to the *Isicholo* hat while answering,

> Sewing and beadwork was taught by my sister; some of the designs were taught by my mother; no classes were taken to learn this skill. The way we learn is by someone sitting us down and showing or telling us what to do. The *Isicholo* hat imitates the hair design; the patterns are connected to match some of the beadwork and shows where you come from. The hat is a part of the attire for traditional women. The bottom of the hat represents respect to the in-laws. We design hats by show and tell; no numbers counted, no traced patterns are involved. Zulu women only wear this hat style.

Data Synthesis

After collecting relevant information and artifacts, I used qualitative techniques to extract emerging patterns related to the social, cultural and mathematical structures underlying the indigenous technology of the *Isichiolo* hat. Furthermore, in each step of the analysis, I correlated data triangulated from the interviews, site visits and field observations, collection of artifacts and informal conversations using the substantivism theory in order to establish a meaningful frame to interpret major ideas. By understanding the process of development of the *Isicholo* hat, I tried to examine the mathematical relevance underlying this object of indigenous technology.

The Story of the *Isicholo* Hat: Juxtaposing Relevance

In an attempt to understand the *Ishicholo* hat as an object of indigenous technology, I mapped several aspects related to its significance in the Zulu community. As such, I interrogated its social, cultural, mathematical, and technological relevance and its contribution to building a sense of community and culture, which are highly cherished values of the traditional Zulu lifestyle. The Zulu people believe that they must continue to share and use the knowledge of making indigenous technological tools in order to keep their culture preserved. Therefore, the *Isicholo* Zulu hat must continue to identify a Zulu woman's marital status in order to keep the importance of traditional Zulu attire alive.

It was obvious that the color patterns and designs embedded in the *Isicholo* hat embody a deep sense of social relevance. The *Isicholo* hat is symbolic of Zulu married women. Originally, Zulu women personalized their hats with bundled locks of hair. Today however, Zulu women personalize the *Isicholo* hat with segments of beadwork and embroidery that compliment themes present in the accompanying skirts and shawls. The *Isicholo* hat continues to portray ancient Zulu tradition. The tradition of receiving this hat as a gift for being a newly married woman continues to exist along with wearing the hat for special occasions. Although simple mathematics is involved in the construction of the hat, Zulu women have not adopted this practice. Rather, Zulu women believe that they decorate the *Isicholo* hat for a purpose. For many, the *Isicholo* hat identifies women and their functional roles within the Zulu society.

The cultural relevance of the female Zulu traditional attire remains abundant in the *Isicholo* hat. The vibrant elaborate placement of patterns throughout the design of each hat demonstrates what the Zulu culture value most. Also, history is preserved with the colors of each hat paying homage to the many original Nations of Zulu women. This hat represents the respected Zulu women and can be used to demonstrate the skills and talent that goes into making the patterns and designs. The fabric, colors and materials of the *Isicholo* hat represent a significant part of the Zulu culture.

The technological relevance persists in assembling the *Isicholo* Zulu hat. The technology involved in making this hat includes utilizing indigenous tools or materials while using skillful talents and techniques to create a design. The design of this technological object reflects the art and science of indigenous knowledge. Technologies used by the Zulu community constitute an important part of its cultural heritage. While new technologies are being adapted in Zulu oral traditions, Zulu traditions are kept preserved using indigenous technologies. For example, the old hat consists of using real hair and earthly products while the new hat is detachable and is designed by weaving yarn on a thin piece of cloth.

The mathematical relevance is also evident in the design of the *Isicholo* hat. A typical *Isicholo* hat consists of an array of lines and colors composing symmetric and geometric art. The lines within each pattern function to relate an African woman to her local tribe. In short, design of the hat utilizes colors, shapes, and patterns to represent and identify authentic African women. Within the *Isicholo* hat there exists many different types of symmetry such as rotational symmetry and translational patterns that are positioned around the border of the hat. For example, the *Isicholo* hat that was examined had rotational symmetry with an angle of rotation equals 80, or an order of rotational symmetry equals 4.5 degrees (360°/80°). The translation of patterns includes 30 units in the horizontal direction and 40 units in the vertical direction. Employing the following equation for ordered pair (x, y): (x+x1, y +y1) after substitution becomes (x+ 30, y+40). Said differently, every x-coordinate will become 30 and every y-coordinate will become 40 for a final mathematical composition of x+30 and y+40.

Implications: Indigenous Technology in Education

For the Zulu community, culture influences and motivates technological developments in response to everyday needs. In order to preserve and sustain their cultural and social heritage, the Zulus continue to use and share indigenous technological tools that were passed on from their ancestors. In the past, the *Isicholo* hat was made of natural material using indigenous technologies. Since new technologies have been adopted, the hats have found their way into contemporary society through oral traditional practices.

The *Isicholo* hat, as an indigenous technology, embodies mathematical ideas that are intertwined with social and cultural significance for the Zulu people. In Westernized society, computers can create such rotationally symmetric shapes and designs that mimic patterns from the Zulu *Isicholo* hat. Relating mathematics to the *Isicholo* hat might provide opportunities to explore ways of how to integrate indigenous technologies in teaching interdisciplinary school subjects. For example, students can use culturally relevant objects to practice with geometry: rotational and translational symmetry. Students and teachers alike can employ several applications of symmetry while sharing information about their hats. Furthermore, teachers and students can extend instruction further by elaborating on the different colors and shapes that reflect different characteristics and personal identities.

The South African Department of Education has adopted indigenous knowledge systems as a part of their curriculum in order to foster student understanding of their local cultures. Educators have realized that when students learn using their local contexts, they have a better understanding of the necessary skills needed to contribute to the growth and development of South Africa. However, the feat of incorporating indigenous technologies into the curricula across countries continues to be a challenge amongst educators. For example, students are only able to explore the contexts in which technologies emerged and developed by reading case studies which could be an indication that indigenous technologies no longer exist. According to Vandeleur (2010) explains "students should be given opportunities in a classroom context to critically engage with technologies this would encourage their reflexive reasoning thus giving them the opportunities to improve and change existing technologies" (p. 86). Gumbo (2015) argues that teachers may find indigenous technology problematic if their training workshops do not include a subject matter which is relevant to their local contexts. Additionally, teachers may not embrace students' various cultures and characters. Gumbo (2016) proposes that in South Africa, the training approach to integrating indigenous technologies should be based on the *Ubuntu* ideology, which alludes to unity, and where learning is seen as a collective social activity. Hence, integrating indigenous technologies in school curricula could be a key to addressing the persistent lack of motivation that students have particularly in STEM subjects and consequently improve student success rate in school and beyond.

REFERENCES

Afọlayan, F. S. (2004). *Culture and customs of South Africa*. Greenwood Press.

Arnoldi, M. J., & Kreamer, C. M. (1995). *Crowning achievements: African arts of dressing the head.* Fowler Museum of Cultural History.

Boyne, R., & Rattansi, A. (Eds.). (1990). *Postmodernism and society*. Macmillan.

Carton, B., Laband, J., & Sithole, J. (2009). *Zulu identities: Being Zulu, past and present.* Columbia University Press.

Chahine, I., & Kinuthia, W. (2013). Juxtaposing form function, and social symbolism: An ethnomathematical analysis of Indigenous technologies in the Zulu culture. *Journal of Mathematics & Culture*, 7(1), 1–30.

Feenberg, A., Robins, K., & Webster, F. (2006). At modernity's limit. *Theory, culture & society*, *23*, 5.

Gumbo, M. T. (2015). Indigenous technology in technology education curricula and teaching. In P. J. Williams, A. Jones, & C. Buntting (Eds.), *Contemporary issues in technology education: The future of technology education* (pp. 57–75). Springer.

Gumbo, M. (2016). An indigenous perspective on technology education. In P. Ngulube (Ed.), *Handbook of research on social, cultural, and educational considerations of indigenous knowledge in developing countries* (p. 137). IGI Global.

Isicholo, Zulu women hat. (n.d.). Retrieved from http://hatsglobally.com

James, V. (1995). *Women and sustainable development in Africa*. Praeger.

Kapff, U. V. (2003). *Zulu, people of heaven*. Holiday African Publications.

Lustig-Arecco, V. (1975). *Technology: Strategies for survival*. Holt, Rinehart and Winston.

Mabunda, R. (2010). Introducing Zulu clothing design. http://www.youtube.com/watch?v=YJc3I0k1ZI0&feature=related

McGovern, S. (1999). *Education, modern development, and indigenous knowledge: An analysis of academic knowledge production*. Routledge.

Misa, T. J., Brey, P., & Feenberg, A. (2003). *Modernity and technology*. The MIT Press.

Ogungbure, A. A. (2011). The possibilities of technological development in Africa: An evaluation of the role of culture. *Journal of Pan African Studies, 4*(3), 86–100.

Pacey, A. (2001). *Meaning in technology*. The MIT Press.

Pacey, A. (1992). *The maze of ingenuity: Ideas and idealism in the development of technology*. The MIT Press.

Precious, O. U. (2010). Globalization and the future of African culture. *Philosophical Chapters and Reviews, 2*(1), 1–8.

Southafrica.info. (2012, November 9). *South Africa: Fast facts*. http://www.southafrica.info/

Spier, R. G. (1970). *From the hand of man: Primitive and preindustrial technologies*. Houghton Mifflin.

Spier, R. G. (1973). *Material culture and technology*. Burgess.

Spector, J. M. (2011). *Foundations of educational technology: Integrative approaches and interdisciplinary perspectives*. Routledge.

Stewart, F. (1979). *International technology transfer: Issues and policy options*. http://documents1.worldbank.org/curated/en/581801468765906402/pdf/multi-page.pdf

Vandeleur, S. (2010). *Indigenous technology and culture in the technology curriculum: Starting the conversation: a case study* (Doctoral dissertation). Rhodes University.

Vandeleur, S., & Schäfer, M. (2011). Indigenous technology and culture. In *International Handbook of Primary Technology Education* (pp. 305–317). Sense.

Vaughan, D. (2011). The importance of capabilities in the sustainability of information and communications technology programs: The case of remote Indigenous Australian communities. *Ethics and Information Technology, 13*(2), 131–150.

White, L. A. (1959). The concept of culture. *American Anthropologist, 61*(2), 227–251.

CHAPTER 19

"MAKING" SCIENCE MEANINGFUL AND EQUITABLE

Colby Tofel-Grehl

ABSTRACT

Often touted as a new and important way for students to access STEM (science, technology, engineering, mathematics) content, maker space technologies offer a unique opportunity to disrupt the traditional perceptions of who can successfully "do" STEM. Electronic textiles (e-textiles) are particularly culturally engaging, because they allow makers to create working circuits using crafting materials atypical of stereotypical STEM pursuits such as conductive thread and sewable LED light bulbs. This chapter describes the experiences of one teacher as he used e-textiles to teach in his middle school science classroom. The author explores the ways in which this teacher's ideas and conceptions of the abilities of his Latino students shifted as he taught science using e-textiles. The discussion highlights the importance of teachers' conceptual changes as a mechanism for diversifying the STEM pipeline.

Evidence-Based Inquiries in Ethno-STEM Research: Investigations in Knowledge Systems Across Disciplines and Transcultural Settings, pp. 361–375

INTRODUCTION

The current national focus on STEM (science, technology, engineering, mathematics) education evolved from reports that the United States' workforce would suffer from a future shortage of skilled STEM participants (President's Council of Advisors on Science and Technology [PCAST], 2010). Recent legislation intended to facilitate better and more effective STEM education, such as the America COMPETES (Creating Opportunities to Meaningfully Promote Excellence in Technology, Education, and Science) Act of 2007, has further promoted the advancement of STEM education. The creation of a national secretary of STEM education exemplified the importance being placed not only on STEM in general, but specifically education within STEM.

Early exposure to meaningfully and personally engaging STEM learning opportunities leads to higher rates of STEM degree attainment (Tai, Liu, Maltese, et al., 2006). While programing proves an essential skill for many STEM professionals, within schools computing is rarely meaningfully integrated within other content classes and tends to lack meaningful application for solving authentic STEM problems students face (Barron et al., 2003). While states and national organizations work to develop standards that address computing as an integrated aspect of STEM learning, the skills instruction for computing and coding remain isolated from the rest of the curriculum. Thus, genuine opportunities to integrate scientific reasoning and computing skills remain unfulfilled and unfulfilling for students (Hofstein & Lunetta, 2004). With a national push toward meaningful integration of STEM learning opportunities in K–12 classrooms, many wonder how to further foster both access and interest in STEM for traditionally underrepresented minority students. In fact, a national epidemic of homogeneity of STEM participants exists, creating a diversity vacuum that leads to the decline of diversification of STEM research agendas (PCAST, 2010). In short, because scientists research the things they have personal connections with, a lack of diversity among scientists leads to less research on diseases that affect underrepresented populations.

Recent enthusiasm for the Maker Movement (Peppler & Bender, 2013) provides a context for exploring the way new activities can engage classroom students in STEM. The Maker Movement uses technology, most frequently in non-formal educational contexts such as camps and after school clubs, to engage students directly with content and skills using a constructionist framework (Vossoughi & Bevan, 2014). Electronic textiles, or e-textiles, emerged as one maker technology that demonstrated a potentially meaningful mechanism for diversifying and improving student interest in STEM activities (Searle et al., 2015). E-textiles projects merge wearable sewn construction projects with programmable microprocessors.

They allow students to control various aspects of the constructed object. The microprocessor can be programmed to do a wide range of tasks, from blinking sewn lights to monitoring a sensor and alerting the wearer when there are changes detected by the sensor. These projects allow students to design and build solutions to personally relevant problems while engaging core scientific knowledge from existing physical and life sciences curricula and developing computer science skills.

The integration of STEM content that is typically included in school science curricula and computing is also a noteworthy aspect of making. While programing proves an essential skill for many STEM professionals, computing is rarely meaningfully integrated with other content classes in schools and tends to lack meaningful application for solving authentic STEM problems (Barron et al., 2003). While states and national organizations work to develop standards that address computing as an integrated aspect of STEM learning, the skills instruction for computing and coding remain isolated from the rest of the curriculum if it is offered at all. Thus, genuine opportunities to integrate scientific reasoning and computing skills remain unfulfilled and unfulfilling for students (Hofstein & Lunetta, 2004).

Traditional science curricula around electricity and circuits use wires, breadboards, batteries, and incandescent light bulbs to allow students to build working models of circuits (see Figure 19.1).

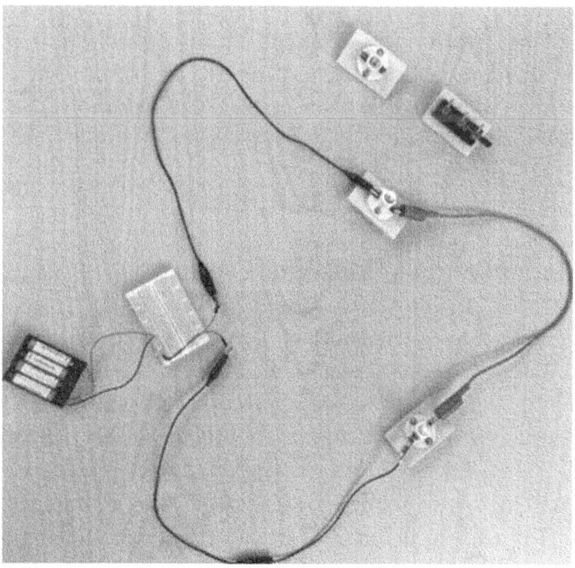

Figure 19.1. A traditional circuit build.

Upon completion of the lesson, these circuits are disassembled to prepare for use by the next group of students. E-textiles projects allow students to make and retain the objects they invest themselves in building. By sewing circuits using new materials (such as conductive fabric, sensors for light, sound, and pressure, and actuators such as LEDs and speakers) students produce wearable items (e.g., texting tee-shirts, force sensing backpacks). Through these projects students engage engineering practices and find solutions to challenging problems while designing something personally meaningful and reflective of their own aesthetic values.

Physics is often the most Western and linear of the sciences. Frequently taught through didactic lectures, these classes exude the mode of the western scientific learning model. This chapter explores the evolution in thinking of one teacher with 10 years' experience teaching eighth grade physical science. A self-identified "innovator" Mr. Von Stuffen taught his core content on electricity and circuits in the traditional fashion using alligator clips and breadboards for most of his career. After experiencing the use of e- textiles circuits, in a professional development workshop, he decided to develop lessons for his students using this technology. Though the e-textiles teaching process his thinking about his students evolved.

This chapter shares the evolution of his perceptions as an educator working in a highly diverse rural district as he engaged in teaching using what he called "culturally applicable technology." Working with students he previously identified as low performing and disengaged from science learning, Mr. Von Stuffen experienced a conceptual shift in his perceptions of both his students' ability and his role in supporting their identity development as scientists within his science classroom.

BACKGROUND

Understanding the ways in which the construction of physical artifacts impacts learning, or constructionism, acts as the framing context for engaging the Maker Movement (Turkel & Papert, 1990) in today's classrooms. With ever changing technology available to teach artifact construction (i.e., 3-D printers, laser cutters), opportunities for students to learn while engaged in modeling and material manipulation (Halverson & Sheridan, 2014). Of additional consideration for teachers is the ever-decreasing cost of these technologies, providing reasonably affordable mechanisms for larger groups of student engagement (Gershenfeld, 2007). By merging traditional crafting techniques with new technology, the opportunity exists to share value for different funds of knowledge. E-textiles has been particularly effective at modeling the values of traditional crafting cultures while engaging students in learning greater content standards. While the technological options ever evolving, a recent interest in merging traditional crafting with modern technologies, leading the Maker Movement to expanded dramatically in past a few years. E-textiles provide students the opportunity to construct objects of personal value) while tackling meaningful problems important to them in ways that are educationally empowering with contemporary technologies (e.g., Blikstein, 2008, 2013).

Within STEM education, motivation and interest are highly correlated; finding activities that promote student engagement while addressing learning standards is essential (Martin & Dixon, 2013). Using e-textiles as a fulcrum, we can embed STEM learning, and science in particular, into culturally meaningful and personally relevant projects that engage students in the Next Generation Science Standards (NGSS) required by districts. By using prior experience or personal relevance as a tool for engaging students, students build knowledge and scaffolds it to interest (Jones, Ruff, Snyder, et al., 2012). However, when student interest is not leveraged and they do not identify their personal identify as engaged in their science learning they develop negative associations with science learning (Basu & Barton, 2007).

Because e-textiles incorporate traditional crafting techniques and materials while engaging users in science and coding, they offer a unique third space from which merged identities can develop. Because of this fusion, e-textiles may disrupt traditional barriers to STEM interest and participation. For example, much potential exists for e-textiles to disrupt historically gendered boundaries to STEM participation through the authentic integration of crafting into more traditionally STEM-related knowledge and skills with electronics and computing (Kafai, Fields, & Searle, 2012). With a focus on hand crafting, e-textiles occupy disruptive space within thee Maker movement. Girls reported being less intimidated to engage in e-textiles based learning due to the crafting elements of e- textiles regardless of prior crafting experience with crafting (Kafai, Fields, & Searle, 2014). Beyond gender, e-textiles offers a valued space for traditional knowledge funds within science and computing. Kafai, Searle, Martinez, and Brayboy (2014) developed a culturally responsive approach that connects community funds of knowledge to computing with Native American students. By showcasing and elevating the value placed by many indigenous communities on the practices of crafting, and connecting it with computational thinking a space evolved for students to own their role as successful stakeholders of the digital age" and tie their community knowledge to their participations in making in an effective way.

RESEARCH QUESTIONS

With the dual goals of better understanding the ways in which e-textiles effected changes on one teacher's perception and attitude towards his Latino students and better understanding the possible impacts e-textiles had on student interest and engagement in science class, I posit the following research questions:

1. What changes in attitude and teacher beliefs occurred over the month-long teaching experience for Mr. Von Stuffen?
2. How did the teacher's instructional practices shift during teaching e-textiles?
3. Did students report any difference in engagement or enthusiasm during their learning unit on e-textiles?

I argue that understanding the way that these three research questions intersect provides insight into the ways that e-textiles engage students in multidisciplinary contexts. Furthermore, this understanding can shed some light on the role of teachers as facilitators reflecting on their own perceptions of their students' capability and positionality in the science classroom.

METHODS

Context

School setting. The School for Understanding and Constructing Knowledge (pseudonym) is a rurally located public school serving a diverse population with 28% of the students on free and reduced lunch program assistance. A public middle school located in the Mountain West of the United States, the School for Understanding and Constructing Knowledge is the only public middle school within the town's city limits. Serving a town of approximately 10,000 the school averages 550 students enrolled annually. The school demographic population is 51% female with 70% of students being Caucasian, 26% being Latino, and 4% of students being from other ethnic backgrounds. Nearly 78% of students score in the proficient or above range in both mathematics and reading standardized testing.

The teacher. Mr. Von Stuffen's professional experience mirrors those frequently seen across rural schools. As a middle school science teacher, he possessed a content area degree in physical science at the bachelor's level. He taught eighth grade science at The School for Understanding and Constructing Knowledge for the entirety of his 12 years teaching.

He was introduced to e-textiles at a district wide professional development workshop and saw connections to his own curricular standards. A Caucasian male in his 30s, Mr. Von Stuffen participated and completed advanced certification as part of his professional training and education.

Data Collected

Multiple data sources were collected to best capture the complex teaching and learning dynamics at play. Observations were made for two months on a daily basis across all treatment and control classes. All observed classes were video recorded and transcribed. Interviews were conducted with 34 students before and after instruction. Additionally, interviews were conducted with the teacher at multiple times during the construction of the curriculum and data collection process.

Introduction to E-Textiles

Several years prior, Mr. Von Stuffen's district invited teachers to attend a STEM professional development workshop. While attending the workshop, Mr. Von Stuffen learned what e-textiles were and had the opportunity to engage in building one of the projects firsthand. He opted to make a class management sweatshirt with lights on the front that blinked faster when he squeezed the conductive patches to indicate his impressions of the students' behavior. Because e-textiles allowed him as a teacher explore electricity and circuitry while exploring computer programing and engaging construction practicesusing nontraditional materials (i.e., sewing, thread, and fabric). Mr. Von Stuffen saw immediate applicability for his own middle school science classes, He reached out to researchers from the university who lead the original professional development workshop. He requested assistance to create a curricular unit comprised of a set of e-textile projects and associated lesson plans designed to meet the learning standards of his district. In talking with the research team, Mr. Von Stuffen described his school as "totally different from a normal high school." He explained his school represented a very divergent population. Half his students he described as "the richest of rich" while the other half he referred to "poor kids from transient families who worked in the booming service industry our town has." As he said they were "two groups a world apart." Further, he went on to state:

> My normal kids do just fine in science. They do the work, they pay attention. I think the e-textiles projects will be great for them because they will

> really enjoy it. My other kids, the ELL kids, need this stuff. They need something that is fun and will work for them. The sewing projects will be great for them. I would expect them to do even better than the regular kids.

Mr. Von Stuffen taught electricity to all eight sections of science for the eighth grade students. He and the other eighth grade science teacher opted to swap students for two units a year; he opted to teach everyone electricity while Ms. Runsalong taught environmental science to all students. In order to understand and explore the ways in which Mr. Von Stuffen taught his students, we arranged for Mr. Von Stuffen to teach half of the sections using e-textiles based projects and curricular materials. The other half of his sections used his traditional curricular materials.

The two curricula were examined and tailored to provide as much equivalency as possible. Both curricula took the same amount of class time, required the same amount of direct lecture from the teacher, and addressed the same standards. With the goal of equivalent learning opportunities in mind, the traditionally taught curricula were amended to include an activity requiring students to debug improperly designed circuits (i.e., short circuits) to provide content parity with the e-textiles classes.

Students learned electricity content within the electronic textiles curriculum through a three-project sequence. The initial project offered students the opportunity to make a paper circuit (see Figure 19.2). The second project presented a bracelet circuit (see Figure 19.3). The third project engaged students in incorporating a preprogrammed microprocessor into e-textiles (see Figure 19.4).

Table 19.1 offers a visual representation of the comparable curricula.

Analysis

To answer the three research questions identified, a qualitative analysis was conducted. Analytic induction (Erickson, 1985) was used as a methodological framework to answer the first research question. Interviews were coded iteratively. Macro level coding focused on teacher attitude and belief indicated two broad level themes of attitudes about students and beliefs about students' ability. In order to unpack what, if any, changes occurred in teaching practices during the teaching of e-textiles observational field notes and transcripts of all lessons were examined for trends instructionally. Lastly, to check the perspectives of the students who participated student interviews were coded and analyzed for common themes of responses to questions and discussions of how they felt about learning with e-textile projects. Understanding the way these three research questions intersect

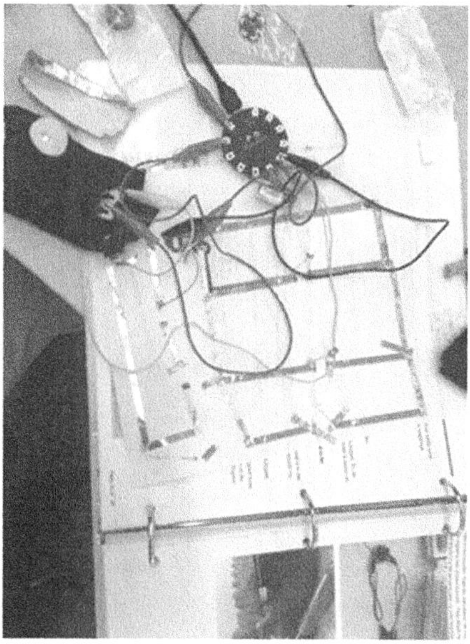

Figure 19.2. A paper circuit.

Figure 19.3. A bracelet circuit.

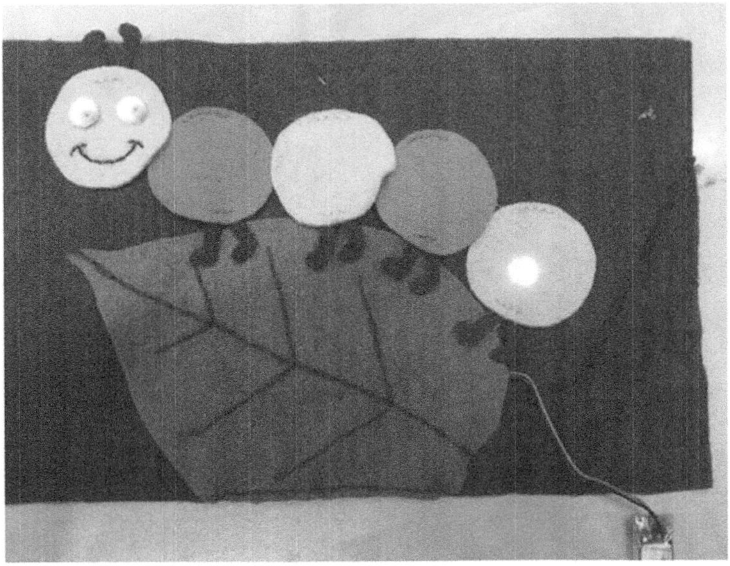

Figure 19.4. Preprogramed microprocessor project.

Table 19.1.
Comparable Curricula Table 1. Comparable Curricula

Traditional	E-textiles
Drawing circuits	Paper circuit
Creating circuits with breadboards	Bracelet with switch
Identifying and debugging short circuits	Preprogrammed circuit

can help understand the ways that e-textiles engages student culture and acts as a facilitator for a teacher to reflect on his own perceptions of his students' ability and place within the science classroom.

FINDINGS

The teaching of the e-textiles curriculum progressed as planned and anticipated. Mr. Von Stuffen taught both the traditional and e-textiles curriculum to the best of his abilities. Upon completing the teaching of both units Mr. Von Stuffen stated, "I definitely want to make e-textiles my core curriculum." Discussions with him about this emphatic shift revealed that he perceived a difference in the behavior of his Latino students. As he explained it during other units the Latino population of his school did not engage in science learning. As he explained:

> We have a very transient Latino group of kids and the fact is that they don't do their work, they don't care about school. Their families don't care about school either. Sometimes, I mean, sometimes the mom cares but she is busy working. So, these kids just float through.

His monolithic beliefs about Latino students was reflected in the observations of his teaching. Because the school opted to lump Latino second language learners into the lowest performing ability group, classes appeared during observation to be fairly homogeneous in their racial make-up. Of the 42 students observed during teaching who identified as Latino, none were in the section tagged as gifted, six were in the middle level classes, and 36 were split across the two remedial sections. While not all Latino students were second language English learners, the school administration and teacher used the terms interchangeably. Mr. Von Stuffen felt that the Latino students he identified as previously disengaged from science learning, engaged more actively during the unit using e-textiles. As he explained "normally I don't get any homework from those kids. During e-textiles I got 100% of work from nearly everyone." His concept of which students did science shifted through e-textiles. Observational data indicates that across race and gender groups enthusiasm and attitude were equally engaged. While changes in affect and attitude toward science learning were uniform across all groups, Mr. Von Stuffen felt the changes for his Latino students were different. He stated that for the first time "Latino students kept pace. You know? They closed the gap." His consistent perception that his Latino students caused us to delve deeper in interviews.

Observational data indicated that Mr. Von Stuffen did change his teaching practices when teaching e-textiles. Specifically, he was observed shifting his classroom discourse practices when teaching e-textiles. It was his questioning technique and linguistics patterns that most notably shifted. During the teaching of the e-textile projects he asked many more questions of his students. Not only was the number of questions higher, but the types of questions shifted. Rather than asking instructionally mundane questions like "did you write down that step" he asked students questions such as "why do you think it did not work?" This instructional shift towards asking more open ended discourse questions lead to greater opportunities for students to engage in classroom discourse and discussion. Because of the problem-solving nature of e-textiles projects, students partnered and discussed projects many more times than during standard curricular learning. When pressed on this during interview, Mr. Von Stuffen reflected

"Yes, I had not thought about that side of it. I guess I do have a lot more control about how we discuss stuff. You just don't think about that when you are teaching." When presented with observational data indicating that this shift was most robust during his remedial sections, he reflected by saying "I didn't even notice I talked differently to those kids. And those classes are most of my ELL kids, right? So basically, I am not talking to them like scientists like I am the other kids. Crap." Beginning to see his role in the dynamic Mr. Von Stuffen invited us to interview students. In talking with students in both traditional and e-textiles, a clear sense of how students perceived e-textiles. As one Latino student clearly articulated "Learning this way was better. I can share this with my mom because she sews. I can't really share a science worksheet with her. But she loved this." When asked about the changes in class he said "it was nice to get to work together and talk. That doesn't happen with worksheets."

In talking with students before and after instruction, a shift in their perceptions became apparent. The students' reflections revolved around two major differences in their perceptions became evident. Firstly, students felt that they were more engaged as learners specifically through e-textiles based instruction. Reflecting on the entire unit, a Latino student shared "If you would have taught this way the whole year, I would have been way more interested in Science." Each student who was interviewed specifically referenced their own individual project in detail. In sharing about what they liked about e-textiles one student noted "my project is all me. It has a Red Sox logo. I am the only kid who like them because I moved here from Boston." Another student, when pressed to share what their project made them think of science as a career noted:

> I never liked science. It's all paper and worksheets, you know? But this was cool because I actually got to do something. If I could do stuff more I would like science a lot I think. But I don't like it a lot.

The students' reported buy in concurred with the teacher's assessment. Mr., Von Stuffen reflected:

> Phillip, like many of my other Latino ELL students did extremely well on their light-up bracelet and their Lily Tiny. However, I would have never predicted why they would be so engaged in this project.... Many of my Latino students have parents and relatives that are expert sewers and expert sewing teachers. For the first time in their academic careers many of my Latino students received instruction and help with their homework from their parents or family members. Many Latino parents in our town do not speak English and often express their frustrations of not understanding their children's homework, not being able to help their children, and not

knowing what homework is due. E-textiles helped generate interest in my students' schoolwork through something as simple as sewing.

This heightened sense of engagement spoke to students' second generalized perception that they could identify more as scientists when doing this kind of science class work. Several students commented that they felt they better understood that scientists don't only do their work in one way or look one way. Summarizing a class discussion one student joked "Not everyone [who does science] wears a lab coat." Another student commented in her interview "No one in my family knows science, but I did this and my parents loved it. I could do this stuff all the time." Harder to convey, but equally important, is the enthusiasm with which students spoke about e-textiles in general and their own projects in particular. When asked about their projects many students spoke rapidly and with enthusiasm. That same tone and affective appearance did not appear when talking about science in general or other units they enjoyed throughout the school year.

CONCLUSION

Between the worlds of maker space technologies and real world classroom content instruction exists a chasm of disconnect. On the one side, maker space technologies, oft touted as innovative and dynamic ways to engage learners in enthusiastic novel approaches to learning, provide students with complex problems surrounding design and function. Seemingly ideal for addressing the multifaceted and deep level goals of the new Next Generation Science Standards (NGSS), these technologies frequently fail to make their way into the classrooms. More often than not, the research indicating the value of these technologies showcases efforts of summer camps, afterschool clubs, or other informal learning environments facilitated by researchers and specialists rather than independent classroom teachers. This research showcases the deep integrated learning possible through maker space technologies. Why then, with the compelling evidence available do we not see widespread adoption of maker space technologies in our schools' classrooms?

On the other side of our chasm are classroom teachers. These teachers provide multiple units of instruction a day to large numbers of children. The groups of children they serve are diverse both in cultural history and educational ability. Students in the classroom do not mirror those of the afterschool programs and camps frequently touted in Maker space programs. Furthermore, teachers are evaluated and strictly monitored based on student performance on standardized tests. They lack the luxury of time needed to provide deep content experiences for their students. Seeking to provide the most effective and safe instruction needed for students to measure proficient on standardized testing, teachers often shy away from novel curricula. Furthermore, maker space technologies require teachers to learn a great deal about the technology before instruction can be identified. Much of the writings and research on these technologies focuses on the skill attainment for students rather than the way in which instruction can be delivered from the teacher side. Few if any of the guides provide teachers with a scaffolded approach for demystifying both the technology and the instructional practices needed to teach with those technologies.

This chapter explored the potential transformative properties of e-textiles for teachers as they learn to navigate their own roles as educators and brokers of educational opportunities. Mr. Von Stuffen engaged deeply in self-reflection about his attitudes and beliefs about who could do science. The process of engaging in e-textiles taught him as much about his students as it did his students about electricity. By engaging students in culturally responsive projects, teachers can affect changes about who perceives themselves as students of science.

REFERENCES

Barron, A. E., Kemker, K., Harmes, C., & Kalaydjian, K. (2003). Large-scale research study on technology in K–12 schools. *Journal of Research on Technology in Education, 35,* 489–507.

Basu, S. J., & Barton, A. C. (2007). *Developing a sustained interest in science among urban minority youth. Journal of Research in Science Teaching, 44*(3), 466–489.

Blikstein, P. (2008). Travels in Troy with Freire: Technology as an agent for emancipation. In P. Noguera & C. A. Torres (Eds.), *Paulo Freire: The possible dream.* Sense.

Blikstein, P. (2013). Digital fabrication and 'Making' in education: The democratization of invention. In J. Walter-Herrmann & C. Büching (Eds.), *FabLabs: Of machines, makers, and inventors.* Transcript.

Erickson, F. (1985). *Qualitative methods in research on teaching* (pp. 119–162). Institute for Research on Teaching.

Gershenfeld, N. (2007). *Fab: The coming revolution on your desktop—from personal computers to personal fabrication:* Basic Books (AZ).

Halverson, E. R. & Sheridan, K. M. (2014). The Maker Movement in education. *Harvard Educational Review, 84*(4), 495–504.

Hofstein, A., & Lunetta, V. N. (2004). The laboratory in science education: Foundation for the 21st century. *Science Education, 88*, 28–54.

Jones, B. D., Ruff, C., Snyder, J. D., Petrich, B., & Koonce, C. (2012). The effects of mind mapping activities on students' motivation. *International Journal for the Scholarship of Teaching and Learning, 6*(1).

Kafai, Y. B., Fields, D. A., & Searle, K. A. (2012). Making learning visible: Connecting crafts, circuitry & coding in e-textile designs. In J. van Aalst,

K. Thompson, M. J. Jacobson, & P. Reimann (Eds.), *The future of learning: Proceedings of the 10th International Conference of the Learning Sciences (ICLS 2012), Volume 1, Full Papers* (pp. 188–195). International Society of the Learning Sciences.

Kafai, Y. B., Fields, D. A., & Searle, K. A. (2014). Electronic textiles as disruptive designs in schools: Supporting and challenging maker activities for learning. *Harvard Educational Review, 84*(4), 532–556.

Kafai, Y. B., Searle, K. A., Martinez, C., & Baryboy, B. M. J. (2014). Ethnocomputing with electronic textiles: Culturally responsive open design to broaden participation in computing in American Indian youth and communities. *SIGCSE '14: Proceedings of the 45th SIGCSE technical symposium on computer science education* (pp. 241–246). Association of Computing Machinery.

Martin, L., & Dixon, C. (2013). *Youth conceptions of making and the Maker Movement.* Paper presented at the 12th International Conference on Interaction Design and Children. New York, New York. https://dl.dropboxusercontent.com/u/422054/DigiFab_IDC2013/Papers/IDC_2013_Martin_Dixon.pdf

President's Council of Advisors on Science and Technology. (2010). *Prepare and inspire: K-12 education in science, technology, engineering, and mathematics (STEM) for America's future.* Report to the President. http://www.whitehouse.gov/sites/default/files/microsites/ostp/pcast-stemed-report.pdf

Peppler, K., & Bender, S. (2013). Maker Movement spreads innovation one project at a time. *Phi Delta Kappan, 95*(3), 22–27.

Searle, K. A., & Kafai, Y. B. (2015). Culturally responsive making with American Indian girls: Bridging the identity gap in crafting and computing with electronic textiles. In *Proceedings of Gender and Information Technology 2015.* Association of Computing Machinery.

Turkle, S., & Papert, S. (1990). Epistemological pluralism: Styles and voices within the computer culture. *Signs: Journal of Women in Culture and Society, 16*(1), 128–157.

Tai, R. H. Liu, C. Q., Maltese, A. V., & Fan, X. (2006). Planning early for careers in science. *Science, 312*(26), 1143–1144.

Vossoughi, S. & Bevan, B. (2014). *Making and tinkering: A review of the literature.* Commissioned paper for Successful Out-of-School STEM Learning: A Consensus Study, Board on Science Education, National Research Council, Washington, DC. Retrieved from http://sites.nationalacademies.org/cs/groups/dbassesite/documents/webpage/dbasse_089888.pdf

CHAPTER 20

DECOLONIZING THROUGH A NEW TRIBALISM

The Recognition of Warriors Through a Re-Evolutionizing Lifespace in Urban Mexico

Sue Kasun and Amita Jyoti Kaneria

It was the end of the academic year at Escuela Raices, and the members were busy preparing a final celebration to mark the closure of a cycle. In order to prepare the final meal, one complete with the celebratory foods of carne asada, grilled meat, and roasted sausage, they needed to make the fire in their hand-made earthen stove, a stove that was part of the curriculum over the last several years at the school.

One of the adult group leaders, Luis, called over five youth to enlist their help. "Hay que hacer el fuego," "The fire has to be made," he explained.

"Yo puedo," a few of the younger children said, "I can do it."

Luis twisted the point of the piercing through his right eyebrow. "Vengan, pues," he told them calmly to, "Come on, then," with a sense of purpose.

Evidence-Based Inquiries in Ethno-STEM Research: Investigations in Knowledge Systems Across Disciplines and Transcultural Settings, pp. 377–396

A few feet away on the cement sidewalk, the children ranging in age from seven to nine gathered their materials, thin, straight wooden sticks of about nine inches in length, hollowed and flat pieces of wood which sat well on the sidewalk, and small bits of dried grass and kindling. They rubbed the long sticks of wood between their palms with the tip in the hollow of the other piece of wood. They rotated the sticks with focus and precision, and within a minute or two—smoke. "Sóplenle," "Blow on it," Luis told them. And just like that—fire. Fire to cook the meal, fire as an essential element of who the school is.

As fire-making has been a pivotal moment of humankind, it was a watershed moment for Escuela Raices when it was "discovered" by independent trial-and-error about five years ago. Since then, the participants have made their own fire to light their stoves at home and regularly as part of daily activities at the school.

The above vignette is from research conducted at a "school," (referred to by those who run it as an *educational project*) designed for "warriors" who survived and thrived despite difficult urban conditions in Mexico. This chapter examines the results of ethnographic observations and interviews with leaders of a school which served historically marginalized youth from summer 2016 and spring 2017 in Cuernavaca, Mexico. Kasun observed the project's work over the last four summers, and she noted that a transformative impact remained durable on participants. The participants created a "new tribalism," or what Chicana feminist/indigenous/postcolonial theorist Anzaldúa refers to as "working together to create new 'stories' of identity and culture, to envision diverse futures. It's about rethinking our narratives of history, ancestry, and even reality itself" (Anzaldúa, 2015, p. 85).

In Escuela Raices, the poorest of Mexico joined together daily for nutrition—both physical and emotional. This project was in operation for over 25 years, but in the last six it shifted its curriculum to create a re-evolution of humans who survived the worst of modern material conditions including poverty, extreme violence, and deep pain. Where Native science engages the deep roots of knowledge of indigenous populations to represent and understand the interconnectedness of all creation (Arrows, 2013; Cajete, 2000), Escuela Raices both reengaged submerged indigenous knowledges and built curricular spaces where participants could make quantum leaps of civilizations (Garcia, 1991) in spans of months toward their growth. The use of the term quantum leaps refers to how many participants transformed dramatically from their original motivators of fear, violence, and depression into empowered, self-aware warriors contributing peacefully, which then provided the space for the whole community at the school (their civilization) to grow together. This is an activation of people who have been described as the "neo-indigenous," or those who may no longer be directly connected to their indigenous pasts but have survived despite genocidal histories (Emdin, 2016). In Escuela Raices's terms, the leaders were merely helping "warriors"—the term ascribed by those who run it to the strong youth who attended daily—find their path both as individuals and in the collective. This is the "rethinking of narratives of history" (Anzaldúa, 2015) which has been reframed successfully in Escuela Raices and exemplified in the chapter through participants' accounts of both current students and of the leaders' lived experiences vis-à-vis participation with the group.

This chapter includes key components of their holistic curriculum and main themes we found, as described by leadership of those who directed the school. It examines the conditions which required such a response in contemporary—dystopian, even—times to situate these "warriors" who could survive and thrive despite the dystopic conditions that worked against students' indigeneity. According to the leaders and youth participants, the situations in which they lived—including the teachers—were of extreme violence, housing insecurity, and lack of income. These conditions are the result of a social system into which they were born, not one which any of the participants directly created (Vitali, Glattfelder, & Battiston, 2011).

We then explore how the school impacted the lives of participants in terms of the significance of community, the practice of gratitude, and their aspirations as part of the evidence section. Specifically, we illustrate and examine how community was built, both by convivencia, sharing together in a meaningful way, and through conflict. For instance, students daily took turns serving food to each other and preparing it while also having a collective reflection about their lives and daily activities in ways that were generative and thoughtful. At the same time, sometimes spaces of reflection were opened for conflict. Next, we show how living gratitude was another key element, one introduced more recently at Escuela Raices, in a way that cultivated deeper community and generosity to both self and others. Students and teachers engaged in lengthy and open discussions about gratitude. Finally, we situate the organization's aspirations as signposts of helping the school situate its efforts, from beginning to more deeply engage participants in arts education—ranging from pre-Hispanic instruments to postcolonial folk protest music—to the art of creating a new society.

BACKGROUNDS

Around 120 youth attended this life space in urban Mexico to be "creative," their only reward for attending. The physical space consisted of a small office and library on two floors, two large open spaces used for a group garden, communal eating, and a well-conditioned outdoor mini soccer field. All this was donated to Escuela Raices, and none of the donations upon which the organization survived were tied to any government's or foundation's expectations or mandates. The youth hailed from all over Mexico, including from many indigenous communities who could no longer support families due to economic constraints and which precipitated their families' migration to the city. In such cases, the children's families had migrated from more rural parts of Mexico in search of work and better lives for their children. This meant, more often than not, leaving land that could not produce agricultural products that could compete with heavily subsidized U.S. agricultural products as a result of NAFTA (Bigelow, 2006) to living in aluminum shacks along railroad tracks, the only places they could afford.

The framework for this chapter directly engages the theoretical underpinnings of the project as crafted by the curriculum designers at Escuela Raices, including creative transformation (Garcia, 1991) and the Four Agreements (Ruiz, 1997) while weaving in the Anzaldúan construct of "new tribalism," referred to above, or the construction of new groups and ways of being who break from old paradigms to form new and more inclusive spaces of belonging. We engage in the same theoretical underpinnings as the project, as we were compelled by their framework based on seeing how the participants both had internalized it and how it seemed to manifest in moment-by-moment practices. We include Anzaldúa's lens as a useful and theoretically aligning complement to enhance our understanding of the impact of Escuela Raices' work. Following, we explore each of these three areas more concretely.

Being creative, the only reward of participating, according to Escuela Raices, is learning how to predict the outcomes of situations and how to act upon their circumstances with information that is true, based largely on both John David Garcia's (1991) foundational work on creative transformation and Don Miguel Ruiz's Four Agreements (1997). By following a conscious return to basic ethics, the participants, including those who were graduates of the program and later worked as teachers, created a life space where participants made quantum leaps in civilizational growth. For instance, when one of a few small groups finally learned to make fire without any machine-made tools after 18 months of day-long efforts, the other groups also learned, within a few weeks and without contact with the initial group, how to also make fire. Similarly, the overall group's cohesion improved toward making its next steps in both values and social cohesiveness, where respect and mutual dialog became increasingly valued.

The Four Agreements were first described in the 1990s by Don Miguel Ruíz. They explore knowledge and ancient Totlec wisdom. The Toltecs were an ancient indigenous group situated in what became Mexico, and like much indigenous wisdom, has persisted, despite the decline of the ancient civilization. Ruiz (1997) explains how in modern times, people tend to have agreements (also understood as guiding self-narratives and values) because they have been given to them by their cultures, and that these agreements are often quite harmful. For instance, he explains how young children learn agreements about who they *are* in ways that are sometimes very harmful (1997), such as being told that a child is "too shy," and thus fulfilling a prophecy about who the child becomes as a "shy" adult. Similarly, children learn oftentimes constricting gender roles and eventually take on these agreements as their own, passing them on from generation to generation. For Ruiz, these agreements are illusions that can only be undone with a different set of agreements, the ones he delineates in his foundational text, *The Four Agreements*. The agreements may appear simple but in reality prove difficult to practice. They include: Be impeccable with your word; do not take anything personally; do not make assumptions; and, do your best. In a vision of a world in which all people accept these agreements, the ethical practice can be imagined to be one wherein people are deeply honest with each other and ready to ask questions about what things mean when they do not understand. People in such a collective could then also trust that the others they are with are trying their hardest, in a way that is free of illusions from other harmful agreements.

The Four Agreements form the ethical basis in Escuela Raices, along with what they call the "ethical axis," which includes a basic version of survival: "The elder takes care of the younger; the one who knows more teaches the one who knows less; the only punishment is not learning; the only reward it to be creative." This axis comes out of an understanding of John David Garcia's study of living organisms, including human animals. Garcia was deeply interested in how humans, especially, could evolve. Garcia (1991) developed formulas which demonstrate an appeal to ethics through creative practice (such as the only reward being to be creative), and that humans can only evolve when the greater good through this model is pursued.

According to the lead curricular developer of the current Escuela Raices project, the basis need not be Garcia's, but it is a strong one that the curricular lead knew well, as she had studied directly with him when pursuing and completing her PhD in psychology years ago. For Garcia, the highest ethic is being creative while not occluding others' abilities to be creative. As such, he explains there are two emotions in conflict with each other—fear and love, emotions found in all peoples universally. For Garcia (1991), love is best exemplified through giving, and it is "the highest form of creativity," (p. 31) and fear is the belief that one cannot create (p. 31). Garcia believes "true information" should be provided to correct those whose efforts work in contradiction of creativity, and that people should embrace negative feedback if it helps them improve. That said, it should be provided in a loving way, he argues. Ultimately, according to Garcia, this feedback will help humans achieve quantum leaps in our evolution, less our civilization, like those that have proceeded ours, fail. In Escuela Raices, the participants thus created their own new civilization based on these ideas toward achieving their own continuing evolution. Within the context of Escuela Raices, their community is in the building itself and the people within; it is also situated in the neighborhood within which they live. Their work strives to see that civilization evolve into one with less violence and less drug addiction towards one with more harmony, empowerment, and the transmission of life skills among the youth. They can thus be equipped to enter the world at large and be *creative*.

Finally, as a complement to the framing of the work at Escuela Raices, we engage Anzaldua's (2002) "new tribalism." The new tribalism suggests that at least momentarily, people can supersede categories and be together, even if just being on the bridge in the work of bridge-building among people who experience differences, toward the goal of greater social transformation (p. 570). The "new tribalism" is rooted in the idea that categories of people are not static, and that identities often intersect in ways that create new spaces of possibility; at the same time, it never negates the situated and often oppressive histories people have lived. These spaces are often temporary, the slash she articulates as a bridge in the Spanish pronoun (given in the feminine version) of *nos/otras*, loosely translated as (us/we/others) (p. 570). Anzaldúa (2000) writes against the oppressions women and non-heteronormative people experienced in social movements, especially those created around ethnic identities, including the Chicano movement, in her case (p. 185). She also explores embracing multiple ways of knowing toward the integration of being able to see the world differently and thus create new ways of seeing among these liminal or in-between spaces (Anzaldúa, 2002, p. 571). For Anzaldúa (2000), this work is critical because we have engaged a global moment where our lives are intertwined, and isolation can no longer become a choice (p. 186).

Ultimately, we must share in this global space, Anzaldúa (2002) says, and that work leads to transformation, both individually and socially:

> By compartiendo [sharing] historias, ideas, las nepantleras [liminal bridge-builders] forge bonds across race, gender, and other lines, thus creating new tribalism. Este quehacer [this doing]—internal work coupled with commitment to struggle for social transformation—changes your relationship to your body, and, in turn, to other bodies and to the world. And when that happens, you change the world. (p. 574)

This new tribalism allows for ways of knowing which echo from the past to the present, allowing individuals and groups to change, in ways that mirror and extend the philosophies of Ruiz and Garcia.

METHODS

Kasun partnered with the school in cross-cultural exchanges, bringing U.S. based university students to learn from Escuela Raices for the last five years, collecting ethnographic and interview data at different points (Kasun & Saavedra, 2016; Kasun & Lee, 2017). Her work with the school has been in efforts to help decolonize (Smith, 2006); specifically, she recognizes her greater audience in the global North can be decolonized by engaging this work, fraught as all contact with North-and-South can be. The global North in this context is specifically the United States and other western, audiences who have historically developed out of European paradigms. The global South includes, generally speaking, those audiences that identify as indigenous, colonized or people of color and who have persisted despite attempts at destroying their histories. Escuela Raices also maintains as part of its mission an outward orientation of wanting to, in their words, "share the fire" far beyond the doors of their school, and the researchers attempt to honor that intention, an indication of their efforts toward decolonizing work.

Kasun focused her research on this question: What is the impact of Escuela Raices's work on students, and how does the school design its work to achieve the impact it does on the participants? In Summer 2016, the lead researcher conducted five focused interviews (Rubin & Rubin, 2012) with the leaders and lead teachers on both the kind of work of the teachers and school leaders as well as the intersections of their life history narratives (Chase, 2005). She conducted them during school hours on the school property in a quiet library space or quiet office space, whichever the participants preferred. These interviews lasted between one hour and two and half hours. Most of the participants had been with Escuela Raices for at least ten years, if not longer. Table 20.1 lists demographic information on key participants.

Marco was trained as a social worker in Mexico City and subsequently established the project with his wife, over twenty years ago. He maintained an openness to improving the project, including the current curricular iteration. Guadalupe was enlisted about six years ago to no longer be a temporary and occasional supporter to being the founding designer of the current curriculum. She held a PhD in psychology and was known as a healer in her local nearby indigenous community. Magdalena, Luis, and Melissa all benefited from the services of Escuela Raices for extended periods of time prior to being offered the chance to be teachers at the school, and they lived in the neighborhood from which the students hailed. These three also shared similar socioeconomic backgrounds from how they were raised, with very limited incomes, and, in the case of Magdalena and Melissa, they had migrated from their indigenous small towns several states

Table 20.1.
Interviewee/Participants

Name (Pseudonyms)	Approximate Age	Role/Important Characteristics
Guadalupe	60	Conceptual designer of curriculum; participated about once a week; held a PhD in psychology
Marco	60	Overall curricular and group leader; lovingly referred to as "el abuelo," the grandfather; social worker
Magdalena	35	Curriculum leader; participant from having her first child with Escuela Raices at 17 and has her children participated there
Luis	29	Youth leader; participant from youth, was studying psychology
Melissa	24	Preschool director; participant from youth days

away. They each commented that Caminando had a profound impact on the people they had become in a deep and meaningful way. Each of the five was deeply committed to their work and the students.

Kasun also collected critical ethnographic data (Foley, 2002; Madison, 2011) through participant observation (Spradley, 1980) of how the school worked over two weeks, during daily observations from 8 A.M. until 2 P.M., when the official work day ended. What made her work critical was her on-going commitment to the school to maintain reciprocal relationships with participants and teachers at Escuela Raices and future students from the U.S. who she would take to visit and learn from and with participants, as well as her reflexive sense about her work and positionality as a researcher. Her daily activities usually included helping serve breakfast, sharing breakfast with the students, talking and playing with youth participants, listening in on group discussions and evaluations, participating in daily workshop activities, and sometimes helping cook food at the end of the day for the final meal served. She and the second author returned in early 2017 to conduct further observations and follow-up interviews with the initial participants. She kept daily notes after each visit and during-visit jottings.

The school opened its doors to about 120 youth, from the smallest of children who attended a small preschool until age five to teenagers. Escuela Raices's daily activities were varied but followed a structured and intentional pattern. First thing in the morning, students sat and waited for breakfast along tables they had helped to clean the previous day, sometimes getting up to play or perhaps complete homework, often with help of teachers at the school or each other. The students helped serve the scratch-cooked and healthy Mexican food, offering a blessing over it collectively. After eating, each student cleaned her or his own plate at a communal sink. Sometimes a guest would arrive and be offered communal and public gestures of gratitude or some kind of discussion. Following, students went with small groups based on interest, such as to making clay objects and tools, weapons as tools (weapons in terms of how they were discovered in human history, such as spears), or agriculture. Teachers served as guides engaging the students on deep emotional levels and allowing the students to learn by discovering. Discovering could and did, at times, include learning how to consult the Internet through computers connected at the school and staff smart phones for problem solving, among, of course trial-and-error with all kinds of natural materials and eventually man-made tools. Next, there was time for reconnecting and large group discussion about what everyone was learning or sometimes to help evaluate activities and even some of the individuals (including directors) in the group. Those who wanted to stay could do so for a scratch-cooked lunch (served around the traditional Mexican meal time of 2 P.M.), and after each one cleaning up, they went home.

Kasun notes her positionality and history in this work as an important lens of interpretation, as are all positionalities (Johnson-Bailey, 2004). She first found the Escuela Raices project riveting the first summer she got to know their work, four years ago. After taking students to partner directly with them for two weeks at a time over two subsequent summers, she reported on the impact of exposure to preservice teachers in terms of the preservice, U.S.-based teachers' dramatic shifts in understanding what is possible in education and in their future work as teachers (Kasun & Saavedra, 2016; Kasun & Lee, 2017). As a White woman who is bilingual in Spanish and English, she knows she could never have "insider" status among the participants. She maintained a mindfulness about attempting to use decolonizing approaches to research, continually reflective of how her positioning and presence could be oppressive despite efforts for her work to be liberatory from colonialism (Smith, 2006).

While collecting data for this research last summer, she found herself in a dramatic moment, one where her mom's suicide had only transpired three months prior (Kasun, 2016). She recognized that the work with Escuela Raices was like so much research a mirror for her, one where she saw people actively and deeply engaged in the critical work of healing despite great difficulty. Like Kasun (2016), other members of Escuela Raices had either lost loved ones to suicide or survived years of abuse, as she had, growing up the daughter of a single mother with mental illness. She realized that the work of Escuela Raices demonstrated the radical possibilities of togetherness and healing, despite some of the most difficult life circumstances. Undoubtedly, all her interpretations are colored by these recognitions and affinities in ways she hopes help her better gauge the school's work. Kaneria recognizes her unique insights as a woman of color from a postcolonial background and with a biracial identity; she is bilingual in Spanish, a non-heritage language for her. She recognizes roots of identity and interethnic solidarity in her work with Escuela Raices while consistently problematizing her positioning as a visitor from the United States, where she resided.

After transcribing all interviews from the original language (all were held in Spanish, a language in which both researchers are fluent, with the exception of one interview conducted in English as offered by the interviewee), the researchers coded and sorted data from the interviews and participant observation notes into first rounds of codes and themes. Similarly, they coded and sorted field notes from the visits as well. The researchers each coded individually in an effort not to influence the other in the first round and allow as many codes as possible to emerge. Initial codes included, for instance: process of maturity, alternate reality, transcendence, group vs. individual work, and respect. Upon reflection and re-analyzing the data, they conducted a second and third round of codes to find the most salient themes from their interview and observation data (Saldaña, 2009). These ultimately led to the three final categories reported in this chapter: community, living gratitude, and aspirations, explored below.

EVIDENCE

By starting with learning how to build fire with no manufactured tools—a project that took 18 months—to eventually determining their curriculum as based in life skills, rooted in communitarian, ethical work, the Escuela Raices achieved a dramatically positive impact on its participants. The project's work is unlike any the researchers had experienced or read about. The work is radical in how it positioned the participants and in its curriculum of exploring indigenous and neo-indigenous knowledges toward truly quantum growth. Historically, the group evaluated their efforts as a coordinating team and among all the youth. They shifted over the last several years as being designated as several "families" who worked in smaller groups to tribes to eventually one large tribe. The school teachers and directors discerned the growth and change and that eventually the students had arrived at a point, along with the teachers, of being able to cooperate as one cohesive group.

Community

Living in community was an integral part of Escuela Raices. In their space, they wove a sense of community amongst the students, within the work of the adults as teachers and in the processes within which they address issues with individual students. To be clear, Escuela Raices was very purposeful about *how* they created community, recognizing that the space they created was more important than fostering friendships among staff—which they also recognized may not occur naturally. Above all, the teachers were committed to living and modeling the Four Agreements and Ethical Axis. The sense of community was also established in regular daily practices as well as purposeful engagement of youth within the ethical axis and the Four Agreements. On a daily level, youth and the adults ate breakfast together—all scratch-cooked, hearty Mexican food. The youth and adults washed their dishes, one by one, and rested until group activities are called shortly afterwards. In group activities, or workshops, they worked together on basic skills, such as the creation of weapons as early human tools—a quantum leap in their evolutionary history—as were the making of jewelry and building a garden, among other evolutionary events in their curriculum.

Sometimes illustrating an exception better shows the general practices and how community building mechanisms are activated. The strong sense of living in community at the school is evident in one instance when a coloring book was taken in summer 2016. As part of the school's process of working within the community when addressing issues such as stealing, group leader Magdalena chose to address the issue of the stolen book with the whole group. The children suggested someone had taken the book. Magdalena was upset by what had happened, yet she took the opportunity to continue to dialogue with all the students about stealing, the consequences for stealing outside the school (which could include a beating by police and/or jail time) and how to correct the action. Magdalena provided background that it is not enough to just say, "Stealing is bad," and that you should not do it because it is bad. She explained that youth might think they can just get away with stealing something without seeing the consequences clearly, something they regularly discussed at Escuela Raices. Generally speaking, they had very little problem with theft.

To deal directly with the theft, Magdalena described the importance of naming the action in front of the children and encouraging them to correct their actions in one of their daily group reflection meeting times. She did this without malice or meanness, just direct and clear delivery of information. A girl from the class eventually told Magdalena she was the one who had the idea of taking the book and together with another boy convinced the third boy to take it. This is when the whole community got involved and the students were encouraged to make right their mistake by admitting in front of an assembly that they had taken the book and that what they did was stealing. This was achieved only after several clear discussions with Magdelena and the three youth involved in a way that helped the children understand the seriousness of stealing and why what had happened was an actual problem. Magdalena explained how it was important for the students to name their actions as one of stealing versus framing it as they were just taking something that they wanted and she further explained the significance of having discussing it seriously with the entire school community:

> Entonces, siempre es como te equivocas, se vale equivocarse pero corriges y sigues hacia delante. Entonces con estos lo que hicimos fue que ellos nombraron que habían agarrado algo que no es suyo enfrente de todo el grupo en un a asamblea. Y entonces entre todos les dijeron, "No. Pues no hay que agarrar cosas que no es suyo porque aquí a lo mejor no te van a mandar a la cárcel, no va a pasar mucho, pero si ustedes lo hacen afuera con otras personas, esas personas van a decir este niño roba y le voy a mandar la policía, o vas quedarte siempre tachado como un niño que roba no, y no es divertido ser alguien que roba, no."
>
> So, always it's that if you make a mistake, it's valid to make a mistake but you correct and you move forward. So with these students, we had them name that they had grabbed something that wasn't theirs in front of the whole group in assembly. and then among all they told them, "No, well, there's no need to take things that aren't one's own because here probably they won't send you to prison, a lot won't happen but if you do this outside with other people, they are going to say, 'This kid steals,' and they're going to send them to the police, or you're going to be forever marked as a kid who steals, and it's not fun to be someone who steals, no."

Naming the action for what it was became part of the educative process for the children to understand their actions and the consequences of their actions, in the context of their community. Magdalena said, "Cuando los niños logran decir yo robe un libro, es saber que el niño esta aceptando algo que hizo," or, "When children arrive at the ability to say that they stole a book, it becomes apparent that he/she is accepting that they have done something." She continued by emphasizing the transformative impact of naming the action of stealing. Magdalena explained to the children about the change and transformation that occurs when the child who took the book decides to return it. Their naming their action illustrates the transformation evident in the children and the process of doing so in community. It also illustrates some basic practices of the community—daily collective discussion, connection and safety among members so that all can speak without fear of retaliation or humiliation, and perhaps most importantly, a deep and genuine care for all those present, one that was reciprocated by the youth who accepted their fault and continued to participate even in spring 2017 when the researchers returned, more integrated in the group than ever.

Living Gratitude

Gratitude practice was a key method of doing in Escuela Raices, albeit one that was drawn into close focus in the last two years as the project continued to evolve. The basis for living in gratitude was to shift from a victim mentality or "poor me" way of being to one of strength, determination and creativity; this finding was echoed by several of the interviewees. Magdalena explained how Guadalupe instructed the teachers to avoid the victim mentality within themselves so that they could live and model a different way of being for the youth,

> En este lugar no hay pobrecitas, en este lugar hay puros guerreros. Quien quiere ser guerrero esta en este lugar, [si], pero no hay pobrecita. Porque ser pobrecita te permite como te creas el derecho de exigir, no? De decir, "Yo no pongo nada de mi parte porque soy pobrecita." No, entonces ella [Guadalupe] dijo, "No, a la chingada con todo esto aquí no hay pobrecitos; aquí hay puro guerreros, ... ustedes son ricos mentalmente, son ricos de corazón, y ustedes tienen mucho que dar, tiene mucha capacidad y se quedan en ese plan nunca van a crecer y se van a quedar allí pobres porque se lo crean. Yo no quiero eso para ustedes la idea es que si ustedes saben mas les pueden ensenar a estos niños pero si ustedes no están bien y no se ven esa capacidad que ustedes tienen no pueden trabajar con los niños, qué les van a ensenar? No hay forma de enseñar algo que tu no puedes hacer, no te puedes comportar, no puedes controlar."
>
> In this place there are no *pobrecitas* (those that say/think "poor me"), in this place there are pure warriors. Whoever wants to be a warrior in this place, [yes], but there's no *pobrecita*. Because being a *pobrecita* allows you to believe you can demand, no? Of saying, "I won't put anything of my own part in here because I'm a *pobrecita*." No, so [Guadalupe] then said, "No, fuck that, here there are no *pobrecitos*; here there are pure warriors ... you all are mentally rich, you are rich in your hearts, and you have much to give, you have so much capability and you stay in that plan then you'll never grow and you'll stay poor because you believe it. I don't want that for you all the idea is that if you know more you can teach these kids but if you are not well and if you don't recognize this capabiltity that you all have you can't work with the kids, what will you teach them? There's no way to teach something that you cannot do, you can't behave well, you can't control yourself."

In an attempt to foster this shift in mentality from "poor me" to *guerrero,* fighting to be different and to see their richness, the youth were instructed to take time specifically to write thank you cards for the gifts and donations they received. On one occasion, a wealthy man and his partner were invited into the space to receive the kids' gratitude for having donated fresh chickens. They sang for him, did cheers, presented a banner with signatures as well as individually signed thank you cards. The donor appeared genuinely pleased and continued donating subsequently. The youth appeared to have succeeded in being reframed as capable and agentic as opposed to "*pobrecitos.*" Indeed, by expressing gratitude, youth were further empowered and positioned to speak in public voices rather than meek recipients of charity.

Another way in which the youth practiced living in gratitude was by making personalized birthday cards. In his interview with Kasun, Marco, one of the founders of the project, shared the cards the youth made for his birthday. He explained that he had taken the time to learn to play the harmonica, to draw and to pray with the youth despite not being a devout Catholic himself. This was all so that he could teach them these same skills for use in their own lives, ideas that had all come directly from the suggestions of the youth themselves throughout the evolution of the program. Marco explained that the cards not only had detailed drawings that showed genuine artistic talent but also contained beautifully written notes of thanks. Some examples from some of the boys/young men included,

> *Example 1*: Tú eres la única persona que me hace dibujar, con amor se hizo el mundo.
> You are the only one who makes me draw. With love the world was made.

> *Example 2:* Gracias Marco, por tus oraciones, gracias por enseñarme a jugar fútbol, por enseñarme el juego, por enseñarme a hacer dibujos. Te voy a extrañar.
> Thanks, Marco, for your prayers. Thanks for teaching me to play soccer, for teaching me the game, for teaching me to draw. I will miss you.

> *Example 3*: Te queremos por orar, enseñar a hacer fuego en todas las cosas. Gracias por enseñarme a orar y creer en Dios.
> We love you for praying, for teaching us to make fire in all things. Thanks for teaching me to pray and believe in God.

Marco explained the gift that it was for him to receive these letters but even more important was the way in which the youth had come to learn to be aware of their feelings and to express themselves, skills he believed transformed their lives. We acknowledge a recognition of the fire, the essential quantum element of the change and yet stability of the group's work. There was also an intense expression of gratitude toward a higher power, something the youth wanted to include and that manifested itself as a Catholic-style altar of gratitude to Mary, mother of Jesus, attended daily with small offerings of flower and candles. Gratitude became, then, transcendent, not just something for unique individual people, but for greater elements of understanding the world.

Aspirations

As mentioned before, Escuela Raices worked with children based on the Four Agreements: "Be impeccable with your word; don't take anything personally; don't make assumptions; and, always do you best." They also followed an Ethical Axis situated in an understanding of how any species survived: the older looked out for the younger, those who knew more taught those who knew less. Using these as guiding principles, Escuela Raices aspired to prepare the youth for integration into society with an ultimate goal of creating a new society, a new tribe.

The youth learned and practiced a different way of being with themselves, their classmates and with their teachers. They learned skills and tools not only by being told about them but by connecting with the rhythm of the others and receiving the transmission of the way of being and doing at Escuela Raices. Luis said at the end of his 2016 interview, "We do it all together, not alone," building on the idea of living in community. Luis illustrated that the transformation that occurred with the youth at Escuela Raices would go with them whereever they would go, guide them to *salir adelante*, evolve and integrate into society. He even described a vision of transmitting their knowledge and their new way of being to others to create a new society and even a new world.

> Si aquí uno los niños pueden salir adelante, que todas las personas pueden salir adelante no solo los niños ni nosotros de aquí que les ensenamos todo … pueden hacer algo pues para salir a delante todos juntos no solo aquí en México.
>
> If here one of the kids can keep moving forward, then all the people can keep moving forward not just the kids nor just us from here who teach them everything… they can do something to move forward all together not just here in Mexico.

Specifically, the founder and "Abuelo" (Grandfather), as he was affectionately referred to ath the school, described a vision that included a depth of appreciation for many things, including the mix of cultures toward a global sense of belonging. Others may or may not have shared all the elements of his vision at Escuela Raices—these are only described and brought into fullness in consensus as a group at Escuela Raices during their communal evaluation sessions. But for Marco, for instance, he wanted the students to learn a love of reading, of all kinds of music, of all kinds of food. A local and well-known Mexican folk music teacher offered a twice-a-week music class for interested youth who learned how to play both pre-Hispanic instruments and contemporary music such as the theme song to the film Titanic on the recorder. Ultimately, though, Marco wanted the neighborhood to become transformed, one that was economically quite diverse, and he wanted the fire of Escuela Raices to be part of a larger transformation of the world. He agreed it required enormous amounts of efforts from all who participate, something Kasun witnessed up close in daily interactions over several years.

For group leader Magdalena, a product herself of the outcomes of Escuela Raices from her later youth years, she expanded on the idea of the tools and skills that were internalized by the students as being theirs to take and teach:

> Entonces ya con herramientas claras de lo que están trabajando en este lugar también les permita sobrevivir alrededor de ellos; ya sea en la casa, la escuela, con los amigos con lo que sea no, porque mucho de lo que le decimos a los niño es que ellos pueden obtener toda las herramientas necesarias para la vida si quieren, porque no lo podemos obligar, no? Cada quien puede tener herramientas de vida se ellos deciden aquí van a estar.
>
> So with the clear tools of what they are working on in this place it also permits them to survive out there—in the home, school, with friends or whatever, because a lot of what we tell the kids is that they can obtain all the necessary tools for life if they want, because we can't oblige them to, no? Each one can have the tools for life if they decide here they stay here.

We were able to observe a new activity for the youth that focused on their aspirations and transformation and became a new tool for their survival. Many of the youth had helped create and decorate small cloth bags. The activity was one where the youth could practice changing and controlling their thoughts while simultaneously creating vision for their personal aspirations. A French visitor who had spent several weeks working closely with the teachers and students instructed the youth to write a positive thought about themselves and a dream they had, both in the present tense as if it were real in the present moment. An example was, "I am creative" or "I am a teacher." The youth then folded the papers on which they wrote their positive affirmations and dreams and placed them in a bag of their choosing.

Once in the bag they placed the bag on the altar to Mary. This French visitor, as were all external visitors, was vetted to be allowed to share in the Escuela Raices space. She brought a new energy and a new challenge, one the school embraced and became part of a new and evolving vision. This evolution in community and of each individual exemplified a connection of all three themes highlighted in this chapter—a gesture of gratitude at the altar, in community, and of each one's vision.

IMPLICATIONS/MOVING FORWARD

Escuela Raices represents an alternative project to the idea of learning in contemporary models, as they so often circulate around the physical institution of a school, complete with mandatory attendance. In Escuela Raices, students attended because they wanted to, not because they were legally obliged. The participants, including interviewees as well as the students, demonstrated a whole other way of knowing and being. Their work was a radical departure from traditional ideas of education, encompassing the development of people—including the teachers and directors—in community and in the individual. This development was understood as evolution (Garcia, 1999), and it was evident in the way the curriculum was created around students' engagement and interests and civilizational movements, such as developing fire and then eventually working toward the creation of weapons. The curriculum design indicated an understanding of more ancient indigenous practices—such as the need to create weapons for hunting and protection, not as a way to create further violence in the neighborhood. Students developed what Western educators might consider "technical" skills, such as succeeding in finding the means and way to create fire and also to create weapons.

Contemporary indigenous practices center on the maintenance of and respect for the community (Cajete, 2000). Community was a foundational element, from the daily practices evidenced in rituals such as sharing the meal and cleaning one's dishes. However, the commitment to the practice of living the values with depth and all together became clear in the example of the stolen coloring book. That something so seemingly small could be understood as important enough to discuss with the larger group is indicative of the importance and effectiveness of their model in their own space. It was evidenced by the participants' commitment to being "impeccable with their word," one of the Four Agreements (Ruiz, 1997), and exemplified ultimately in how the youth accepted their error in the larger group evaluation activity. The leaders and other community members indeed "did their best" by weaving them into the community through understanding their behavior and the seriousness of it, also as a way to protect them from such errors on the outside of the community.

The school also incorporated the idea of gratitude as one of its newest evolutions over the two previous years. It helped bring about an increased evolution, the objective of Garcia's (1991) creative transformation. Students and teachers agreed they were more united at the school, and it seemed evident in a genuine sense of transcendence evidenced in the ways they expressed gratitude through letters, speech, and in the spirit with which they did these activities. It also helped address a key underlying problem that Guadalupe identified, that of the potential of framing oneself as *pobrecito* or "poor little one." By expressing gratitude, the Escuela Raices participants developed a confraternity with those who also donated, thus creating a co-being instead of mere recipients of charity. This can also be understood as part of the creation of a "new tribalism" wherein donors are, albeit temporarily, a very real part of the Escuela Raices community. Similarly, despite the differences in religious experiences of participants, the sense of gratitude and completion of an altar helped create a unity among the different within the space, the way *nos/otras* are connected along the slant in Anzaldúa's (2002) term which underpins her sense of the new tribalism.

The aspirations and vision of participants in Escuela Raices closely parallel the visionary understanding of a new tribalism. The leaders rethought pedagogies in ways that were both radical and situated in the realities toward creating "new tribalisms" where participants understood and recrafted histories, literally through evolutions—and created new ways of being as guerreros. The participants collectively re-did history through their own evolutionary practices, with a full awareness of these evolutions. These guerreros, warriors, were warriors who could face their realities with new tools of understanding, including complex principles by which to live and by which to elevate each other. Finally, they exhibited a deep understanding of a vision which included the creation of a newer, bigger tribe, beyond the walls of the school. Former participants were, in many ways, living it as teachers in the school and with one as a student of psychology as well. Students became Anzaldúan bridges between Escuela Raices and their home community, often then bringing in other siblings or parents to participate more deeply. The students at the time of research also embodied and lived by the principles and vision of creating a new life space.

In the field of education, we have much to learn from this project. A more Western stance might observe a code of ethics and several daily workshops as modes of learning, perhaps the only modes. This chapter surely could also have identified the Western senses of mathematics and science involved in jewelry making, weapons making, and the creation of fire. However, such principles have already been described in countless monographs. We selectively attempted to recount the substantive senses of major themes of Escuela Raices where the project stood when we visited, highlighting connections to their core underlying guiding principles and the successful invention of a "new tribalism." We see how literal historic stages of human evolution were centered as guiding principles for curriculum development and a deep engagement of all the students in this work. We also see how a new tribalism can be created out of linking into what we know of human development through history by literally recreating it.

We are left with many questions from this model of education. What would happen if urban schools were so deeply connected in following a well-grounded ethical model that linked indigenous knowledges on a large scale? What would happen if these themes of community, living gratitude, and aspirational vision undergirded all our work in education? We challenge other researchers to be both inspired by these practices and to consider what evolutionary moves we might need more largely in the field of education beyond more static conceptualizations of education from the past.

REFERENCES

Anzaldúa, G. E. (2000). Quincentennial: From victimhood to active resistance, an interview with Inés Hernández-Avila. In A. Keating (Ed.), *Interviews/Entrevistas/Gloria Anzaldúa* (pp. 177–194). Routledge.

Anzaldúa, G. E. (2002). now let us shift ... the path of conocimiento ... inner work, public acts. In G. E. Anzaldúa & A. Keating (Eds.), *This bridge we call home: Radical visions for transformation* (pp. 540–578). Routledge.

Anzaldúa, G. E. (2015). Geographies of selves--reimagining identity: Nos/Otras (Us/Others), las Nepantleras, and the new tribalism. In A. Keating (Ed.), *Light in the dark Luz en lo oscuro: Rewriting identity, spirituality, reality* (pp. 65–94). Duke University Press.

Arrows, F. (2013). *Teaching truly: A curriculum to indigenize mainstream education* (Vol. 3). Peter Lang.

Bigelow, B. (2006). *The line between us: Teaching about the border and Mexican immigration*. Rethinking Schools.

Cajete, G. (2000). *Native science: Natural laws of interdependence*. Clear Light.

Chase, S. E. (2005). Narrative inquiry: Multiple lenses, approaches, voices. In N. K. Denzin & Y. S. Lincoln (Eds.), *The SAGE handbook of qualitative research* (3rd ed., pp. 651–679). SAGE.

Emdin, C. (2016). *For white folks who teach in the hood ... and the rest of y'all too: Reality pedagogy and urban education*. Beacon Press.

Foley, D. E. (2002). Critical ethnography: The reflexive turn. *Qualitative Studies in Education, 15*(5), 469–490.

Garcia, J. D. (1991). *Creative transformation: A practical guide for maximizing creativity*. Noetic Press.

Johnson-Bailey, J. (2004). Enjoining positionality and power in narrative work: Balancing contentious and modulating forces. In K. B. DeMarrais & S. D. Lapin (Eds.), *Perspectives and approaches for research in education and the social sciences*. Lawrence Erlbaum Associates.

Kasun, G. S. (2016). Breaking the taboo: What my mother's suicide might teach us in critical social justice and faith work, and perhaps beyond. *The Journal of Faith, Education, and Community, 1*(2), 1–9.

Kasun, G. S., & Lee, H. K. (2017). From "blissfully unaware" to "another perspective on hope": An Indigenous knowledge study abroad program's impacts on the ways of knowing of pre-service transnational English language learner teachers. *Teacher Education & Practice 30*(3), 425–442.

Kasun, G. S., & Saavedra, C. M. (2016). Disrupting ELL teacher candidates' identities: Indigenizing teacher education in one study abroad program. *TESOL Quarterly*, *50*(3), 684–707.

Madison, S. D. (2011). Introduction to critical ethnography: Theory and method. In *Critical ethnography: Methods, ethics, and performance* (pp. 1–18). SAGE.

Rubin, H. J., & Rubin, I. S. (2012). *Qualitative interviewing: The art of hearing data* (3rd ed.). SAGE.

Ruiz, D. M. (1997). *The four agreements: A Toltec wisdom book*. Amber-Allen.

Saldaña, J. (2009). *The coding manual for qualitative researchers*. SAGE.

Smith, L. T. (2006). *Decolonizing methodologies: Research and indigenous peoples*. Zed Books.

Spradley, J. P. (1980). *Participant observation*. Holt, Rinehart, and Winston.

Vitali, S., Glattfelder, J. B., & Battiston, S. (2011). The network of global corporate control. *PLoS ONE*, *6*(10).

ABOUT THE AUTHORS

EDITORS

Iman C Chahine is an Associate Professor of Mathematics Education at the College of Education at University of Massachusetts Lowell and an appointed Extra-Ordinary Professor at North West University in Potchefstroom, South Africa. She is also a coeditor of the *Journal of Mathematics and Culture* and a 2019–2020 Fulbright Scholar in South Africa. Dr. Chahine has published in peer-reviewed premier journals and newsletters including *Educational Studies in Mathematics* (1999, 2002); *Journal of Humanistic Mathematics* (2013, 2018), *Journal of Mathematics Teaching in Middle Schools(2017), International Journal for Studies in Mathematics Education* (2011); *Teaching Children Mathematics* (2007); *Journal of Urban Mathematics Education* (2010); *Near and Middle Eastern Journal of Research in Education* (2014); *Journal of Mathematics Education* (2015); *The Journal of Mathematical Behavior* (2013), *Latin American Journal of Ethnomathematics* (2013), *Indilinga: African Journal of Indigenous Knowledge Systems* (2013) *and International Study Group on Ethnomathematics* (2011). Her research and teaching interests include modeling indigenous knowledge systems; ethnomathematics; embodied and situated cognition; neurocognitive and multimodal learning; and problem solving in nonconventional settings.

Josef de Beer is a Research Professor in the Research Unit-Self Directed Learning Faculty of Education North West University, Potchefstroom–South Africa. His PhD (obtained in 1994) was in science education, and his recent MSc was carried out in ethnobotany, on the plant use of the Khoi-San people. He is currently doing a second PhD focusing on anthropomorphic and ethnobotanical perspectives in KHOISAN indigenous knowledge. He had been involved in teacher education for most of his life, and this remains his first love. In 2009 Josef was awarded the Higher Education Learning and Teaching Association of South Africa (HELTASA) national award for teaching excellence. In 2012 the American National Association of Biology Teachers honored Josef with the Research in Life Sciences Education award. Josef was the recipient of the 2019 NRF Excellence in Science Engagement Award. He is the editor of the 2019 AOSIS publication, *The decolonization of the curriculum project: the affordances of indigenous knowledge for self-directed learning.*

AUTHORS

Piet Ankiewicz is full Professor of Technology Education in the Faculty of Education at the University of Johannesburg. His research interests include the affordances of the philosophy of technology for technology classroom pedagogy, teacher education, indigenous technology knowledge systems, and STEM education. He also has an interest in students' attitudes towards technology. He has authored a chapter on the perceptions and attitudes of pupils toward technology in the *Springer International Handbook of Technology Education*, as well as a chapter on the implications of Andrew Feenberg's critical theory for technology education in the book, *Reflections on Technology for Educational Practitioners: Philosophers of Technology Pnspiring Technology Education.* He is a member of the Editorial Board of the *International Journal of Technology and Design Education* and a member of the Editorial Board of the Brill/Sense International Technology Education Studies Series. He has been rated by the National Research Foundation as an established researcher with international recognition.

Erica Bass-Flimmons holds a PhD from the Learning Technologies Division at Georgia State University. Her research interest includes studying mobile devices in developing countries with an interest in international education. Erica has presented her research at the American Education Communication Technology Conference (2014–2017), Comparative International Education Society (2015), The African Virtual University Transform Research Colloquium (2015) ELearning Africa Conference (2015). Her research interest includes analyzing the mobile technologies in culturally sensitive settings, which could result in the development of culturally specific content and training tools necessary to support various cultures.

Iwan R. Elstak was an Associate Professor of Mathematics (now retired) at the Department of Mathematics at Valdosta State University (GA). Dr. Elstak has published in peer-reviewed premier journals such as: *Georgia Journal of Science* (2013, 2015), *International Journal of Mathematical Education in Science and Technology* (2010, 2012, 2013, 2017), *American Mathematical Association* (2016), *Mediterranean Journal for Research in Mathematics Education (2011) and Educational Studies in Mathematics (2009)*. He has also published in peer-reviewed conference proceedings: SIGMAA on RUME's Conference on Research in Undergraduate Mathematics Education (2015), Proceedings of the 32nd annual meeting of the North American Chapter of the International Group for the Psychology of Mathematics Education (2009, 2010) and Proceedings from the Sixth Conference of European Research in Mathematics Education (CERME) (2009). His research and teaching interests include geometry, mathematical education, and history of math.

Jennifer K. Henderson has been a secondary mathematics teacher in Marietta Georgia for over 10 years. Dr. Henderson completed her doctoral studies at Georgia State University focusing her research on the effect of research-based professional development on teacher professional dialogue. Her research and teaching interests include teacher professional development, teacher dialogue, and equitable teaching using ambitious teaching practices.

Divan Jagals is a Senior Lecturer in the Curriculum Studies subject group at the Faculty of Education, North-West University, Potchefstroom campus. His main research focus is on facilitating metacognitive awareness to promote and understand self-directed learning. The most recent project involvements include the facilitation of metacognitive awareness through a personalized and adaptive online learning platform, as well as a project set towards developing a self-transcendent philosophy of self-directed learning. He is one of three research fellows to complete a 4-month research fellowship under the guidance of the UNESCO Chair for Personalized and Adaptive Distance Education at the Swiss Distance University of Applied Sciences.

Jyoti Kaneria is a doctoral student in the Teaching and Learning/ Language and Literacy PhD Program at Georgia State University. She has worked with and taught emerging bilinguals, both adults and youth, in the Southwestern U.S., Southeastern U.S., and Mexico. She has also conducted research in the U.S. and Mexico and is interested in continuing her research focused on emerging bilinguals with critical approaches to her work.

Sue Kasun is an Associate Professor of Language Education in the Department of Middle and Secondary Education and the founding Director of the GSU Center for Transnational and Multilingual Education. She is also an affiliate faculty member at the Center for Latin American and Latino/a Studies at Georgia State University. She researches issues related to ways of knowing, transnationalism, language, and education, having more recently recognized the need to engage indigenous knowledges in the field of education. Her work has appeared in the following publications, among others: *Teachers College Record, TESOL Quarterly, Anthropology & Education Quarterly,* and *Equity & Excellence in Education.*

Anteneh Kibret is a Lead Math Tutor at Georgia State University Perimeter College Learning and Tutoring Center, and he is a trainer and member of Georgia Tutor Association. He also develops math handouts for the Student Success Learning and Tutoring Center. His training focuses on successful tutoring. His tutoring includes college-level math and physics.

Kimbeni Mansion is a doctoral student in the Teaching and Learning with a concentration in Mathematics Education degree program at Georgia State University. She received her M.Ed. in Mathematics at Georgia State University and her B.S. in Mathematics from Southern University. Kimbeni hopes to specialize in ethnomathematics research, integrating cultural practices in the mathematics instruction. She has also presented in various arenas on the topics of STEAM education and ethnomathematics.

Kori L. H. Maxwell is the Director of Examinations at LOMA. He is responsible for leading efforts to monitor examination development, quality, and security through the use of statistical applications such as item analysis, item response theory, and excessive similarity software. He has undertaken a leadership role in providing professional and nontraditional students with the knowledge required to excel at their jobs in the insurance, financial services, and retirement industries, which also helps to prepare them to take on more challenging roles within their organizations. Dr. Maxwell is a reviewer for the *International Journal for the Scholarship of Teaching and Learning* (IJ-SoTL) and a member of the Editorial Board for the *Journal of Mathematics and Culture* (JMC). He has also published his research in peer-reviewed premier journals including *Journal of Humanistic Mathematics* (2013); *Indilinga African Journal of Indigenous Knowledge Systems* (2013); *and International Journal of Science and Mathematics Education* (2016). His area of interests includes statistics research; statistics education; and multicultural education.

Alesia Mickle Moldavan is an Assistant Professor of Mathematics Education in the Graduate School of Education at Fordham University in New York City. She received her Ph.D. in Teaching and Learning in Mathematics Education from Georgia State University. She spent several years teaching high school mathematics and served as a STEM Saturday Enrichment program instructor. Her area of interest and expertise is in equity-based teaching practices, ethnomathematics, and the use of digital technology in teacher preparation. She is a section editor for the *Journal of Urban Mathematics Education* and has published papers in peer-reviewed journals and conference proceedings, including *Georgia Educational Researcher* (2018), the North American Chapter of the International Group for the Psychology of Mathematics Education (2016, 2019), and the Georgia Association of Mathematics Teacher Educators (2015, 2017). She has also presented at over twenty local, national, and international conferences. Her recent research projects extend from research on culturally responsive pedagogy to create innovative teaching practices that promote cultural awareness, elicit mathematical engagement, and support social justice and advocacy in STEM classrooms.

Nirmala Naresh is a Senior Lecturer in the Department of Mathematics at the University of North Texas. Dr. Naresh has published in peer-reviewed premier journals including *International Journal of Science and Mathematics Education* (2014), *Mathematics teaching in the Middle School* (2018, 2013), *Mathematics Teacher* (2014), *Bolema, Mathematics Education Bulletin* (2014), *Latin American Journal of Ethnomathematics* (2013), *Fields Mathematics Education Journal (2013)*, and *Journal of Mathematics and Culture* (2011). Her research and teaching interests lie in the domains of ethnomathematics, teaching mathematics for social justice and the teaching and learning of probability and statistics.

Hannah Oldham is a high school mathematics teacher and STEM scholar academy coordinator in Marietta, GA. Dr. Oldham's primary research focuses on self-efficacy and goal setting and she is the creator of the self-efficacy goal spectrum, a means of targeting goals in order to maximize mathematical efficacy. Dr. Oldham has presented at multiple national conferences to discuss her research and is passionate about showing how mathematics can be used by anyone, anywhere at any level.

Daniel Clark Orey is Professor Emeritus at California State Univeristy, Sacramento and is currently a professor of mathematics education in the *Departamento de Educação Matemática* in the *Instituto de Ciências Exatas e Biológicas* at the *Universidade Federal de Ouro Preto*, where he works with distance education and master's degree students in the mathematics education program. In 1998, at the invitation of Professor Ubiratan D'Ambrosio, Dr. Orey served as a Fulbright Scholar at the Pontifícia Universidade Católica de Campinas in Brasil, where he conducted research in classrooms and taught courses in ethnomathematics and mathematical modeling. He has published articles in numerous Brazilian and international journals, chapter books, and books in English, Portuguese, and Spanish. He has also participated in Brazilian and international conferences by giving lectures, speeches, and participated in round tables about ethnomathematics. Currently, Prof. Orey serves as the coordinator of the Ethnomathematics Research Group at the Universidade Federal de Ouro Preto and participates in: (a) International Study Group of Ethnomathematics (ISGEm), (b) North America Study Group of Ethnomathematics (NASGEm), (c) Red Latinoamericana de Etnomatemática (RED), (d) Grupo de Estudo e Pesquisa em Etnomatemática (Study and Research Group on Ethnomathematics in Brazil, and e) Topic Study Group on Ethnomathematics (TSG35) from the International Conference on Ethnomathematics (ICME).

Elijah Porter II is a doctoral candidate at Georgia State University in the College of Education & Human Development. His research interests focus on activity theory, the teaching experiment, theory-building in mathematics education. Elijah's major publications include: *Critical Book Review: Lipman, P.* (2011), *The New Political Economy of Urban Education: Neoliberalism, Race, and the Right to the City*. Elijah has 15 years of business and Electrical Engineering experience, and another 15 years as a Mathematics Educator in the K–12 environment.

Erika Potgieter is a research assistant for the Self-Directed Learning Research Unit at North West University, Potchefstroom in South Africa. She completed her master's degree focusing on the use of different pedagogies (e.g., puppetry and indigenous games) for the enhancement of mathematics teachers' metacognitive awareness and pedagogical content knowledge. Her other research interests include the development of cases based on teachers' experiences in the mathematics classroom.

Umesh Ramnarain is a Professor of Science Education, and head of Department of Science and Technology Education at the University of Johannesburg, South Africa. His main research interest is on inquiry-based science education, with a particular focus on its uptake in South African classrooms. His research has been published in top-tier journals such as *International Journal of Science Education, Research in Science Education, Journal of Research in Science Teaching, Chemistry Education Research and Practice,* and

Teaching and Teacher Education. His work has also been disseminated at prominent international conferences such as NARST, ESERA and IOSTE. He has received awards for best paper presentations at conferences in Singapore and Thailand. He is associate editor of *Research in Science Education* and has served on the editorial board of *Journal of Research in Science Teaching*.

Milton Rosa is Professor of Mathematics Education in the *Departamento de Educação Matemática* in the *Instituto de Ciências Exatas e Biológicas* at the *Universidade Federal de Ouro Preto*, where he works with distance education and master's degree students in the mathematics education program. He has experience in mathematics education research areas such as ethnomathematics, mathematical modeling, ethnomodeling, curriculum and instruction, educational leadership, and distance education. Dr. Rosa has published articles, chapter books, and books mainly in ethnomathematics and ethnomodeling in Portuguese, Spanish, and English. His works appear in Brazilian and international journals. He has participated in conferences and congresses in Brazil and other countries; with contributions of plenaries, speeches, and round tables. He also participates in: (a) International Study Group of Ethnomathematics (ISGEm), (b) North America Study Group of Ethnomathematics (NASGEm), (c) Red Latinoamericana de Etnomatemática (RED), (d) Grupo de Estudo e Pesquisa em Etnomatemática (Study and Research Group on Ethnomathematics in Brazil, and (e) Topic Study Group on Ethnomathematics (TSG35) from the International Conference on Ethnomathematics (ICME).

Karen Ratajczak is a Secondary School Math Teacher in Atlanta, Georgia. She graduated from Trinity College with a BA in history and later earned her MAT in mathematics education from Georgia State University. Her interests include interdisciplinary instruction, applicable math skills in daily life, math history, folklore, and music.

Alexandra Starke is currently a doctoral student at Georgia State University. Her research interests include ethno-mathematics and the effects of school culture on high school mathematics classroom environments. Alexandra currently teaches mathematics to high school students in Gainesville, Georgia.

Jessica Steele received her master's in mathematics education from Georgia State University. Her research interests include ethno-mathematics and the stages of mathematical development at the elementary level. Jessica Steele currently teaches mathematics to high school students in Decatur, Georgia.

Erica Dorethea Spangenberg (University of Johannesburg) is a Deputy Head of the Department of Science and Technology Education and a Senior Lecturer in Mathematics Education. Her research mainly focuses on socio-psychological imperatives in the teaching and learning of mathematics; as well as affective constructs, ethnomathematics, and technological pedagogical content knowledge. (ORCID: orcid.org/0000-0003-3073-9239). She has published in peer-reviewed accredited journals, such as *Pythagoras, Independent Journal of Teaching and Learning, Journal of Education and South African Journal of Science and Technology, Gender & Behaviour, and EURASIA.*

Colby Tofel-Grehl is an Associate Professor of Science Education in the School of Teacher Education and Leadership at Utah State University. She is principle investigator of multiple National Science Foundation grants including a CAREER grant focused on developing novel models of teacher professional development for rural teachers. Her published research can be found in journals such as *The Journal of Advanced Academics* (2014), *Gifted Child Quarterly (2017),* and *Roeper Review (2018).* Her research focuses on the instructional affordances afforded teachers as they engage with novel technologies during their normal classroom instruction. By identifying the ways in which professional learning can influence teacher instructional choices, Dr. Tofel-Grehl's research seeks to improve access and STEM identity outcomes for groups historically excluded from those arenas.

Marthie Van der Walt is an Associate Professor of Mathematics Education at the Faculty of Education, North West University, South Africa. She has been working in the field of Mathematics teacher education for the past 20 years, and her research niche encompasses metacognition as reflective methodology for mathematics teachers' professional development and the affordances of contextualizing mathematics topics in indigenous knowledge (games, music, and puppetry) to make mathematics more meaningful and accessible to all learners. She developed the Tshimologo Mathematics Shoestring Kit to support the contextualizing of mathematics in IK (games, music and using puppetry) and presented to participant teachers in 2018.

Patrice Parker Waller is an Assistant Professor of Mathematics Education in the Department of Secondary Education at California State University, Fullerton. Her research interests include teaching and learning mathematics abroad, improving teaching and learning in K–12 and undergraduate mathematics, and in-service teacher development. Dr. Waller has travelled to several countries, including Brazil, South Africa, Morocco, and Ghana, to explore the teaching of mathematics and to examine instances of mathematics embedded in various cultures. In 2014 Dr. Waller was awarded a Fulbright-Hayes Scholarship to investigate mathematics, women, and culture in Belize. Dr. Waller has published her global research in *the Journal for Multicultural Education, International Journal of Science and Mathematics Education* and in the Government Report for the Office of International and Foreign Language Education of the U.S. Department of Education Group Projects Abroad.

Printed and bound by CPI Group (UK) Ltd, Croydon, CR0 4YY

15/09/2026

14962336-0006